D0301588

PRINCIPLES OF MEAT SCIENCE

A SERIES OF BOOKS IN FOOD AND NUTRITION

Editor: B. S. Schweigert

PRINCIPLES OF MEAT SCIENCE

John C. Forrest
PURDUE UNIVERSITY

Elton D. Aberle
PURDUE UNIVERSITY

Harold B. Hedrick
UNIVERSITY OF MISSOURI

Max D. Judge
PURDUE UNIVERSITY

Robert A. Merkel
MICHIGAN STATE UNIVERSITY

W. H. FREEMAN AND COMPANY
San Francisco

Library of Congress Cataloging in Publication Data

Principles of meat science.

Includes bibliographies.
1. Meat. 2. Meat industry and trade. I. Forrest, John C.
TX373.P74 664'.9 75-8543
ISBN 0-7167-0743-8

Printed in the United States of America

Contents

Preface

Meat-science courses at many universities serve students having a broad range of interests and backgrounds. Our goal in preparing this book was to provide a text for sophomore level college students enrolled in their first meat-science course. Although many excellent reference works were available for use in our own introductory meat-science courses, no single up-to-date text covered the bulk of the information included in our courses. *Principles of Meat Science* was generated to fill this void. *The Science of Meat and Meat Products,* a companion text to *Principles of Meat Science,* has served upper-division and graduate students well for many years. It was not our intent to duplicate the material in that book, but to supplement it with material written at an introductory level.

This text has been five years in preparation, and most of the chapters have been used for several semesters in the courses taught by the authors. In our quest for completeness we have attempted to provide information on meat produced by cattle, swine, sheep, poultry, fish, and other species. Since many of the basic principles of meat science apply to all meat-producing species, specific references to species were often considered unnecessary. Properties of meat that are unique to an individual species have been included where appropriate, but it was nearly impossible to be totally complete in the coverage of all species. This text is also oriented toward the meat industry as it exists in the United States;

there is a wide variation in much of the specific information that is presented here when consideration is given to the industry in other areas of the world. This is particularly true with regard to meat-inspection regulations, standards and grades, and the nomenclature of meat cuts.

Since this is a multi-authored text, some variation in writing style exists among the chapters. We attempted to minimize this variation by assigning chapters to each author for initial writing, and then having each chapter reviewed by all five authors. Authors of the individual chapters are not identified because each chapter received major input from all authors.

The authors wish to acknowledge the help of many colleagues across the country who have been willing reviewers of various parts of the book. We also wish to extend special thanks to Dr. C. Eugene Allen, Dr. B. S. Schweigert, and Dr. D. G. Topel, each of whom reviewed the entire manuscript and offered many valuable suggestions.

April 1975

JOHN C. FORREST
ELTON D. ABERLE
HAROLD B. HEDRICK
MAX D. JUDGE
ROBERT A. MERKEL

PRINCIPLES OF MEAT SCIENCE

PART I

INTRODUCTION

CHAPTER 1

Meat as a Food

Eating is a process essential to the maintenance of life itself, and few foods can quiet the pangs of hunger and satisfy the appetite so quickly and completely as meat. In Western society, foods of equal nutritional value are hard pressed to compete with the great pleasure derived from the consumption of sizzling steaks, succulent roasts, spicy sausage, crisp friend chickens, juicy lobster tails, or any of the hundreds of other meat dishes available as the main attraction for a meal. An entire industry has developed to provide the nutritious, high quality meat supply sought after by much of the world's human population. You are beginning a study of that industry and the unique food products it prepares for the hungry masses.

WHAT IS MEAT?

Were you surprised to see lobster tails and fried chickens listed along with steaks and roasts? If so, it is time to broaden your thinking before you begin the study of Meat Science. *Meat* is defined as those animal tissues which are suitable for use as food. All processed or manufactured products which might be prepared from these tissues are included in this definition. While nearly every species of animal can be used as meat, the majority of the meat consumed by humans comes from domestic animals and aquatic organisms.

Meat as an entity can be subdivided into several general categories. The largest category, in terms of volume of consumption, is "red" meat. Beef, pork, lamb or mutton, and veal are the most common "red" meats. However, horse, goat, eland, llama, camel, water buffalo, and rabbit meats are commonly used for human consumption in many countries. Poultry meat is the flesh of domestic birds, and includes that of chickens, turkeys, ducks, geese, and guinea fowl. Sea foods are the flesh of aquatic organisms, of which the bulk are fish. However, the flesh of clams, lobsters, oysters, crabs, and many other species are also included in this category. A fourth category is that of game meat, which consists of the flesh of all nondomesticated animals.

MUSCLE AS MEAT

It should be immediately emphasized that any discussion of meat must be directed primarily toward muscle. Although meat is composed of numerous types of tissue, such as those found in nerves, fat, and blood vessels, the major component of meat is muscle. The chemical and physical properties of muscle tissue and the associated connective tissues are of paramount importance in dictating the usefulness of meat as a food.

Since muscles are organs whose unique structure and function primarily serves to provide locomotion, they possess several characteristics associated with this function that require modification when they are utilized as a food. For example, because of the particular function they perform in the body, certain muscles contain relatively large quantities of connective tissue. Connective tissue is associated with toughness. However, these muscles can be as palatable as any other, provided a cooking method is used which will degrade the connective tissues. Had muscle been engineered and designed to function primarily as a food product, it would undoubtedly have differed in several characteristics. With the advent of space age technology, many foods are being engineered to specification, and attempts are being made to substitute other protein sources for the muscle in meat. Surprisingly enough, it is extremely difficult to imitate the texture and flavor imparted to meat by muscle and other animal tissues. Attempts to prepare meat-like products without using any animal tissue have met with only limited success. However, the partial replacement of muscle with other protein sources in various meat products has a great potential for making a meat supply available to the world's poor.

WHAT IS MEAT SCIENCE?

Meat Science is a broad field of study, an important part of which is the basic study of the unique characteristics of muscle and the other animal tissues that are used as meat. A complete understanding of the basic properties of the tissues used in meat can lead to improved utilization and better meat products.

Meat Science is not limited to the study of tissues. It includes all facets of the meat industry, beginning with animal production and ending with final preparation for consumption. Breeding, feeding, and management are extremely important parts of the food chain because quality control actually starts on the farm, ranch, or feedlot.

The market system, through which livestock move from feedlots to packing plants, has a profound influence on the industry. The market system is the only line of communication between the consumer and producer. The language is money, and producers adjust their production to meet the demands of the market, which (ideally) should reflect the desires of the consumer. Whether or not the market system accurately communicates the consumers wishes to the producers, it is a major determinant of the type of livestock and poultry being produced.

The meat packers, processors, and purveyors make up that segment of the industry which converts the live animals into food products and then distributes them to retail stores, hotels, restaurants, and institutions. It is in this segment of the industry that many of the aspects of meat technology are applied, both to maintain product quality and wholesomeness and to develop new and different meat products.

Retail meat markets and the hotels, restaurants, and institutions (HRI) group are also a part of the market system. Retailers are the meat industry's representative to the consumer. They prepare many of the fresh meat cuts for the consumer. If they maintain quality and wholesomeness, and display all meat products in an attractive manner, the entire industry benefits. The HRI group carries food processing to its ultimate end, and places the food on the plate of the consumer. The final cooking and preparation of meat is just as important as any segment of the complex industry that brings meat from grazing lands and feedlots to the consumer. Part of the reason for the increasing complexity of the livestock and meat business is that new competitive food products are continually being developed. These new competitors might seek to entice the consumer with modifications in convenience, price, quality, uniformity, nutritional value, or even with novelty. If the meat industry is to maintain its present position of importance in the food production chain and maintain a dynamic and growing market, it must strive to produce the highest quality products with the greatest possible efficiency, as well as to develop innovative new products and employ more sophisticated advertising and promotion programs.

Students who plan to be associated with any part of the livestock and meat industry during their working careers must not be satisfied with learning only the status of the industry today, for this knowledge will soon be sadly out of date. Instead, they must learn basic concepts and be prepared to apply these to changing situations—indeed, they should prepare to be the initiators of changes.

It is not the purpose of this book to describe the intricacies of all phases of the complex meat industry. This would provide only a historical background, and would not prepare students for involvement in a dynamic industry. In this text emphasis is placed upon the basic principles and characteristics that govern the operations of the meat industry.

The characteristics of the animals which produce the tissues used for meat is considered, with emphasis being placed on the mechanisms responsible for their growth and development. The structure, composition, and function of muscle tissue components are examined in detail, and the contribution that each makes to the characteristics of the final product is stressed.

The subjects of meat processing and preservation are approached from the standpoint of the principles behind the processes used to prepare and preserve hundreds of different meat products. Recipes and detailed instructions for manufacture of specific products are available from other sources, if they are needed for reference. Many other topics, such as inspection, grading and standardization, and by-products are included, providing an in-depth coverage of key aspects of the total meat industry activity.

MEAT AND THE ECONOMY OF A NATION

Throughout recorded history, the consumption of meat has indicated a position of social and economic prestige. It is noteworthy that meat consumption is often an indicator of the economic status of a country or individual. As a nation industrializes and improves its economic position, its meat consumption increases. Moreover, as persons raise their social or economic status, they tend to demand a greater and higher quality of meat products.

Meat is one of the most nutritious foods used for human or animal consumption. Meat is an excellent source of high quality protein, and it also contains large amounts of minerals and essential B vitamins. In this latter respect, it is one of the few good sources of vitamin B_{12}. Adequate protein nutrition is essential. By adequate protein nutrition, we mean a sufficient quantity of *high quality protein*; that is, protein with all of the essential amino acids necessary for physical well-being, and for proper mental and intellectual development. It has been postulated that if the people of the underdeveloped nations of the world could be adequately supplied with meat or other high quality protein foods, their capability for rapid industrial, social, political, and intellectual development would increase many times.

The breeding and rearing of livestock and poultry for meat production constitutes a significant proportion of our total agricultural production. If we also consider the production of forages and grains used as feed for livestock and poultry, the proportion of our total agricultural resources devoted to meat production exceeds 60 percent. In the United States, more than one-half of total farm income is derived from the sale of livestock (including poultry) and livestock products, and approximately 40 percent of total farm income is derived from the sale of livestock and poultry for meat production.

The food industry is the largest industry in the private sector of the United States economy, in terms of total assets, employment, and gross business receipts. If the meat packing and processing industry is compared to other segments of the food industry, we find that the meat industry ranks either first or second in total assets, value added to product by processing, and total employment. More important, however, the meat industry easily ranks first in total business receipts, earning approximately 30 percent of the total food industry receipts. The next largest is the dairy industry with about 15 percent of total food receipts. These figures reflect the proportion of the consumer's food dollar spent for meat and meat products and emphasizes the important position of meat in our food economy.

THE MEAT INDUSTRY IN THE UNITED STATES

The meat industry has had a long and colorful history that includes a nostalgic era of development referred to as the "good old days," as well as rougher times marked with scandals and labor problems similar to those experienced by other industries. The unsanitary practices of a few unscrupulous meat packing companies rocked the entire industry when they were given nationwide publicity at the turn of the century. However, a highly sophisticated industry that employs modern computer age business practices and applies scientific knowledge to food manufacturing, has emerged from this background to provide this country with one of the safest, highest quality meat supplies in the world.

The U. S. meat industry began when colonial butchers, most of whom had learned their trade in Europe, began slaughtering and dressing animals for others beyond their own immediate family. Colonial meat shops became the first retail meat markets. The term "meat packing" arose with the first commercial attempts at meat preservation in this country—the salting and packing of pork in wooden barrels for storage, or for shipment to Europe.

As cities grew, small packing plants were established. These plants generally expanded in size as the population of the city grew. The major packing companies originated in the large cities because of the availability of the labor force needed to operate the large plants. Animals were often driven on hoof from the production areas to rail heads, then moved by rail to the large terminal livestock markets that had developed in conjunction with the growth of the packing industry.

Refrigeration and transportation have both played major roles in the development of the meat industry. In the early days, slaughter plants had to be located in the large cities so that the highly perishable product could be moved quickly to the consumer. With slow transportation and no refrigeration, meat could best be moved from the grazing lands and feedlots on the hoof. Before the advent of mechanical refrigeration, slaughter and processing was limited to the cooler seasons, except in areas where ice could be obtained from rivers and lakes and stored for use during the summer months.

Large meat packing plants capable of producing a full line of fresh and processed meats were established in all of the major cities in the U. S. Chicago, with its geographical location in the center of the corn belt, became immortalized as the "hog butcher of the world" by poet Carl Sandburg. Many factors led to the rise of the Chicago meat packing industry, including the availability of all the necessary resources, raw material (livestock), labor, transportation (both water and rail), and a ready market for the products. The meat industry in Chicago and other major cities reached its peak in the early 1950s. The large plants which had started in the 1890s or earlier and grown piecemeal into huge units, had become obsolete. Automation had reduced the requirement for a large labor force, and the development of the interstate highway system combined with the perfection of refrigerated rail and truck transportation made possible the movement of perishable foods over long distances. In the 1960s, irrigation changed the livestock industry in the southwest and large cattle feedlots were established in areas formerly considered as being fit only for grazing. Consequently, the finishing of livestock for market is no longer concentrated in the cornbelt, but has spread over a much wider area of the midwest and southwest.

When new meat packing plants were constructed to replace the obsolete ones, they were not built in the big cities but rather were placed close to areas of livestock production. It became more feasible and economical to ship meat and dressed carcasses, rather than live animals.

These modern plants are very different in character from the old ones. They are usually attractive in appearance and often located in open country, rather than in crowded industrial areas. On the inside they are bright, well lighted, and easy to clean, and the comfort and safety of the employees have been considered. Modern meat packing plants are often very specialized, perhaps producing only dressed carcasses from one species, or slaughtering one species and manufacturing a limited line of processed meat items. Other plants can produce only manufactured meat products, receiving wholesale cuts, carcasses, or boneless meat from outside sources.

The retail meat business has also changed. Many of the retail chains have established large central facilities where carcasses are broken into retail cuts that can be distributed to local supermarkets in the area on a daily basis. This centralized processing reduces the labor required to operate meat counters in the supermarkets.

The foregoing is but a brief account of the recent history of a dynamic industry. Technological advances in meat processing, packaging, preservation, and transportation did not cause the recent changes in the meat industry, but they made change possible. The meat industry is still in an evolutionary stage, and more changes are in sight. Frozen meat is being used extensively by institutional consumers, and it is becoming technologically feasible to produce frozen meat that will have an appearance similar to unfrozen fresh meat, with all parts of the cut completely visible. Will this be the next change? The answer is only a few short years away.

SCIENCE AND MEAT

Once, the manufacture of meat products was strictly an art handed down from generation to generation. Today, the development of new products and the improvement of old ones is a science. Many major meat packing companies have extensive research facilities and employ highly trained basic food scientists to solve problems and develop new products. Major research programs in Meat Science are being carried on at universities, research institutes, and government research facilities in this country and abroad. Because of this research effort, the past decade has seen a tremendous expansion of the basic science of meat as a food. Much of the information in this book is the result of this research. Since many basic questions remain unanswered, and new technical problems frequently arise, there is still much fertile ground to be plowed by the imaginative scientist.

REFERENCES

Boorstin, Daniel J., *The Americans: The Democratic Experience* (Random House, New York), 1973.

Mayer, O., *America's Meat Packing Industry* (Princeton University Press, Princeton, New Jersey), 1939.

Sandburg, Carl, *Chicago Poems* (Henry Holt & Co., New York), 1916.

Sinclair, Upton, *The Jungle* (The New American Library Inc., New York), 1905.

CHAPTER 2

Animal Production Goals

Animal agriculture includes the production of many different species of domestic animals, including cattle, sheep, swine, poultry, rabbits, horses, goats, mink, and many others. Throughout its long and colorful history, the economic uses to which animals were put shaped the character of animal agriculture at each stage of its development. At one time the production of power for agriculture and transportation was a major goal of animal agriculture in the U. S., as it still is today in many areas of the world. As petroleum and other power sources were developed, the resources of animal agriculture were shifted more directly to food production. Thus, a major change from the production of horses to that of domestic meat animals has occurred, resulting in a steady increase in the per capita consumption of meat up to the present. As we approach the 21st century, we find that animal agriculture is taking on an added dimension; a contribution to leisure time enjoyment, through a growing pet, zoo, and pleasure horse industry. While the portion of the animal industry associated with leisure time continues to increase in economic and social importance, and utilizes an increasing amount of our animal production resources, the production of food and fiber will remain the first priority for several decades.

The food produced by animal agriculture includes meat, milk, eggs, and their associated processed products. (We acknowledge the importance of milk and eggs as products of animal agriculture, but will confine the remainder of the discussion in this text to meat from the various economically important species.)

ESTABLISHING GOALS

The short term goal of most individuals engaged in animal agriculture is to earn a reasonable profit return for their labor and management skills. In this chapter however, we are concerned more with the long term goals that must be established by the entire livestock industry in order to maintain its viable competitive position with other systems of food production, in order to assure an adequate supply of meat for the portion of the world population desiring to consume it.

The pressures of increasing world population and associated ecological problems dictate that a higher efficiency of food production must take high priority. Animal agriculture, more than any other type of food production, will feel the pressure to improve efficiency, because many of the resources used in animal food production have the potential for being processed and used directly as human food more efficiently than if they were to be converted to food indirectly by an animal. An acre of land could produce up to 10 times as much plant protein as animal food protein. However, when consumers can afford the luxury, most of them prefer the more succulent and sometimes more nutritious animal products over plant food products. This is one reason that vigorous attempts are being made to process plant foods in such a way as to imitate the more popular animal products, rather than develop entirely new and different products.

If the efficiency of animal production is not improved, economic and population pressures could force the abandonment of a major portion of animal agriculture as it is known in the U. S. today; particularly the feeding of cereal grains to livestock. This could restrict meat animal production to grazing lands which are not suitable for other types of agriculture. Even without this development, there is little doubt that the utilization of grazing lands for meat production will increase in significance. A potential exists for the improvement of efficiency of cereal grains utilization, as well as that of grazing lands, in meat production. The realization of this potential should be one of the major goals of the animal industry.

Food must not only be produced efficiently, but its quality must be maintained if it is intended for human consumption. In recent years, experience has shown that good nutrition alone is not enough. Even people suffering from malnutrition often refuse to eat foods which are unpalatable to them even though they may be nutritionally adequate.

The goal now becomes three-fold, the *efficient* production of high *quality* food that is also *acceptable* to the people being fed. The techniques and tools available for achieving that goal are of particular significance to animal agriculture. Efficiency and quality improvement are not always compatible goals. Producers who achieve an improvement in production efficiency often find that they lose ground with respect to quality. Three examples will serve to illustrate this point. (1) In swine, some strains within breeds have made tremendous improvement in muscularity, and lean to fat ratio, in recent years; but, in the process, some of the animals became extremely susceptible to stress, and the meat from these animals was often pale, soft and watery. This reduced consumer acceptability, and in extreme cases, animal death losses were high during the growing and marketing phases of production. (2) Very often, beef cattle having a large mature size combined with exceptional growth rates and high lean to fat ratios fail to meet grading standards for the present day market. (3) Certain strains of turkeys have undergone selection to the extent that their extreme muscularity hinders their ability to move. Similar examples could be given for other species of meat animals.

EFFICIENCY IN MEAT PRODUCTION

Every aspect of meat production from conception to consumption must be carefully evaluated in seeking ways to improve efficiency. The following discussion is intended only to suggest ways and means of improvement in the areas of animal production, conversion of animals to food, marketing and distribution, and utilization at the point of consumption. Future technological developments will, in all probability, extend the possibilities for improvement far beyond those points covered in this discussion. All segments of the industry must work together to achieve efficiency. No longer can the producer, packer, processor, and retailer consider themselves as separate industries which must compete with one another. They must function as parts of a single system that efficiently moves meat from the feedlot to the table.

Animal Production

Continuous attention to both efficiency and quality is necessary if the meat animal industry is to meet the challenges it faces. One of the prime requisites for animal and product improvement is information. In order to make intelligent decisions in the selection of animals for breeding, information relative to the performance of the parent stock must be available. Techniques are available for giving much of the needed information, and these are the tools which must be used to meet the goals of improved efficiency and quality.

REPRODUCTIVE EFFICIENCY. Improved reproductive efficiency can reduce the number of animal units in the breeding herd or flock that must be maintained solely for the purpose of producing the animals to be used for meat. Embryonic, prenatal, and early postnatal death losses are major contributors to the present low reproductive efficiency. A concerted research effort will be required for a major improvement in reproductive efficiency. However, some progress can be made through breeding and selection. Longevity of production should also be considered in selection of breeding animals.

FEED CONVERSION EFFICIENCY. In most meat producing domestic animal species there is still room for improvement in the conversion of cereal grains to meat. This improvement must be achieved if cereal grains are to continue to be a significant nutrient source for meat production. Improved efficiencies in feed conversion must not be achieved at the expense of the quality of the final meat product.

Forages and pastures which cannot be easily harvested and processed for human consumption offer an opportunity for meat production without competition with humans for nutrients. Development and improvement of many of the rough lands of the world could do much to assure an adequate meat supply for future generations.

Animals may offer a solution to a potentially serious problem, since many waste materials in this country and the world which pollute the environment can also be utilized as nutrients for animals. The recycling of these materials through animals offers an opportunity for their conversion to a palatable food. Many by-products in the food industry are suitable for animal feed with very little processing. Other by-products and waste materials will require considerable processing before they can be used as animal feed. The development and utilization of new and different nutrient sources of this nature offer a tremendous challenge to the animal industry.

GROWTH. Some efficiency in animal production can be achieved through an improved rate of growth. This can increase meat production per unit of labor and capital invested, when combined with efficient management. Increase in weight of the live animal is a measure of growth. The most common way of expressing growth in terms of weight is by absolute gain in weight, per unit of time, which can be expressed as average daily gain, average growth rate, or rate of gain. Average growth rate, though often used in meat animal evaluation programs, gives no accurate indication of the changes occurring in body or carcass composition. It only gives a measure of gross change in weight of the whole body. Growth rate can be a valuable measure of efficiency, when used in combination with the carcass measurements, described in the next section, that can be obtained at slaughter.

CARCASS COMPOSITION. Several linear measurements are indicators of carcass composition. Measurement of the thickness of the subcutaneous fat layer on the carcass has been demonstrated to be a simple, yet reliable, indication of fatness. As subcutaneous fat thickness increases, the yield of muscle or lean cuts decreases. Fat measurements are directly related to total carcass fatness, and indirectly related to the total amount of muscle. Obtaining measurements of the fat layer on the carcass is a very simple procedure, and requires only a small ruler. In swine, fat thickness may be evaluated in the live animal by probing with a small steel ruler. More expensive ultrasonic sound techniques called "sonoray" have also been used, with some success, in estimating fat thickness in live animals.

Area measurements of muscle cross-sections, particularly of the *longissimus* muscle in the thoracic–lumbar region of the carcass, have been shown to be directly related to the total amount of muscle in the carcass. Measurements of the *longissimus* muscle are included in most carcass selection and evaluation procedures. Area measurements are usually accomplished by tracing the muscle outline on special paper and determining the area with a compensating polar planimeter, or by using a transparent plastic grid which is placed on the muscle surface in order to estimate the area directly. Muscle area is sometimes estimated in the living animal using "sonoray" techniques, but the accuracy is highly dependent upon the skill of the equipment operator.

Carcass length is used in most current pork carcass evaluation procedures. Although frequently utilized, this measurement is poorly related to natural muscling. The relationship between carcass length and the total amount of muscle in the carcass is influenced by the muscular development of the animal. For instance, comparatively short animals could have large thick muscles and longer animals could have rather thin muscles. Therefore, the total amount of muscle would be similar in both cases. The value of any added length on a carcass is entirely dependent upon a "standard" degree of muscularity.

In general, fat measurements are more useful in the prediction of carcass composition than are muscle measurements. No doubt this is due to the wider variation in the quantity of fat, than muscle, per unit of body or carcass weight. Since excess carcass fat is one of the major problems faced by the meat industry, a reduction in fat alone would solve one of the industry's biggest problems.

The use of cutting tests to appraise carcasses is widely used. Improved accuracy in the evaluation of carcass composition is attainable by these techniques, which require more equipment, skill, and training than the measurements just discussed. The cutting tests may be based on the four lean cuts of pork, or on specified boneless trimmed retail cuts of beef and lamb. For these tests to be meaningful, available standardized procedures for cutting and trimming must be followed. Although cutting tests give a rather accurate and practical measure of saleable or edible meat in the carcass, based on present day marketing practices, they do not give an accurate measure of the chemical composition of the carcass.

Dissection methods consist of physically separating the carcass or selected wholesale cuts into component parts, i.e., into muscle, fat, and bone. The method is tedious and time consuming, and a certain amount of subjectivity is involved in the separation of the tissues. Although the method has these disadvantages, it is one of the more precise methods available for the evaluation of carcasses. Detailed chemical analysis of the entire animal or carcass is the most accurate and reliable method of assessing body composition, if proper sampling procedures are used. Chemical analysis generally consists of moisture, fat, and protein determinations that destroy the carcass and are consequently very costly. For these reasons other methods are often used to estimate body or carcass composition.

When information on body or carcass composition is combined with growth rate, feed efficiency, and other production factors, it is possible to determine the efficiency with which animals of a given genetic strain can convert raw materials into edible tissues. Today's consumer demand indicates that the emphasis should be on the production of edible lean with a minimum of excess fat.

QUALITY EVALUATION. Carcass quality must be considered hand in hand with the tools for quantitative evaluation just presented. For the purposes of the present discussion, the evaluation of *quality* is an attempt to predict the palatability, processing, and cooking characteristics of the meat. This is by no means an easy task. As you progress through the study of Meat Science, you will learn that the producer has little or no control over many factors that affect the palatability of meat. The goal of the producer should be to provide animals that have the potential for being converted to highly palatable meat products. Proper breeding, feeding, and management can achieve this goal. The following examples of meat quality factors are known to be affected to some degree by animal production practices.

Meat tenderness is one of the major palatability factors that must be maintained or improved in most of the meat producing species. Meat tenderness is affected greatly by the age of the animal. Meat from the carcasses of relatively young animals is more tender than that from older animals mainly because the connective tissues of young animals are more easily broken down during cooking than are the connective tissues from older animals. (Other reasons are discussed in Chapters 4 and 14.) Producers can influence tenderness by selecting breeding stock with the ability to reach the desired market weight at a young age, and then they must manage their feeding operations to be sure that the animals realize their full potential in gaining ability. They must also see to it that animals are marketed at the proper age and weight.

The physiological maturity of an animal can be evaluated from its carcass and used as an indicator of potential tenderness. Bone and skeletal characteristics, and muscle color, are used in this evaluation. A more detailed discussion of maturity evaluation is found in Chapter 14.

Marbling, or the interspersing of fat within the lean, has often been discussed as a factor associated with tenderness in meat of certain species. However, research over the years has failed to show that marbling has any consistent effect upon tenderness. Marbling may contribute more to the juiciness and flavor of fresh meat products. Marbling has long been used as a quality indicating factor in carcass evaluation, and is of considerable economic importance because of its use in determining USDA grades.

The color of fresh meat has a major influence on the consumer when meat is purchased at retail. Most consumers have a concept of the proper appearance of meat from any given species, and any significant deviation from that color will be discriminated against. They might associate darker than normal color with dryness, toughness, or even with an off flavor. For this reason and others which will be discussed later, color is an important quality factor which can be evaluated in the carcass.

Quality evaluations can also be made using mechanical shearing or penetration devices to determine tenderness, chemical analyses of fat in muscle to quantitate marbling, and spectrophotometers or colorimeters for evaluating color. The ultimate in quality evaluation is the use of a trained taste panel or a consumer panel, but even these methods are not without substantial error. Some variation in quality assessment obviously stems from the variation in preferences among the panel members.

DUAL PURPOSE ANIMALS. A small but significant portion of our meat supply comes from animals that were grown initially for other purposes and then converted to meat when their initial productive phase ended. Dairy animals in particular are dual purpose animals, even though only a few of the dairy breeds are specifically identified as such. Each year nearly all of the male calf crop is converted to either veal or beef. Also, when it is time for a milk cow to be replaced in a dairy herd, she will most likely be converted to meat. But, it is generally true that as the amount of muscling improves in dairy breeds, their milk production declines. Since each dairy animal that is produced is a potential contributor to the meat supply, perhaps more attention should be given to their meat producing characteristics. Some European countries have found that economic conditions are such that they can afford a small sacrifice of milk production to get more meat production from their dairy animals. It is possible that some quality improvement could be achieved without sacrificing milk production.

Poultry layers are also dual purpose animals. In the large commercial flocks, hens are in egg production for no more than 1–1.5 years before they become a part of the meat supply. These hens are used primarily in chicken soup or sausage products. The possibility of improving some of the meat producing characteristics of these birds should be explored.

Certainly there are other animals which could fit into the dual purpose category. In other countries the horse is a significant contributor to the meat supply. In the United States only a small number of horses are normally slaughtered for human consumption, but as other meat prices increase rapidly, consumer interest in horse meat increases for economic reasons, and more people are finding that its consumption is not objectionable.

MANAGEMENT. If any progress is made toward the goals of efficient meat production, it will depend upon wise management of livestock enterprises. Managers of these enterprises must develop production systems which efficiently utilize the resources available for meat production. This includes the utilization and application of information and techniques that will bring about improvements in carcass traits, both qualitative and quantitative, as well as in reproductive efficiency, growth rates, and efficiency of feed conversion. Much of the technology for accomplishing the task of improvement in efficiency is presently available but it still remains to be applied on a broad basis.

There are many other ways in which improved management practices can contribute to efficiency, such as by developing the philosophy among livestock producers that they are in reality handling food products, and therefore, that sanitation, disease control, and careful handling are as important on the farm and in the feedlot as in the packing plant. If such a philosophy were instilled in all personnel handling live animals, a considerable reduction in losses from condemnation could occur.

Marketing

Livestock marketing systems of the future must continue to become more responsive to consumer needs and desires. Some live marketing systems now in use, effectively insulate the livestock producer from the consumer. This is because livestock buyers must make value judgments based on averages, or without adequate information, and the tendency is for the best animals to be purchased at a price below their true value while poor animals are purchased at a price above their true value. This system averages out differences, so that even though the total livestock purchase reflects the total true value, the average value derived from it does not hold on an individual animal (or small group) basis. To encourage rapid improvement or change when required by consumer demand, the system must reflect the true value of each animal with regard to its potential for utilization by the consumer.

Livestock marketing systems are being modernized to eliminate the guesswork in the trading of meat animals. Live animals are a unique food raw material, and market systems are now being designed specifically to fit the characteristics of the animals. The technology is available to design a modern responsive system for marketing meat animals, and this should be a high priority goal of the livestock and meat industry, in order to speed its quest for efficiency.

Conversion of Animals to Food

The meat packing and processing industry has acquired a reputation for using "everything but the squeal," and some contend that even that has been done. This reputation is well deserved, because the industry has done a good job of converting tissues that are not used for food into useful by-products. Economic pressure has forced by-products to be utilized with great efficiency. However, ecological concern and the ever growing need for food protein demand even greater efficiency in the future. The potential exists for converting some of the animal tissues that now find their way into inedible by-products into high protein food products for human consumption. For example, blood is a source of nutritious food protein; but, because of general prejudice among Americans, little of it is used for human food in the United States. However, in other parts of the world, as well as among many ethnic groups in the United States, meat products composed of blood are often considered delicacies, and certain African tribes utilize blood taken from their cattle as their main source of protein. Furthermore, the concept of "no waste" will likely be adopted for the future, and economically profitable uses will be found for the small amounts of waste now generated by packing and processing plants.

The continued application of science and technology to the slaughter of animals and processing of meat is essential. Modern slaughter and processing plants have already achieved considerable reduction in their labor requirements. However, the potential for further automation exists. The industry will probably continue on the path toward total automation. Not only will greater automation allow for greater speed and efficiency, but it will allow a shift of personnel to quality control and research and development.

The trend toward centralized preparation of retail meat products calls for improved meat distribution techniques. There are two aspects of distribution that must be improved. First, special equipment and technology are required to maintain freshness and extend shelf life. This will reduce spoilage, another major source of loss in the meat industry. Secondly, along with the distribution system, a market system should be established which will assure that each individual product goes into the market where the consumer demand for that product is greatest. This will assure greater consumer satisfaction and at the same time, increase profits to the total industry.

Retailing

The retailing of meat will continue to undergo considerable change as the breakdown of carcasses into retail cuts is moved to centralized locations, whether they are in the meat processing plants or in plants established by the retail chains themselves. Handling of meat at the retail level must improve, in order to assure that contamination and loss of shelf life does not occur in the retail meat cases. Meat markets of the future will handle greater volumes of meat with reduced numbers of personnel, but those people must be expert in inventory management, and in maintaining the specialized display equipment.

Final Preparation

The very last step before consumption is as important as any other in the entire meat chain. Whether prepared by a homemaker, French chef, backyard barbecue buff, short order cook, or precooked during processing by industry personnel, if the meat is not properly prepared, all of the previous effort to efficiently produce quality and wholesomeness will be for naught. Proper utilization at the point of consumption is extremely important, and any improvement in this area requires education of the persons responsible for final preparation. Educating such a diverse group of people is a difficult task. Mass educational techniques are required. The producers of agricultural products are finding that they are having to do this job through state and national commodity organizations and the various livestock species groups. Both education and promotion are accomplished by the programs of these organizations, and both are important to the future of the livestock and meat industry.

Students of Meat Science who enter any phase of the livestock and meat industry should realize that they have a mission to fulfill in helping to educate their fellow consumers, either on a personal basis or through vigorous support of the research and educational organizations and institutions that have been established to do the job on a mass basis.

EFFICIENT PRODUCTION OF HIGH QUALITY FOOD

The industry that puts meat on tables around the world is truly a complex production chain whose every link must function dependably. All segments of the meat and livestock industry must establish the *efficient production of high quality food* as the basic precept from which goals and policy decisions can be based. Each segment must look at the entire industry when determining goals and setting policy, and consider the overall effect of their own decisions. A few examples have been presented in this chapter to suggest possible ways in which the different phases of the industry might contribute to this basic goal. These examples are by no means intended to be all inclusive, and so, as you study Meat Science, try to keep the basic goal in mind and develop your own ideas as to how that goal can best be achieved. We feel that this will be the best way to prepare yourself either for a career in the meat production industry, or to be a well informed consumer of meat products.

REFERENCES

American Society of Animal Science, Symposium—Prenatal and Perinatal Development of Swine, Journal of Animal Science, 38, 977 (1974).

American Society of Animal Science, Symposium—Protein Synthesis and Muscle Growth, Journal of Animal Science, 38, 1050 (1974).

Dovring, F., "Soybeans," Scientific American 230, 14 (1974).

Hodgson, R. E., "Place of Animals in World Agriculture," Journal of Dairy Science, 54, 442 (1972).

Van Horn, H. H., T. J. Cunha and R. H. Harms, "The Role of Livestock in Meeting Human Food Needs," Bio Science, 22, 710 (1972).

PART II

MUSCLE AND ASSOCIATED TISSUES

CHAPTER 3

Structure and Composition of Muscle and Associated Tissues

Meat is composed primarily of muscle, plus variable quantities of all the types of connective tissues, as well as some epithelial and nervous tissues. Skeletal muscle is the principal source of muscle tissue in meat. However, a small amount of smooth muscle is also present in meat. Although all of the connective tissue types are present in meat, adipose tissue (fat), bone, cartilage, and connective tissue proper predominate. The muscle and connective tissues are the gross compositional components (muscle, fat, and bone) of the meat animal's carcass, and they contribute, almost exclusively, to the qualitative and quantitative characteristics of meat. It can now be appreciated that meat is not synonymous with muscle, even though it is made up mostly of muscle. Because of the importance of muscle and connective tissue in meat, their structure and composition will be discussed in considerable detail. A brief discussion of epithelial and nervous tissues will also be included.

MUSCLE TISSUE

With the exception of excessively fat animals, the skeletal muscles constitute the bulk (35–65 percent) of the carcass weight of meat animals. A knowledge of the structure, composition, and function of muscle

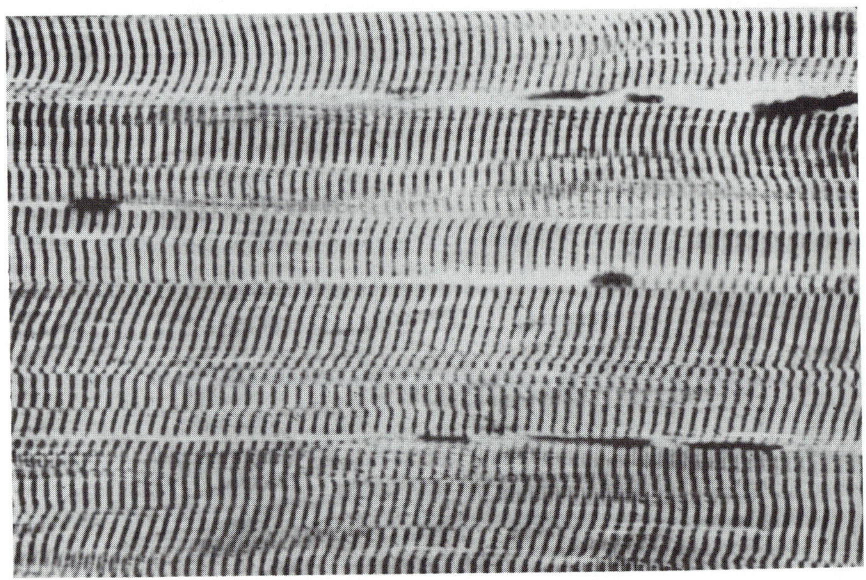

FIGURE 3-1
Photomicrograph of skeletal muscle fibers (× 630).

in the living animal is necessary to understand the postmortem changes associated with its conversion to meat, and its properties and utility as meat.

In addition to skeletal muscle, meat contains some *smooth* muscle, primarily as a component of blood vessels. Another specialized form of muscle tissue, called *cardiac* muscle, is confined solely to the heart. Skeletal and cardiac muscle are also referred to as *striated* muscle because of the transverse banding pattern observed microscopically, as shown in Figure 3-1. Skeletal muscle is sometimes referred to as voluntary muscle, while smooth and cardiac muscles are called involuntary muscle.

Skeletal Muscle

Most skeletal muscles are attached directly to bones, but some of them are attached to ligaments, fascia, cartilage, and skin, and therefore only indirectly to bones. There are more than 600 muscles in the animal body, and they vary widely in shape, size, and activity. The specific characteristics of a given muscle are related to its function. Each muscle is covered with a thin connective tissue sheath, which is continuous with connective tissue constituents that course the interior of the muscle.

Nerve fibers and blood vessels enter and exit the muscle, with the connective tissue networks providing the muscle within an innervating system as well as a vascular bed for nutrient supply and waste removal. Many of the above features of muscle are shown in Figures 3-2 and 3-3. The structural unit of skeletal muscle tissue is the highly specialized cell which is usually referred to as a *muscle fiber*. Muscle fibers constitute 75–92 percent of the total muscle volume. The connective tissues, blood vessels, nerve fibers, and extracellular fluid make up the remaining volume, with the extracellular fluid comprising the major proportion of this volume. The ultrastructure of muscle fibers and their association with connective tissues in the formation of muscles is the subject of the following discussion.

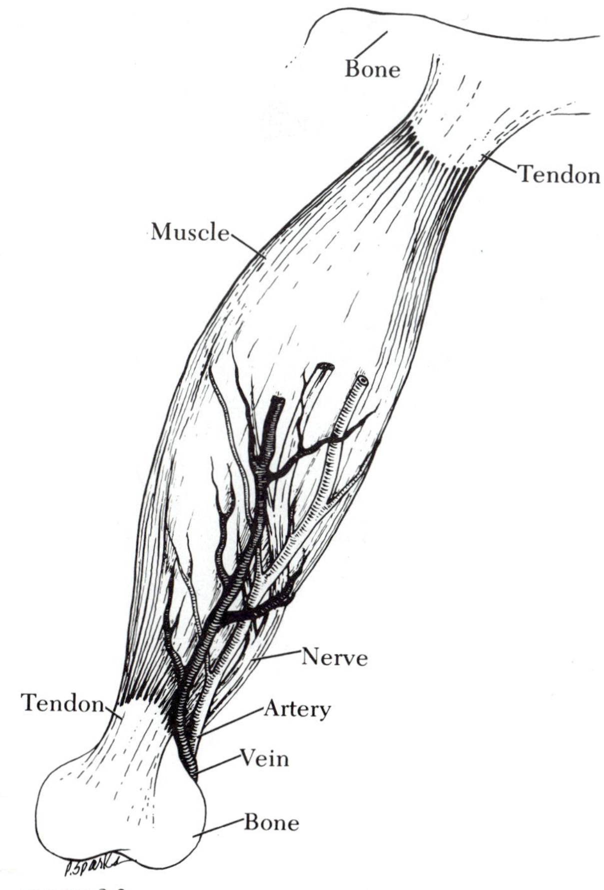

FIGURE 3-2
Drawing of a skeletal muscle showing the structural relationships among tendons, blood vessels, and nerves.

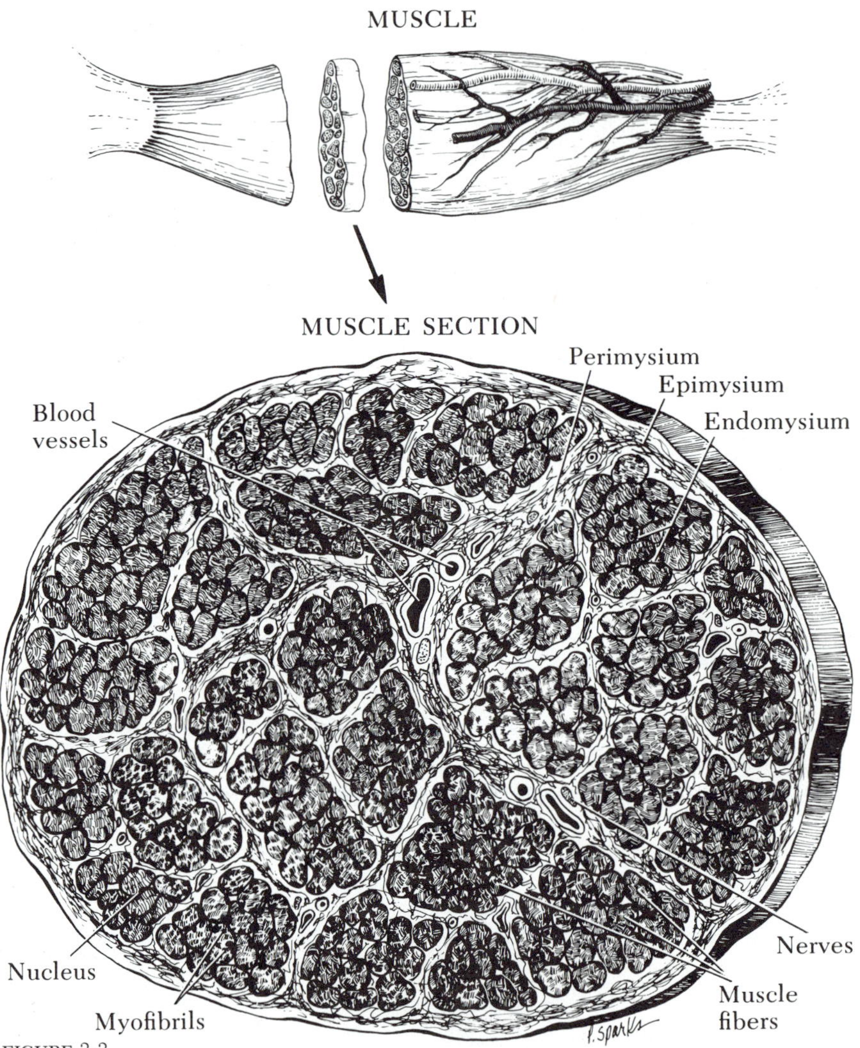

FIGURE 3-3
Drawing of a skeletal muscle in cross section showing muscle fibers, bundle arrangement, pervading connective tissues, nerves, and blood vessels. [Modified from J. E. Crouch, *Functional Human Anatomy*, 2nd ed. 1972. Lea & Febiger, Philadelphia.]

Skeletal Muscle Fiber

Mammalian and avian skeletal muscle fibers are long, unbranched, threadlike cells which taper slightly at both ends, as shown in Figure 3-4. Although fibers may attain a length of many centimeters, they generally do not extend the length of the entire muscle. They vary con-

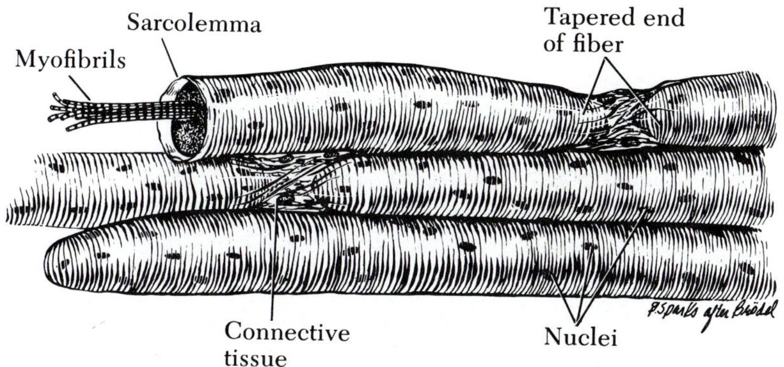

FIGURE 3-4
Drawing of skeletal muscle fibers showing structural features, and their longitudinal orientation. [After M. Brödel, Johns Hopkins Hosp. Bull. 61:295, 1937; © The Johns Hopkins University Press.]

siderably in diameter, ranging from 10 μm to more than 100 μm (1 μm [micrometer] equals one millionth of a meter), within the same species, and even within the same muscle.

SARCOLEMMA. The membrane surrounding the muscle fiber is called the *sarcolemma.* The sarcolemma is diagrammatically illustrated in Figure 3-4. The prefix *sarco-* is derived from the Greek words *sarx* or *sarkos*, which mean flesh, and the suffix *-lemma*, which is the Greek word for husk. It is composed of protein and lipid material. The sarcolemma is relatively elastic, which should be readily apparent when the tremendous distortion it endures during contraction, relaxation, and stretching is considered. The structure, composition, and properties of the sarcolemma are thought to be identical to the *plasmalemma* (cell membrane) of other cells of the body. Periodically, along the length of the fiber, and around its entire circumference, invaginations of the sarcolemma form a network of tubules, that are called *transverse tubules* (Figure 3-5). The transverse tubules are usually referred to as the *T system* or as *T tubules.*

Motor nerve fiber endings terminate on the sarcolemma at the *myoneural* junction, as shown in Figure 3-17. The prefix *myo-* is derived from the Greek word *mys*, meaning muscle. The myoneural junction is the site where the endings of the motor nerve fiber are implanted in small invaginations of the sarcolemma, as shown in Figure 5-4. The structures present at the myoneural junction form a small mound on the surface of the muscle fiber. The entire complex is called the *motor end plate.* A more detailed description of the structure of the myoneural junction is presented in Chapter 5.

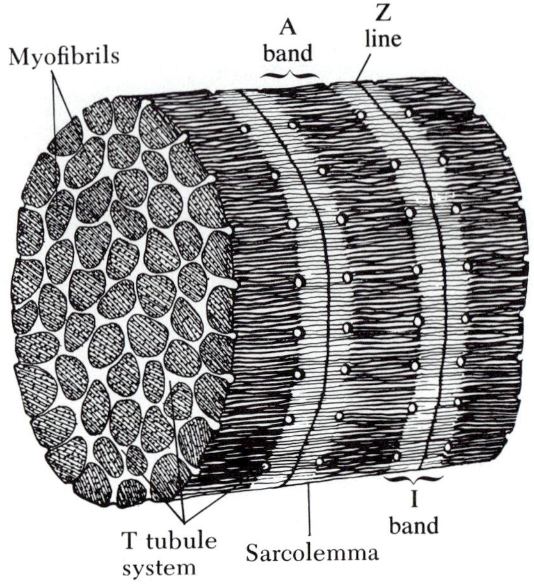

FIGURE 3-5
Drawing of a cross section through a muscle fiber at the A band–I band junction, showing the transverse tubule system in mammalian skeletal muscle. Invaginations of the sarcolemma are shown along the fiber and around its entire circumference. [Modified after L. D. Peachey, *In*: Briskey, Cassens, Marsh, eds. The Physiology and Biochemistry of Muscle As a Food. (The University of Wisconsin Press, Madison; © 1970 by the Regents of the University of Wisconsin), p. 307, 1970.]

SARCOPLASM. The cytoplasm of muscle fibers is called *sarcoplasm*. It is the intracellular colloidal substance in which all the organelles are suspended. Water constitutes about 75–80 percent of the sarcoplasm. In addition to water, the sarcoplasm of skeletal muscle contains lipid droplets, variable quantities of glycogen granules, ribosomes, numerous proteins, nonprotein nitrogenous compounds, and a number of inorganic constituents.

NUCLEI. Skeletal muscle fibers are multinucleated but, because of the tremendous variation in their length, the number of nuclei per fiber is not constant. A fiber several centimeters long might have several hundred nuclei with a regular distribution, about every 5 μm along its length, except near tendinous attachments where they increase in num-

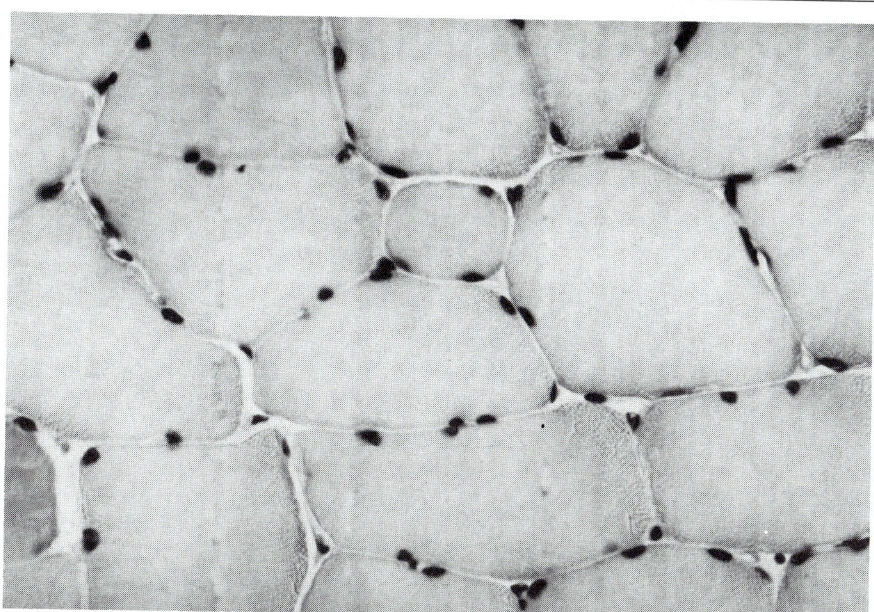

FIGURE 3-6
Photomicrograph of a cross section of skeletal muscle fibers of the pig showing their polygonal shape, and the peripheral position of nuclei (× 630).

ber and are more irregularly distributed. The number of nuclei also increases in the vicinity of the myoneural junction. In mammalian muscle the nuclei are located at the periphery of the fiber, just beneath the sarcolemma, as shown in Figure 3-6. They are ellipsoidal in shape, with their longest axis oriented parallel to the long axis of the fiber. These features of muscle fiber nuclei can best be visualized by examining the transverse and longitudinal diagrammatic illustrations in Figure 3-3. (In contrast, the nuclei of fish skeletal muscle are usually centrally located within the fiber.)

MYOFIBRILS. The *myofibril* is an organelle unique to muscle tissue, as is implied by the prefix. Myofibrils are long, thin, cylindrical rods, usually 1–2 μm in diameter. In most muscles, and in all mammalian species, their long axis is parallel to the long axis of the fiber. The myofibrils are bathed by the sarcoplasm, and extend the entire length of the muscle fiber. A diagrammatic illustration and electron photomicrographs of the myofibril are shown in Figure 3-7d, and e, and in Figure 3-8, respectively. A muscle fiber with a diameter of 50 μm from meat animals will have at least 1000 and could have as many as 2000 (or more) myofibrils.

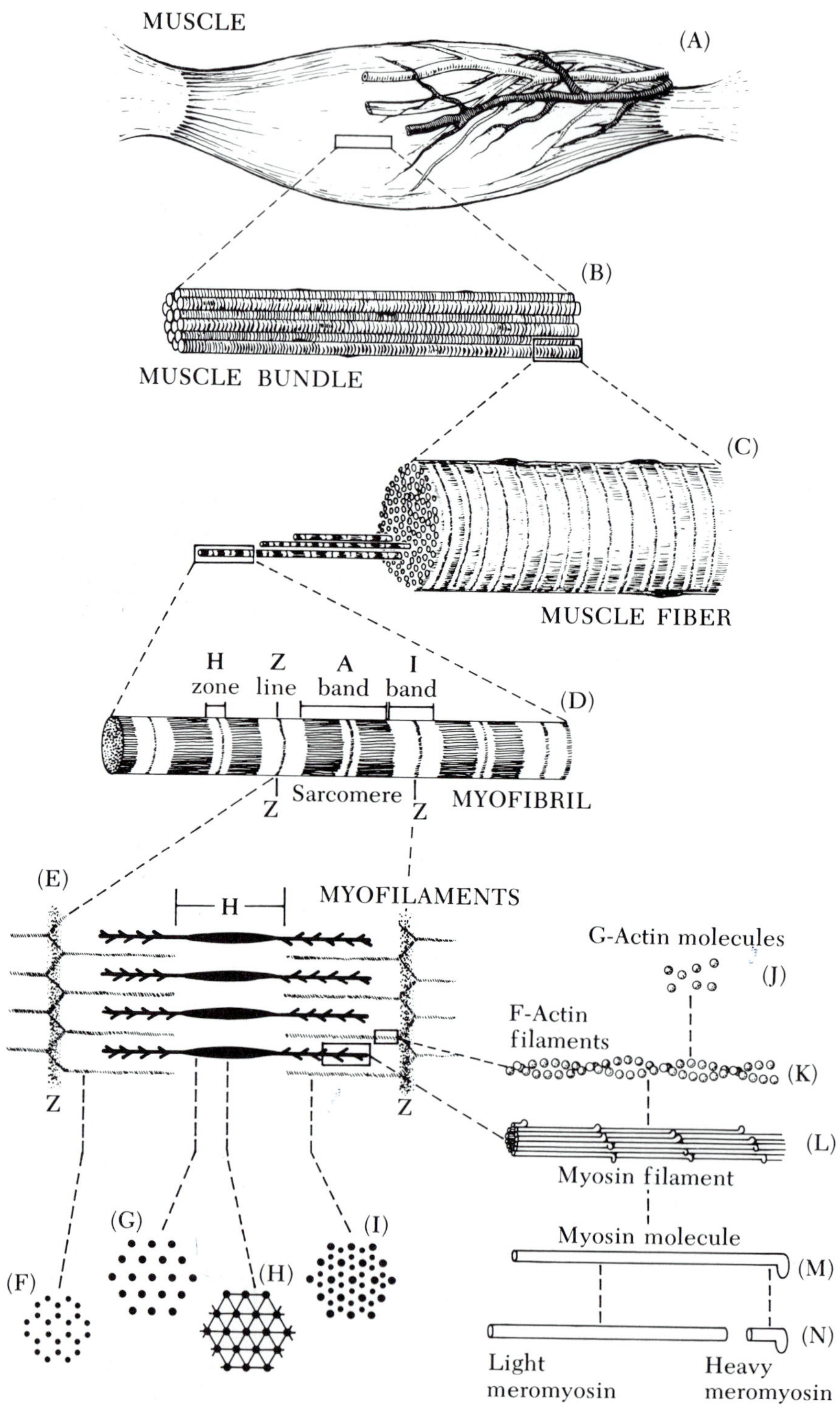
MUSCLE
(A)
(B)
MUSCLE BUNDLE
(C)
MUSCLE FIBER
H zone
Z line
A band
I band
(D)
Z
Sarcomere
Z
MYOFIBRIL
(E)
H
MYOFILAMENTS
G-Actin molecules
(J)
F-Actin filaments
(K)
Z
Z
(L)
Myosin filament
Myosin molecule
(M)
(N)
Light meromyosin
Heavy meromyosin
(F)
(G)
(H)
(I)

FIGURE 3-7 (*opposite*)
Diagram of the organization of skeletal muscle from the gross structure to the molecular level. (A) skeletal muscle, (B) a bundle of muscle fibers, (C) a muscle fiber, showing the myofibrils, (D) a myofibril, showing the sarcomere and its various bands and lines, (E) a sarcomere, showing the position of the myofilaments in the myofibril, (F–I) cross sections showing the arrangement of the myofilaments at various locations in the sarcomere, (J) G-actin molecules, (K) an actin filament, composed of two F-actin chains coiled about each other, (L) a myosin filament, showing the relationship of the heads to the filament, (M) a myosin filament showing the head and tail regions and, (N) the light meromyosin (LMM) and heavy meromyosin (HMM) portions of the myosin molecule. [Modified after Bloom and Fawcett, *A Textbook of Histology*, 9th ed., W. B. Saunders Company, Philadelphia, p. 273, 1968.]

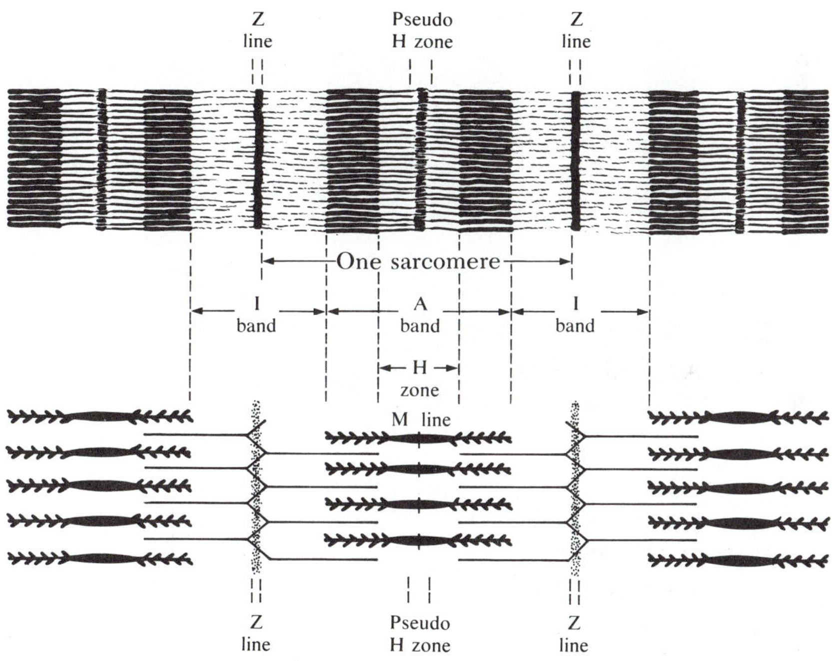

FIGURE 3-8
A drawing, adapted from an electron photomicrograph, showing portions of two myofibrils and a sarcomere (× 15,333) and a diagram corresponding to the sarcomere, identifying its various bands, zones, and lines. [Modified from H. E. Huxley, "The Mechanism of Muscular Contraction." Copyright © 1965 by Scientific American, Inc. All rights reserved.]

Cross-sections of myofibrils show a well-ordered array of dots that have two distinct sizes (Figure 3-9). These dots are actually the *myofilaments* (within the myofibrils), shown in longitudinal section in Figure 3-7d, and e, and in Figure 3-8. The myofilaments are commonly referred to as the *thick* and *thin filaments* of the myofibril. In longitudinal

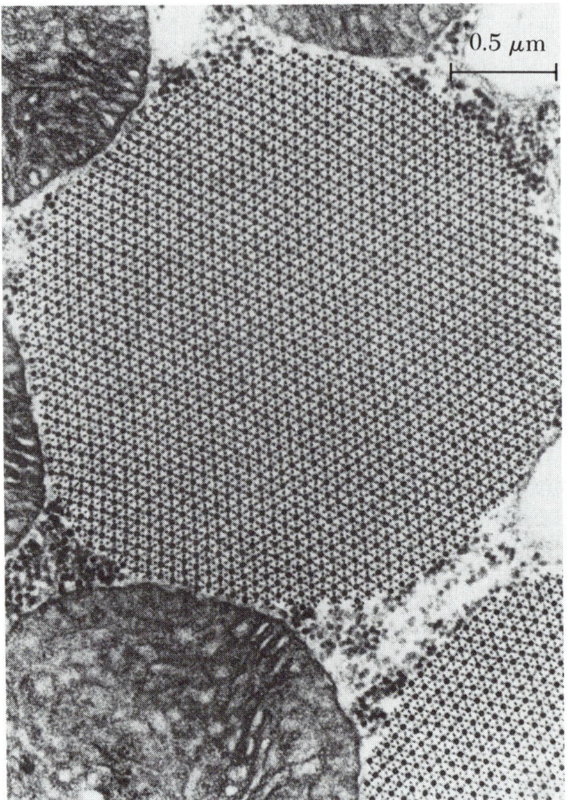

FIGURE 3-9
Electron photomicrograph of a cross section of a myofibril, showing the thick and thin filaments, and their hexagonal arrangement. [Adapted from D. S. Smith, *Muscle,* Academic Press, Inc., New York, p. 14, 1972.]

section, the thick filaments are aligned parallel to each other and are arranged in exact alignment across the entire myofibril. Likewise, the thin filaments are exactly aligned across the myofibril, parallel to each other and to the thick filaments. This arrangement of the myofilaments, and the fact that the thick and thin filaments overlap in certain regions along their longitudinal axes, accounts for the characteristic banding or striated appearance of the myofibril (Figure 3-8). In turn, the bands of each myofibril are aligned across the entire muscle fiber giving the fiber a striated appearance (Figure 3-1). This banding effect, which takes the form of alternating light and dark areas, explains the use of the term *striated muscle* in reference to skeletal muscle.

Areas of different density are visible within the light and dark bands of the myofibrils. Because the light band is singly refractive when viewed with polarized light, it is described as being isotropic and is called the *I band*; whereas, the broad dark band is doubly refractive, or anistotropic, in polarized light and thus designated the *A band*. The A band is much denser than the I band. Both the A and I bands are bisected by relatively thin lines. The I band is bisected by a dark thin band called the *Z line*. The unit of the myofibril between two adjacent Z lines is called a *sarcomere*. The sarcomere includes both an A band and the two half I bands located on either side of the A band. The sarcomere is the repeating structural unit of the myofibril, and it is also the basic unit in which the events of the muscle's contraction–relaxation cycle occur. Sarcomere length is not constant and its dimensions, as well as those of the I band, are dependent upon the state of contraction at the time the muscle is examined. In mammalian muscle at rest, a sarcomere length of 2.5 μm is rather typical. In the central region of the A band, there is an area that has slightly less density than the remainder of the band. This lighter region is called the *H zone*. Additionally, a narrow dense band, known as the *M line*, bisects the center of the A band. A narrow region of relatively low density appears within the H zone on either side of the M line. This low density region is referred to as the *psuedo H zone*. These features of the myofibril are shown in Figures 3-7d, e, and 3-8. The physical significance of these bands will become apparent from our later discussions.

MYOFILAMENTS. The thick and thin filaments of the myofibril not only differ in dimensions, but in their chemical composition, properties, and position within the sarcomere. The thick filaments of vertebrate muscles are approximately 14–16 nm (nanometers) in diameter (1 nm equals one-billionth of a meter) and 1.5 μm long. The thick filaments constitute the A band of the sarcomere, as illustrated in Figures 3-7e and 3-8. Since the thick filaments consist almost entirely of the protein *myosin* they are referred to as *myosin filaments*. The myosin filaments are believed to be held in transverse and longitudinal register by thin cross bands located periodically along their length, and by cross connections which are aligned across the center of the A band. The alignment of these cross connections in the center of the A band corresponds to the transverse density characteristics of the M line.

The thin filaments are about 6–8 nm in diameter, and they extend approximately 1.0 μm on either side of the Z line. These filaments constitute the I band of the sarcomere, although the filaments themselves extend beyond the I band and into the A band between the thick myosin filaments as seen diagrammatically in Figure 3-7e. They consist primarily of the protein *actin*, and are referred to as the *actin filaments*.

The H zone is less dense than the rest of the A band because it is the center region between the ends of the opposing actin filaments (from each half sarcomere). Therefore it contains only myosin filaments, as can be seen from the cross-sections in Figures 3-7h and 3-8. The width of the H zone varies with the state of contraction of the muscle. The densest area of the A band is on either side of the H zone, where both the actin and myosin filaments are present. Since the I band contains only the thin actin filaments, it is the least dense band of the entire myofibril. These features are evident from cross-sections of a sarcomere through the H zone, I band, and that portion of the A band where the actin and myosin filaments overlap, as shown in Figure 3-7f, g, h, and i.

The distribution of the myofilaments in cross-section (Figures 3-7 and 3-9) shows the orderly arrangement of thick (myosin) and thin (actin) filaments. The myosin filaments in the H zone region of the sarcomere are oriented in a definite hexagonal pattern (Figure 3-7g, and h). A cross-section through the A band where the actin and myosin filaments overlap (Figure 3-7i), shows that six thin filaments surround each thick filament.

Z LINE ULTRASTRUCTURE. In longitudinal section, an actin filament on one size of the Z line lies between two actin filaments on the opposite side of the Z line (Figures 3-7e and 3-8). This arrangement indicates that the actin filaments *per se* do not pass through the Z line. The actin filaments are believed to terminate at the Z line. Ultra-thin filaments, called *Z filaments*, constitute the material of the Z line, and they connect with actin filaments on either side of it. Near the Z line, each actin filament connects to four Z filaments that pass obliquely through the Z line. Each of the four Z filaments then connects with an actin filament in the adjacent sarcomere, as shown diagrammatically in Figure 3-10. This structural arrangement of the Z line shows each actin filament of one sarcomere oriented in the center of four actin filaments from the next sarcomere. In longitudinal sections, this offset (or oblique) arrangement of the Z filaments results in the characteristic zigzag pattern of the Z line, as shown in Figures 3-7e and 3-10, and also explains why an actin filament on one side of the Z line appears to lie between two actin filaments on either side of it.

PROTEINS OF THE MYOFILAMENTS. The proteins actin and myosin constitute approximately 75–80 percent of the protein in the myofibril, and the remaining fraction consists of the *regulatory proteins.* These latter proteins are so named because of their direct or indirect regulatory functions on the adenosine triphosphate–actin–myosin complex. Included among the regulatory proteins are tropomyosin, troponin, two M proteins, α-actinin, C protein, and β-actinin (listed in decreasing order of concentration in the myofibril). The proteins of the thick and thin filaments, as well as the regulatory proteins, are referred to as *myofibrillar proteins.* Recent findings suggest that the regulatory proteins are associated with various elements of the myofibril. Tropomyosin,

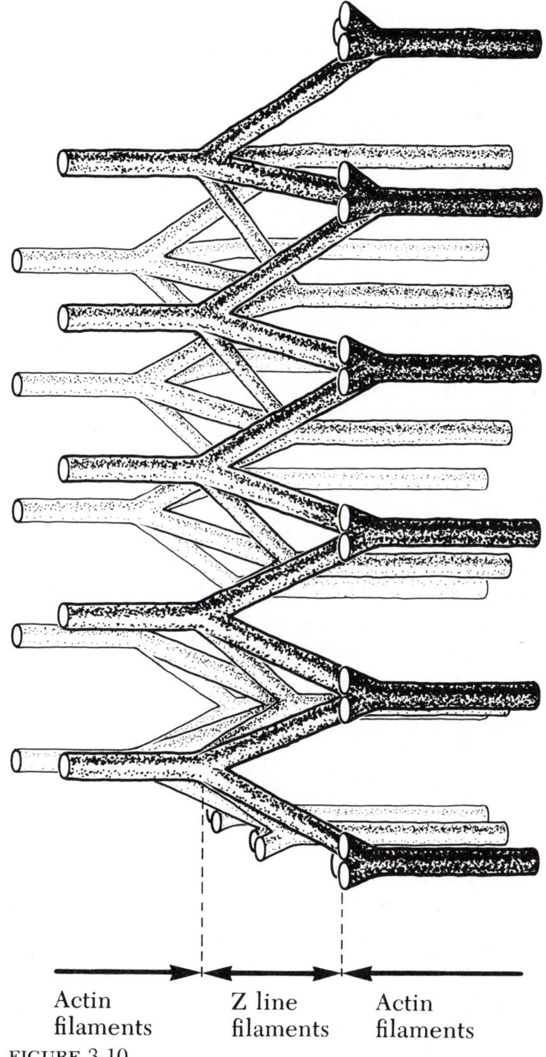

FIGURE 3-10
Diagrammatic representation of Z filaments and their attachment to actin filaments. [From Knappeis and Carlsen, *J. Cell Biol.* 13:332, 1962.]

troponin, and β-actinin are associated with the actin filament; C protein is present in the myosin filament; α-actinin is a component of the Z line; and, M proteins are believed to be the substances composing the M line.

Actin constitutes about 20–25 percent of the myofibrillar proteins. The actin molecule is rich in the amino acid proline. This amino acid, by

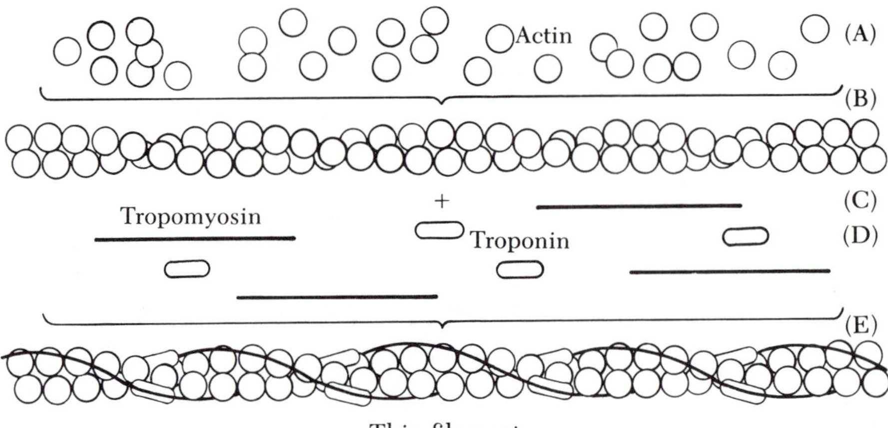

FIGURE 3-11
Diagrammatic representation of the construction and fine structure of an actin filament, showing the relation of tropomyosin and troponin to its (A) G-actin molecules, (B) two chains of polymerized actin (F-actin) coiled about each other, (C) long thin molecules of tropomyosin, (D) pellet shaped troponin molecules and, (E) an actin filament, showing how two coiled chains of F-actin and polymerized tropomyosin molecules form two long strands of tropomyosin, one strand being associated with each of the F-actin chains. The periodic location of troponin along the actin filament every seven G-actin molecules, and its association with the tropomyosin strand are also illustrated. [Modified from J. M. Murray and A. Weber, "The Cooperative Action of Muscle Proteins." Copyright © 1974 by Scientific American, Inc. All rights reserved.]

virtue of its imino group (>N—H), contributes to the folding among the polypeptide chains that results in the formation of a globular (spherical) shaped molecule approximately 5.5 nm in diameter. This globular molecule is referred to as *G-actin* (for globular actin) and as such constitutes the monomeric (single molecule) form of actin. The fibrous nature of the actin filament occurs because of the longitudinal polymerization (linking) of G-actin monomers to form *F-actin* (fibrous actin), as shown in Figure 3-11b and Figure 3-7j and k. In F-actin, the G-actin monomers are linked together in strands, much like the beads on a string of pearls. Two strands of F-actin are spirally coiled around one another, as shown in Figures 3-11b and 3-7k, to form a "super helix" that is characteristic of the actin filament. From this helical arrangement of G-actin monomers, the actin filament's overall diameter of approximately 6–8 nm can be readily visualized (Figure 3-7k, and Figure 3-11b and e). The *isoelectric pH* (pH of minimum electrical charge and solubility) of actin is approximately 4.7.

Myosin constitutes approximately 50–55 percent of the myofibrillar protein and is characterized by a high proportion of basic and acidic amino acids, making it a highly charged molecule. Actin, on the other hand, possesses a relatively low charge. The isoelectric pH of myosin is approximately 5.4. Myosin, with a lower proline content than actin, has a more fibrous nature. The structure of the myosin molecule is an elongated rod shape, with a thickened portion at one end. The thickened end of the myosin molecule is usually referred to as the *head region*, and the long rod-like portion that forms the backbone of the thick filaments is called the *tail region*. The portion of the molecule between the head and the tail regions is called the *neck*. The head region of the molecule is double headed, and it projects laterally from the long axis of the filament (Figures 3-12b and 3-7m). When myosin is subjected to the proteolytic (protein breaking down) action of the enzyme trypsin, it is split near the neck into two fractions that differ in molecular weight; *light meromyosin* and *heavy meromyosin* (Figures 3-12a and 3-7n).

In the center of the A band, on either side of the M line, the myosin filament contains the tail portion of the myosin molecules without any of the globular heads. This region within the H zone, on either side of the M line, is called the *psuedo H zone*. The polarity of the myosin filaments is such that the heads on either side of the bare central region of the A band are oriented at an oblique angle away from the M line (Figures 3-7e and 3-8). The protruding heads are the functionally active

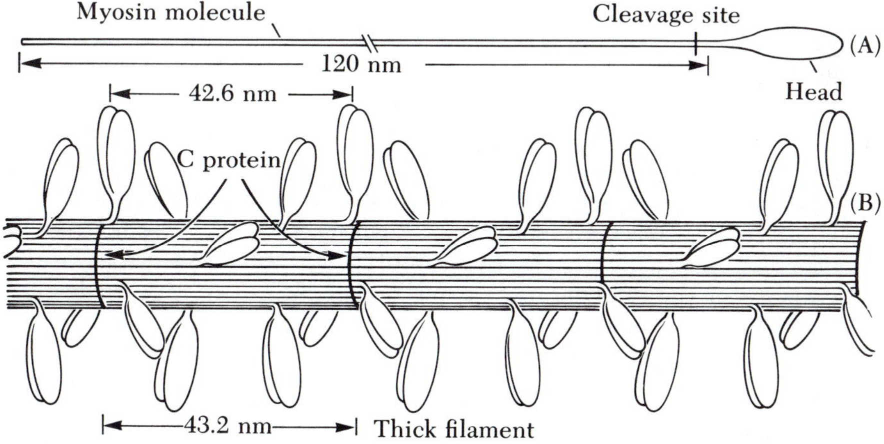

FIGURE 3-12
Diagrammatic representation of the construction and fine structure of the myosin filament. (A) The myosin molecule, showing the head and tail, and the cleavage site and, (B) a myosin filament, showing the projecting double heads and periodic (every 43.2 nm) encircling band of C protein. [Modified from J. M. Murray and A. Weber, "The Cooperative Action of Muscle Proteins." Copyright © 1974 by Scientific American, Inc. All rights reserved.]

sites of the thick filaments during muscle contraction, since the myosin heads form crossbridges with actin filaments. During muscle contractions each myosin head attaches to a G-actin molecule of the actin filament. The formation of crossbridges through this interaction of actin and myosin filaments produces the chemical complex known as *actomyosin*. The formation of actomyosin results in a rigid and relatively inextensible condition in the muscle. Actomyosin is the major form of the myofibrillar proteins that is found in postmortem muscle, and the rigidity that develops following death (rigor mortis) is largely due to this complex. It is a transient compound in the living animal, since the crossbridges between the actin and myosin filaments are broken during the relaxation phase of the contraction cycle. (Crossbridges are almost nonexistent in muscle when it is at rest.)

Another protein, the C protein, is found in the myosin filament. It constitutes about 2–2.5 percent of the myofibrillar proteins. A narrow band of C protein encircles the myosin filament and binds the myosin molecules together into the bundle that forms the thick filament. Eighteen C protein bands 43.2 nm apart, encircle each myosin filament. There are nine bands on each side of the H zone. A few C protein bands encircling a myosin filament are shown diagrammatically in Figure 3-12b.

Tropomyosin constitues 8–10 percent of the myofibrillar protein and, like myosin, is a highly charged molecule with a high content of acidic and basic amino acids. The isoelectric point of tropomyosin occurs at a pH of about 5.1. Tropomyosin also has a very low proline content that contributes to its fibrous nature. Tropomyosin molecules, consisting of two coiled peptide chains, attach end to end to one another, and thus form long, thin, filamentous strands. In the actin filament, one such tropomyosin strand lies on the surface of each of the two coiled chains of F-actin. The tropomyosin strands lie alongside each groove of the actin super helix. A single tropomyosin molecule extends the length of 7 G-actin molecules in the actin filament. The relationship of tropomyosin to the actin filament is diagrammatically illustrated in Figure 3-11d and e.

Troponin, a globular protein with a relatively high proline content, also constitutes 8–10 percent of the myofibrillar protein. Like tropomyosin, troponin is present in the grooves of the actin filament where it lies astride the tropomyosin strands. It is also probably present near the end of the tropomyosin molecules. The troponin units show a periodic repetitiveness along the length of the actin filament, as shown in Figure 3-11e. There is one molecule of troponin for every 7 or 8 G-actin molecules along the actin filament. The structural relationship of F-actin, tropomyosin, and troponin in the actin filament is diagrammatically illustrated in Figure 3-11e. Troponin is a calcium-ion-receptive protein, and calcium ion (Ca^{2+}) sensitivity is its major function in the actomyosin-tropomyosin complex.

α-actinin has a proline content comparable to that of actin, and it too is a globular molecule. α-actinin is present in the Z line and constitutes about 2–2.5 percent of the myofibrillar protein. It has been suggested that α-actinin functions as the cementing substance in the Z filaments.

β-actinin, which is also a globular protein, is located at the ends of actin filaments and is believed to regulate their length by maintaining a constant length of about 1 μm in each half sarcomere. In the absence of β-actinin, actin filaments *in vitro* attain lengths of 3–4 μm or more.

Very little is known about the M proteins. They are believed to constitute the substances of the M line that bind the myosin tails together in mammalian muscle, and thus they maintain the arrangement of myosin filaments. The M proteins comprise about 4 percent of the myofibrillar protein.

SARCOPLASMIC RETICULUM AND T TUBULES. In embryological origin, the *sarcoplasmic reticulum* (SR) corresponds to the endoplasmic reticulum of other cell types. The SR is a membranous system of tubules and cisternae (flattened reservoirs for Ca^{2+}) that forms a closely meshed network around each myofibril. The SR and transverse tubules (T tubules), although usually discussed together, are two separate and distinct membrane systems. The T tubules are associated with the sarcolemma, as mentioned earlier in the discussion of the sarcolemma. The SR, on the other hand, is intracellular in nature. The reticulum membranes of the SR are the storage site of Ca^{2+} in resting muscle fibers.

The SR consists of several distinct elements, and the structural features of each of these elements are discussed with reference to a single sarcomere. Relatively thin tubules, oriented in the direction of the myofibrillar axis, constitute the *longitudinal tubules* of the reticulum. In the H zone region of the sarcomere the longitudinal tubules converge, forming a perforated sheet that is called a *fenestrated* (window-like opening) *collar*. These features of the SR are shown in Figure 3-13. At the junction of the A and I bands the longitudinal tubules converge and join with a pair of larger, transversely oriented, tubular elements called *terminal cisternae*. The longitudinal tubules extend in both directions from the fenestrated collar to the terminal cisternae. The two tubular elements comprising the terminal cisternae lie parallel to each other with one tubule of the pair transversing the A band, and the other transversing the adjacent I band, of the sarcomere. A T tubule also runs transversely across the sarcomere, at the A–I band junction and lies between the two tubular elements of the terminal cisternae pair. The central T tubule and the two tubular elements of the terminal cisternae

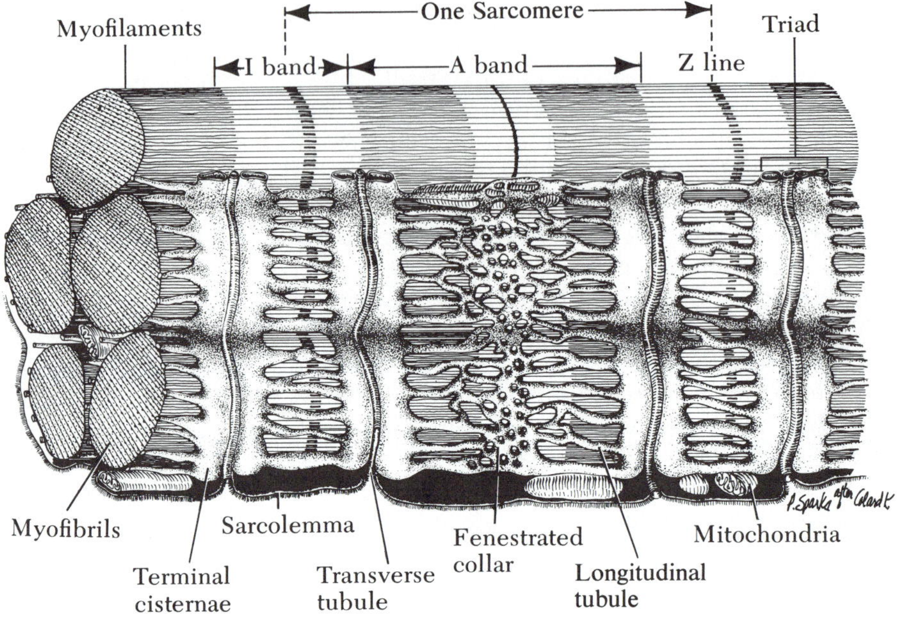

FIGURE 3-13
Diagrammatic representation of the sarcoplasmic reticulum and T tubules, and their relation to the myofibrils of mammalian skeletal muscle. [Modified after L. D. Peachey, *J. Cell Biol.* 25:209, 1965, from Bloom and Fawcett, *A Textbook of Histology*, 9th ed. W. B. Saunders Company, Philadelphia, p. 281, 1968.]

collectively form a structure known as the *triad.* The structure and orientation of the SR, T tubules, and these triads, with reference to the sarcomere, are also diagrammatically illustrated in Figure 3-13. Each sarcomere has two triads, one in each half sarcomere at the A–I band junction. The triads encircle each myofibril at the A–I band junction of mammals, birds, and in some fish. In some species the triads are located at the Z lines.

The extensiveness of the SR can be appreciated if you consider the fact that these structures are associated with each sarcomere along the entire length of the myofibril, and that the muscle fiber contains at least 1000 myofibrils. SR volume varies from one muscle fiber to another, but it is estimated as constituting approximately 13 percent of the total fiber volume. The T tubules, on the other hand, comprise only about 0.3 percent of the fiber volume.

MITOCHONDRIA. Mitochondria are oblong organelles (Figure 3-13) located in the sarcoplasm. They are frequently referred to as the "powerhouse of the cell" because they "capture" the energy derived from carbohydrate, lipid, and protein metabolism, and then provide the cell with a source of chemical energy. They contain the enzymes that the cell uses in the process of oxidative metabolism. There exists a considerable variation in mitochondrial numbers and size in muscle fibers. Skeletal muscle mitochondria are relatively abundant at the periphery of the fiber near the poles of the nuclei, and are especially abundant at the myoneural junctions. Additionally, a number of mitochondria are located between the myofibrils, adjacent either to the Z lines, the I bands, or to the A–I band junctions, (Figure 3-13).

LYSOSOMES. Lysosomes are small vesicles, located in the sarcoplasm, that contain a number of enzymes collectively capable of digesting the cell and its contents. Included among these lysosomal enzymes are a group of proteolytic enzymes known as the *cathepsins*. Several cathepsins have proteolytic effects on some of the muscle proteins, that might contribute to meat tenderization during postmortem aging.

GOLGI COMPLEX. Structures called the Golgi complex are located in the sarcoplasm near the nuclei. They consist of flattened vesicles which apparently function as the "concentrating" and "packaging" apparatus for the products from the metabolic "production line" of the cell. The muscle fiber, being multinucleated, has numerous Golgi complexes. The vesicles of the Golgi complex resemble the membranes of the sarcoplasmic reticulum.

Smooth Muscle

Although smooth muscle comprises only a small proportion of meat, a brief description of its ultrastructure is included and the principal differences from skeletal muscle are discussed. Smooth muscle is present in the greatest quantity in the walls of arteries, lymph vessels, and in the gastrointestinal and reproductive tracts.

Smooth muscle fibers vary in size and shape, depending upon their location. They are not always spindle shaped, as is usually depicted, but can be tremendously uneven in contour along their length. In cross-section, smooth muscle fibers vary from extremely flattened elipsoids to triangular and polyhedral shapes. Photomicrographs of smooth muscle fibers are shown in Figure 3-14.

The sarcolemma of smooth muscle fibers forms membrane to membrane contact bridges with neighboring fibers. The smooth muscle fiber has a single nucleus that is usually centrally located within the cell; although, in large fibers, it may be displaced slightly from the center. The sarcoplasmic reticulum (SR) is much less well developed than in

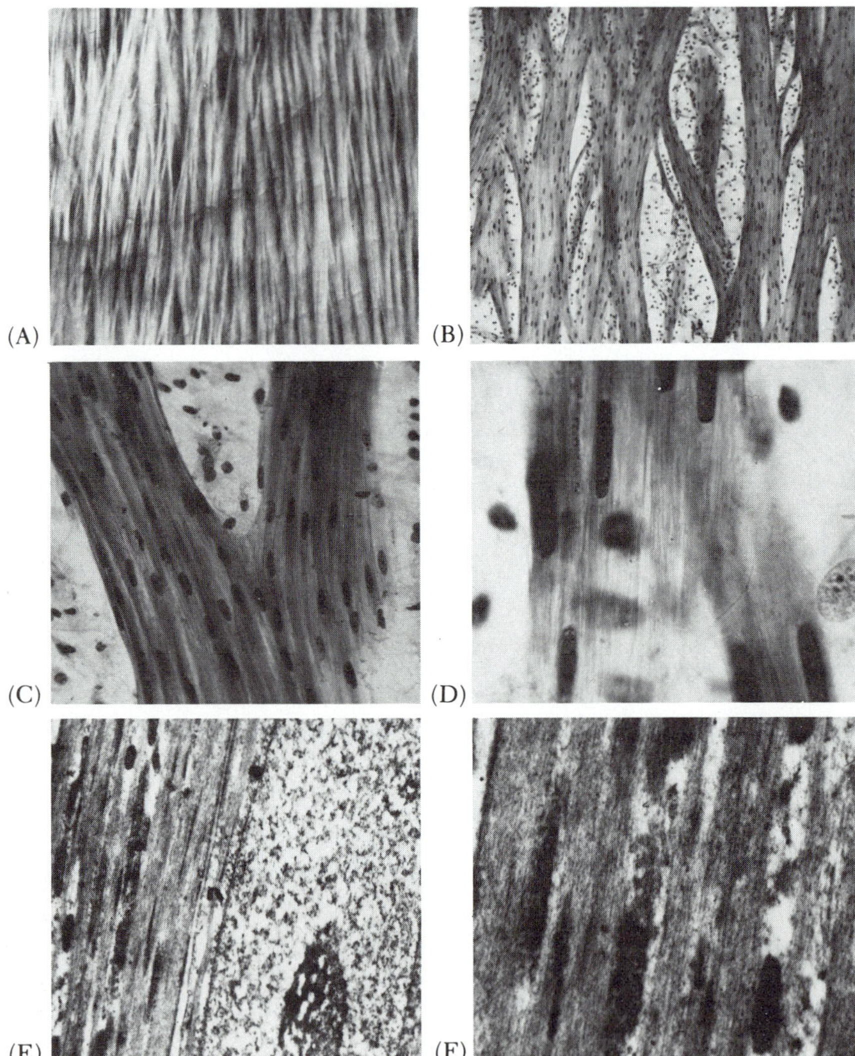

FIGURE 3-14
Photomicrographs of smooth muscle fibers from the longitudinal muscle layer of a pregnant rabbit's uterus at term. Photomicrographs (A–D) light microscopy (× 12, × 75, × 300, and × 1000, respectively) and, (E, F) electron photomicrographs (× 10,000 and × 20,000, respectively). [From, The Structure and Function of Muscle, Vol. 1. 1st ed. G. H. Bourne, ed. Academic Press, Inc., New York, p. 233, 1960; Courtesy of Dr. Arpad I. Csapo.]

skeletal muscle. The myofilaments of smooth muscle are also less well ordered than in skeletal muscle; they do not appear to be arranged in groups to form myofibrils. The myofilaments are, however, arranged in pairs that run parallel to the longitudinal axis of the fiber. In contrast to skeletal muscle, the ends of the fibers contain the densest aggregation of filaments. Among these are actin filaments that resemble skeletal muscle F-actin. Additionally, tropomyosin, troponin, and α-actinin are also associated with the actin filaments. On the other hand, myosin filaments are not visible, even though myosin is extractable from smooth muscle. Actin and myosin are present in the same proportions as in skeletal muscle, yet there are no striations.

The myofilaments of smooth muscle are attached to *dark zones* on the sarcolemma. These dark zones are evident along the surface of the sarcolemma, but they are most numerous at the ends of the fibers where myofilaments are also most abundant. These dark zones are described as being analogous to the Z lines of skeletal muscle, since they move with the contractile action. They might be responsible for transmitting the myofilamental contraction force to the sarcolemma.

Smooth muscle fibers occur either singly or in bundles, but whatever the arrangement, each is surrounded by a delicate network of reticular fibers that support and bind them in place. Scattered collagenous and elastin filaments, and an associated ground substance, occupies the narrow space between the fibers. Blood vessels and nerve fibers are associated with these connective tissues. However, smooth muscle, compared to skeletal muscle, is poorly supplied with blood.

Cardiac Muscle

Cardiac muscle possesses the unique property of rhythmic contractility which continues ceaselessly from early embryonic life until death. Cardiac muscle has properties that resemble characteristic properties of both skeletal and smooth muscle. Like smooth muscle, it generally has a single centrally placed nucleus and is innervated by the autonomic nervous system. The fibers of cardiac muscle are branched, with the main trunk being smaller in diameter and shorter than the fibers of skeletal muscle. These branches are smaller in diameter than the main fiber trunks. The sarcoplasm of cardiac muscle contains numerous glycogen granules. The mitochondria of cardiac muscle are especially large and numerous. These and other features of cardiac muscle can be seen in Figure 3-15. The myofilaments of cardiac muscle are not organized into discrete myofibrils as in skeletal muscle. Instead, aggregates of myofilaments form fibrils of extremely variable size, and the dimensions vary along the longitudinal axis of the fibrils. However, thick

(A)

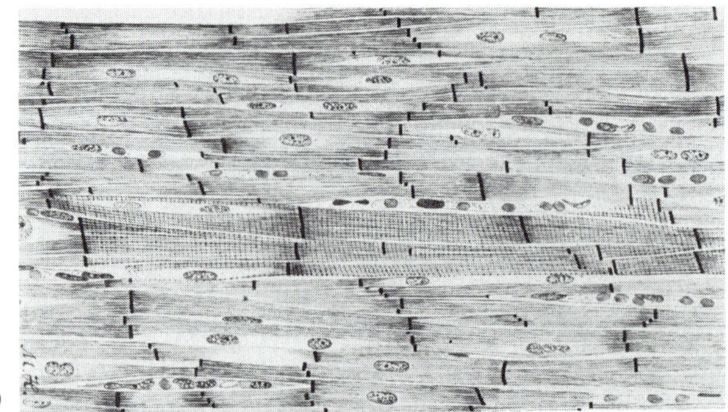

(B)

FIGURE 3-15
Photomicrographs of cardiac muscle. A longitudinal section (A), showing fiber arrangements and nuclei locations and, (B) an electron photomicrograph showing the banding pattern, intercalated disks, mitochondria, and glycogen granules. [From D. W. Fawcett, *A Textbook of Histology*, 9th ed. W. B. Saunders Company, Philadelphia, p. 285, 289, 1968.]

(myosin) and thin (actin) filaments are readily apparent under the microscope and they are aligned to give a striated appearance that is identical to that of skeletal muscle (Figure 3-15).

The T tubules of cardiac muscle are larger in diameter and occur at the Z lines, in contrast to their occurrence at the A–I band junction in mammalian and avian skeletal muscle. The elements of the SR are not only less well developed in cardiac muscle, but structures comparable to the terminal cisternae are absent.

At regular intervals along the longitudinal axis of cardiac muscle, dense lines called *intercalated disks* transect the fiber. In mammalian cardiac muscle, the intercalated disks transect the myofilaments at offset intervals across I bands of the fiber, giving a stepwise appearance to the disks (Figure 3-15). The intercalated disks are continuous across the entire fiber, and their paired membranes represent the cell membranes of adjoining muscle fibers. Thus, these disks provide a cohesive link between the fibers of the myocardium (heart muscle) as well as facilitating the transmission of the contractile force (in the direction of the fiber axis) from one fiber to another.

The myocardium is the contractile layer of the heart, and contains the bulk of the cardiac muscle. The fibers of the myocardium are held in place by reticular and collagenous fibers, which are continous with connective tissue sheaths that group them into fascicles (bundles) of fibers. Blood and lymph vessels, and nerve fibers, enter and exit the myocardium via the connective tissue between muscle bundles. Cardiac muscle is endowed principally with a capacity for oxidative metabolism; consequently, it has an extensive blood capillary network.

EPITHELIAL TISSUE

Of the four tissue types present in the animal body, quantitatively, epithelial tissue contributes the least to meat. However, the characteristic flavor and crispness of fried chicken is partly due to the properties of epithelial tissue and its underlying connective tissue. It forms the linings of the external and internal surfaces of the body and some of its organ systems, and is usually removed during the slaughter and processing operations. (Some of these tissues, such as the hide, become important by-products.) Most of the epithelial tissue remaining in meat is associated with blood and lymph vessels, and edible organs, such as the kidney and liver.

Epithelial tissue is characterized as having little intracellular material, and is classified according to cell shape and the number of layers forming the epithelium. The cell shapes vary from thin flat cells to tall columnar cells (Figure 3-16). Some cells form single layers, and others form stratified layers of cells. In either case, they maintain extensive intercellular

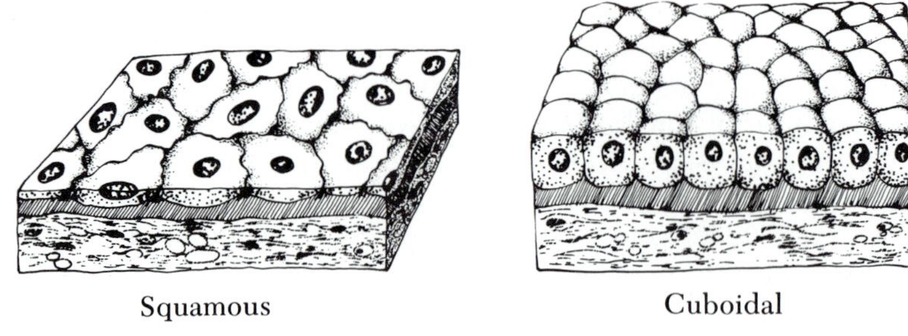

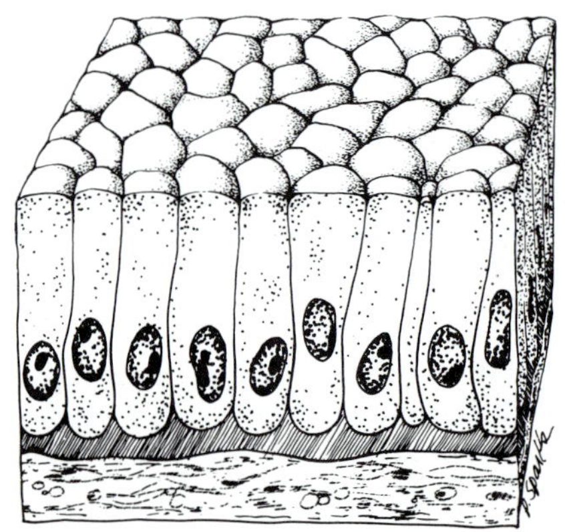

FIGURE 3-16
Diagrammatic illustration of squamous, cuboidal, and columnar epithelial tissue. [Modified from J. E. Crouch, *Functional Human Anatomy*, 2nd ed. Lea & Febiger, Philadelphia, p. 50, 1972.]

contacts, thus forming cohesive cellular sheets that cover surfaces and line cavities. Additionally, their free surfaces may be modified to increase their efficiency in performing such functions as protection, secretion, excretion, transport, absorption, and sense perception.

NERVOUS TISSUE

Nervous tissue constitutes a small proportion of meat (less than 1 percent) but its function in the period immediately prior to and during stunning and exsanguination (bleeding) during slaughter may have an important influence on meat quality. The interrelationship of nervous and muscle tissues is described in Chapters 5 and 6.

Nervous tissue is generally categorized as being part of either the *peripheral* or *central nervous systems* for the purpose of describing its structure and function. The central nervous system consists of the brain and spinal cord, and the peripheral nervous system consists primarily of nerve fibers in other parts of the body. The *neuron* comprises the bulk of nervous tissue. It consists of a polyhedrally shaped cell body and a long cylindrical structure called the *axon.* Encased within the neuron is the *neuroplasm* (cytoplasm). The neuroplasm in the axon of the neuron is frequently called *axoplasm.* Embedded in the neuroplasm of the cell body is a centrally placed nucleus, and radiating from the cell body are several short branched structures called *dendrites.* The axon terminates in a fine multi-branched structure called the *axon ending.* These features of the neuron are shown in Figure 3-17. Occasionally the axon will have one or more *collateral* branches. Each of the collateral branches also terminates in fine multi-branched axon endings.

Nerve fibers are composed of groups of the neuronal axons, and the assembly of groups of fibers into fascicles results in the formation of *nerve trunks* (Figure 3-18). In nerve fibers, neurons are oriented longitudinally in relation to each other; the axon ending of one cell interdigitates with the dendrite of another cell. The interdigitating areas of these neurons are called *synapses.* The neuron structures are not actually joined at the synapses, but they are close enough to permit chemical substances released by one neuron to influence the adjacent neuron which, in turn, influences the next, and so on. Fascicles of nerve fibers are held together by sheaths of connective tissue, and the nerve trunk itself is ensheathed in a connective tissue covering. All peripheral nerve fibers are ensheathed by *Schwann cells* and, in addition, the larger fibers are enveloped in a *myelin sheath* within the sheath of Schwann cells (Figures 3-17 and 3-18). While small peripheral nerve fibers contain a Schwann cell sheath, they are devoid of a myelin sheath. Thus, nerve fibers are frequently referred to as either *myelinated* or *unmyelinated.*

CONNECTIVE TISSUE

As the name implies, connective tissue literally connects and holds the various parts of the body together. Connective tissues are distributed throughout the body as components of the skeleton, in the framework of organs, blood and lymph vessels, as well as in sheaths that surround structures such as tendons, muscle, nerve trunks, muscle fibers, and nerve fibers. The skin or hide is attached to the body by connective tissues. Connective tissues also provide the body with a barrier against infective agents, and they are vitally important in wound healing. The *connective tissue proper* envelopes muscle fibers and bundles, and finally, the muscles themselves. The connective tissue proper and adipose tissue contribute to qualitative and quantitative muscle properties.

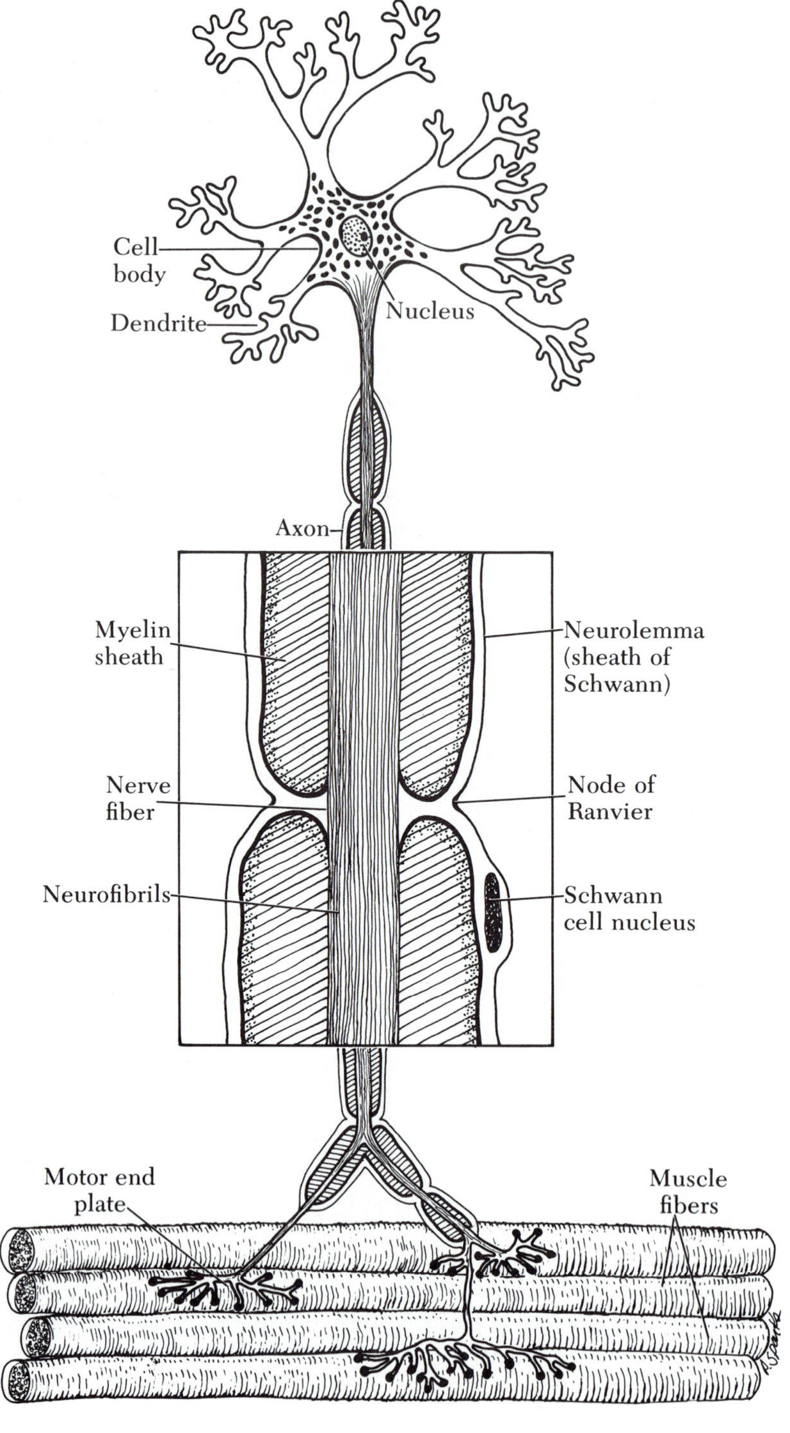
Cell body
Nucleus
Dendrite
Axon
Myelin sheath
Neurolemma (sheath of Schwann)
Nerve fiber
Node of Ranvier
Neurofibrils
Schwann cell nucleus
Motor end plate
Muscle fibers

FIGURE 3-17 (*opposite*)
Diagrammatic illustration of a neuron and the motor end plates associated with it. The magnification of the axon (center) shows details of the nerve fiber, neurofibrils, myelin sheath, Schwann cell nucleus, and node of Ranvier. [Modified from J. E. Crouch, *Functional Human Anatomy.* 2nd ed. Lea & Febiger, Philadelphia, p. 480, 1972.]

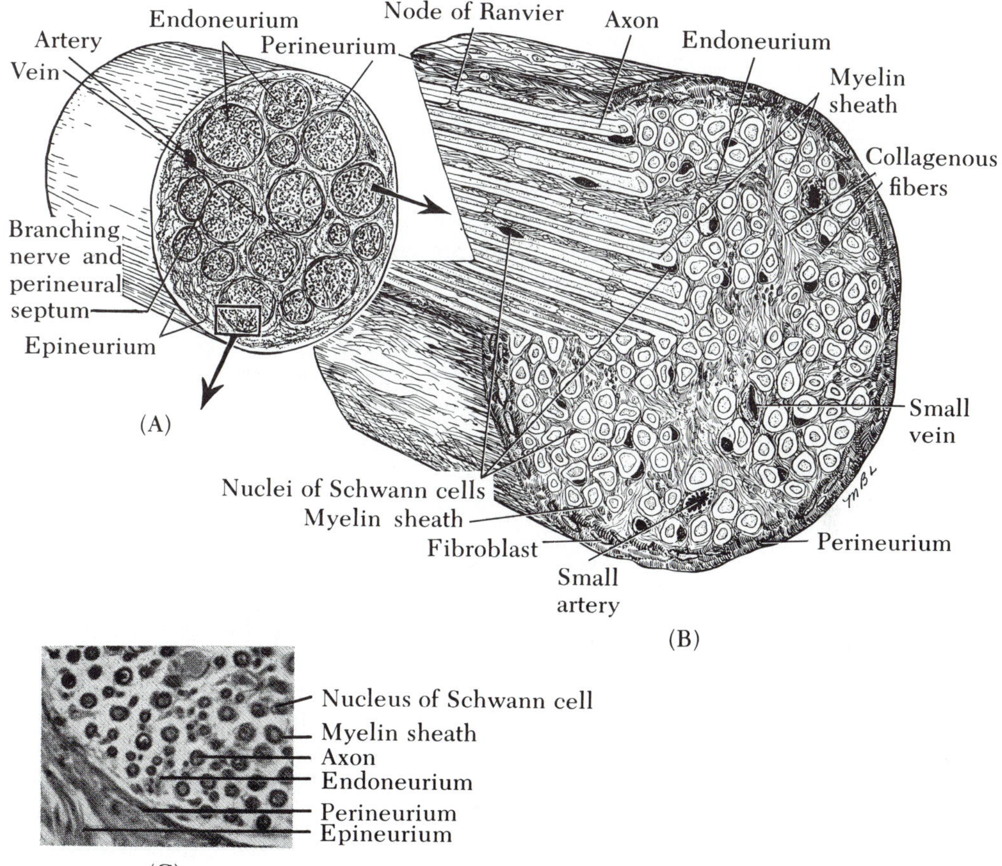

FIGURE 3-18
Microscopic anatomy of a peripheral nerve. (A) Cross section of a nerve (× 40). (B) A nerve fascicle, shown in longitudinal and cross section. Myelin sheaths are shown in white. (C) Photomicrograph of a cross section of a fascicle (bundle) of a myelinated nerve (× 600). (Staining with osmic acid has made the myelin sheath black.) Greatly magnified. [From J. E. Crouch, *Functional Human Anatomy.* 2nd ed. Lea & Febiger, Philadelphia, p. 485; 1972.]

The connective tissues are generally characterized as having relatively few cells, and considerable extracellular substance. This *extracellular substance* of the connective tissue proper varies from a soft jelly to a tough fibrous mass, but it always contains fibers embedded in it that provide the structural elements of the connective tissues. In cartilage, the intercellular substance has a rubbery consistency, but in bone it is much tougher and is impregnated with calcium salts. Both cartilage and bone contain fibers, but they are masked by the intercellular substance. In blood and lymph, the intercellular substance is a fluid, and there are no fibers.

In contrast to the connective tissue proper, which has few cells, the bulk of adipose tissue is composed mainly of cells. In blood, lymph, cartilage, and bone, cells constitute an intermediate proportion of the total tissue volume.

Since the connective tissue proper is characterized by the presence of distinct fibers, it is usually referred to as *fibrous connective tissue*. Bone and cartilage, on the other hand, are considered *supportive connective tissues* because other tissues are attached to them, thus providing the body with structural support. Although the preceding generalizations about connective tissues might indicate that they are quite different from each other, they have many similarities in composition and function. Fibers of the connective tissue proper are sometimes continuous with bone and cartilage, and some chemical compounds are common to a number of connective tissues.

Connective Tissue Proper

The connective tissue proper consists of a structureless mass, called the *ground substance*, in which the cells and extracellular fibers are embedded. The extracellular fibers include those of *reticulin*, *collagen*, and *elastin*. (The detailed structure of each is described later in this section.) The connective tissue proper contains two different populations of cells, which are referred to as either *fixed* or *wandering cells*. Fixed cells include fibroblasts, undifferentiated mesenchyme cells, and specialized adipose (fat storage) cells that are derived from the latter. Wandering cells are primarily involved in the injury reaction; they include eosinophils, plasma cells, mast cells, lymph cells, and free macrophages. The extracellular components (ground substance and fibers) are produced and maintained by the cellular component, the fibroblasts (Figure 3-20), and are defined in the next two sections.

GROUND SUBSTANCE. The ground substance is a viscous solution containing soluble glycoproteins (carbohydrate containing proteins). These glycoproteins are usually referred to as *mucoproteins* or *muco-*

polysaccharides. The ground substance also contains the substrates and end products of connective tissue metabolism. Included among these are the precursors of collagen and elastin, *tropocollagen* and *tropoelastin,* respectively. Notable among the mucopolysaccharides are *hyaluronic acid* and the *chondroitin sulfates.* Hyaluronic acid is a very viscous substance found in joints (synovial fluid) and between connective tissue fibers. The chondroitin sulfates are found in cartilage, tendons, and adult bone. These two mucopolysaccharides and associated proteins function as lubricants, intercellular cementing substances, and structural matter in cartilage and bone. They also provide a barrier against infectious agents.

EXTRACELLULAR FIBERS. The arrangement of the extracellular fibers in densely packed structures is referred to as *dense connective tissue*, and those forming a loosely woven network are called *loose connective tissue.* Dense connective tissue may be further characterized by fiber arrangement. In *dense irregular connective tissue* the fibers are densely interwoven, but in a random arrangement. However, in *dense regular connective tissue* the fibers are arranged in bundles lying parallel to each other, such as in tendons; or in flat sheets, such as those present in *aponeuroses* (tendinous extensions of the connective tissues surrounding muscle), as shown in Figure 3-28. The extracellular fibers include *collagen*, *elastin*, and *reticulin.*

Collagen is the most abundant protein in the animal body and significantly influences meat tenderness. In most mammalian species it constitutes 20–25 percent of the total protein. Collagen is the principal structural protein of connective tissue, and is a major component of tendons, ligaments, and to a lesser extent that of bones and cartilage. Additionally, networks of collagen fibers are present in essentially all tissues and organs, including muscle. The distribution of collagen is not uniform among skeletal muscles, but the amount present generally parallels their physical activity. Muscles of the limbs contain more collagen than those of the back and, consequently, the former are less tender than the latter.

Collagen is a glycoprotein that contains small quantities of the sugars galactose and glucose. Glycine is the most abundant amino acid in collagen, and comprises about one-third of the total amino acid content. Hydroxyproline and proline account for another one-third of the amino acid content of collagen, as shown in Figure 3-19. The hydroxyproline content is a relatively constant component of collagen (13–14 percent) and does not occur (to any significant extent) in other animal proteins. Thus, chemical assays for it are commonly used to determine the amount of collagen in tissues.

The *tropocollagen* molecule, the structural unit of the collagen fibril, is formed by the alignment of tropocollagen molecules in an overlapping

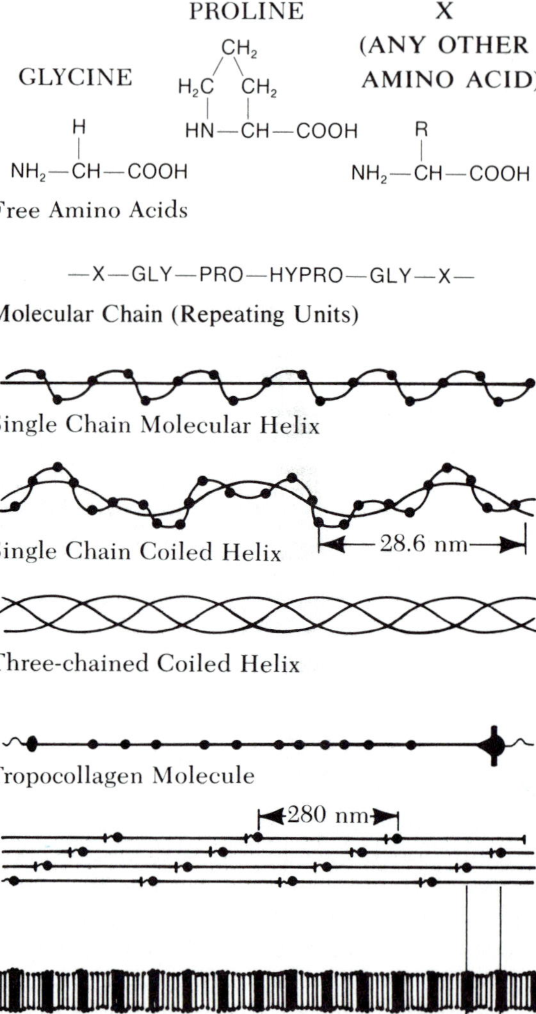

FIGURE 3-19
Diagrammatic illustration of the amino acid sequence and molecular structure of collagen and tropocollagen molecules, and of collagen fibril formation. [From J. Gross "Collagen." Copyright © 1961 by Scientific American, Inc. All rights reserved.]

fashion. As shown in Figure 3-19, each tropocollagen molecule overlaps its lateral counterpart by one-fourth of its length, thus explaining the fibril's striated appearance. The sequence of events in collagen fibril formation is diagrammatically illustrated in Figure 3-20. Collagen fibers are composed of a variable number of collagen fibrils aligned in a parallel array. The size of these fibers depends upon their source. They are almost completely inextensible. Individual fibers are colorless, but aggregations, such as those in muscle sheaths or in tendons, are white in color.

The relative insolubility and high tensile strength of collagen fibers result from *intermolecular cross linkages.* These cross linkages are fewer in number, and more easily broken, in young animals. As the animal grows older the number of cross linkages increases, and the easily broken linkages are converted to stable linkages. Coincidentally, collagen is more soluble in young animals, and becomes less soluble as the animal ages.

Elastin is a much less abundant connective tissue protein than collagen, and its ultrastructural characteristics are not as well known. Elastin is a rather rubbery protein that is present throughout the body in ligaments and the walls of arteries, as well as in the framework of a

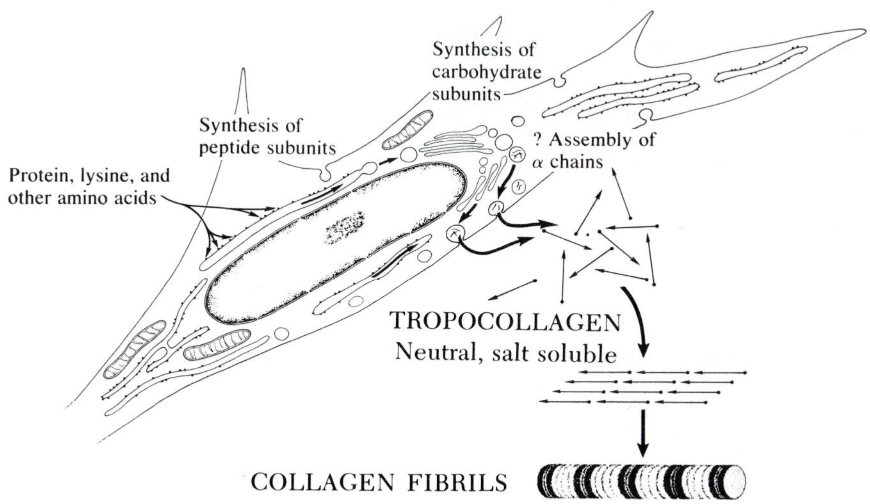

FIGURE 3-20
Diagrammatic illustration of the fibroblast, including the major intracellular and extracellular events in the synthesis of tropocollagen and its extracellular formation into the collagen fibril. As indicated in the diagram, protein and lysine and other amino acids are associated with the ribosomes of the endoplasmic reticulum. [From D. W. Fawcett, *A Textbook of Histology*, 9th ed. W. B. Saunders Company, Philadelphia, p. 145, 1968.]

number of organs (including muscle). The *cervical ligament* (*ligamentum nuchae*) supporting the neck of ruminants is a large strand of elastin fibers arranged essentially in a parallel fashion. Elastin fibers are easily stretched and, when the tension is released, they return to their original length.

Although the amino acid composition of elastin differs from that of collagen, glycine is still present in the greatest quantity and at a comparable percentage. Elastin contains about 8 amino acid residues of each of two unique amino acids, desmosine and isodesmosine, per 1000 residues. These two amino acids are synthesized from four lysine molecules. The extreme insolubility of elastin is largely attributable to its high content (about 90 percent) of nonpolar amino acids and its desmosine cross links. The desmosine linkage binds 4 neighboring elastin chains together by intermolecular cross-linkages between their lysine molecules.

Aggregations of elastin fibers, such as in the *ligamentum nuchae*, have a characteristic yellow color. Elastin is highly resistant to digestive enzymes, and thus contributes little or nothing to the nutritive value of meat. As indicated above, elastin is extremely insoluble, and no method of cookery has any appreciable solubilizing effect upon it.

Reticulin is composed of small fibers that form delicate networks around cells, blood vessels, neural structures, and epithelium which hold them in place. During embryonic development, reticular fibers are the first to appear in the differentiation of loose connective tissue, but eventually they are numerically overwhelmed by increasing numbers of collagenous fibers. It should be emphasized, however, that reticular fibers are present, and continue to serve the same function in the adult animal as described above.

Connective Tissue Cells

Of the several kinds of cells found in connective tissue, only the fibroblasts, undifferentiated mesenchymal cells, and adipose cells are discussed in this chapter because of their immediate relevance to the properties of meat.

FIBROBLASTS. Fibroblasts vary from spindle to star shapes, but, in general, they are long and slender. Fibroblasts synthesize the precursors of the extracellular components of connective tissue, namely tropocollagen, tropoelastin, and the ground substance. These connective tissue proteins are synthesized within the fibroblast and released into the extracellular matrix. The construction of collagen and elastin fibers from tropocollagen and tropoelastin subunits, respectively, occurs extracellularly in the connective tissue matrix. The synthesis of tropocollagen within the fibroblast, and the extracellular formation of fibrils, is diagrammatically illustrated in Figure 3-20. The construction of elastin probably follows a sequence similar to that of collagen.

UNDIFFERENTIATED MESENCHYMAL CELLS. The mesenchymal cells are spindle shaped cells that are slightly smaller in size than fibroblasts. They are undifferentiated cells that can become one of several different cell types, depending upon the specific stimulus. Those which accumulate lipids are the precursors of adipose cells. It should be emphasized that fibroblasts *per se* do not give rise to adipose cells. The mesenchymal cells that are the precursors of adipose cells are located in the loose connective tissue matrix close to blood vessels. When these primitive cells begin to accumulate lipids they are known as *adipoblasts*; as they continue to accumulate fat they eventually become adipose cells (*adipocytes*). These events are shown diagrammatically in Figure 3-21.

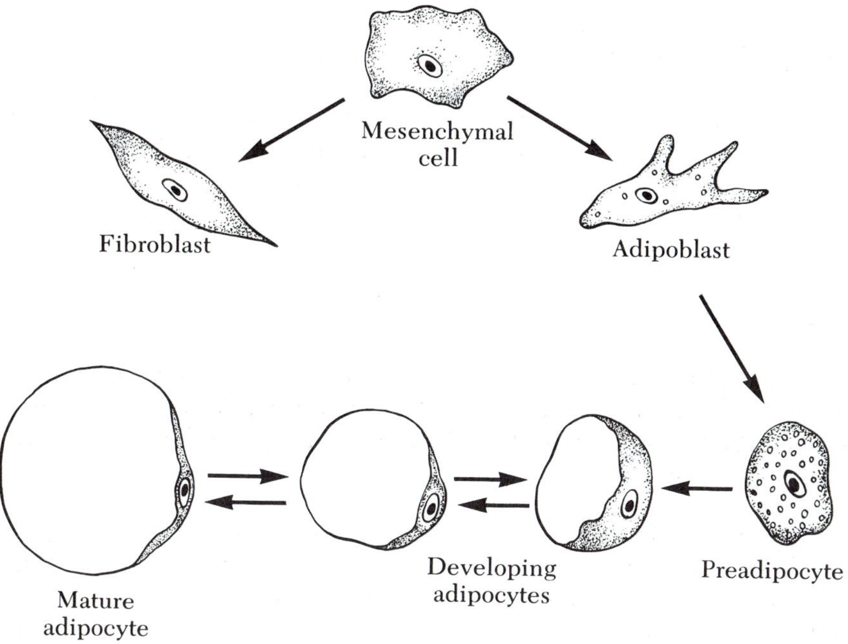

FIGURE 3-21
Diagrammatic representation of adipocyte development.

Adipose Tissue

The accumulation of numerous adipocytes, to the extent that they are the predominant cell type, results in the formation of *adipose tissue*, also known as fat. Many such sites, known as adipose tissue depots, are present in the animal body; certain specific sites are discussed in subsequent chapters. The composition of several adipose tissue depots of meat animals, and the approximate fat content of beef and sheep carcasses at various levels of fatness, are presented in Tables 3-3 and 3-5, respectively.

In many animal species the body contains two types of adipose tissue; *white* and *brown fat*. Most of the adipose tissue in the bodies of meat animals is of the white type, but brown fat is present in all of them at birth, especially around the kidneys, and persists in some mammals, even in the adult stage. Brown adipocytes are smaller in size than those of white fat. The color of brown fat is due to the high content of cytochrome in the mitochondria of these cells. Most brown fat disappears within a few weeks after birth, or it may possibly be converted to white fat.

Cartilage

The specialized connective tissues, cartilage and bone, constitute the supportive elements and the skeleton, respectively, of the animal body. During embryonic development, most of the skeleton is formed as cartilage and is converted to bone later. Much bone growth occurs through the cartilage intermediary. However, not all cartilage is converted to bone, and some, such as the articular disks of bone joints, persists throughout the life of the animal.

Cartilage is composed of cells and extracellular fibers that are embedded in a gel-like matrix. Cartilage cells are called *chondrocytes* and they are isolated in small cavities within the extracellular matrix. The chondrocytes tend to cluster in small groups in the matrix that fill the cavities they occupy. These features of cartilage are shown in Figure 3-22. Interlacing collagen fibrils form a fine network in the cartilage matrix. These fibrils constitute as much as 40 percent of the cartilage; but, they are normally not visible since they are completely surrounded by the matrix.

Cartilage differs in the relative amount of collagenous and elastin fibers, and in the abundance of the matrix in which the fibers are embedded. These differences allow cartilage to be categorized as *hyaline*, *elastic*, or *fibrocartilage*. Hyaline cartilage is the most abundant type in the body, and the other two types are merely modifications of it. Hyaline

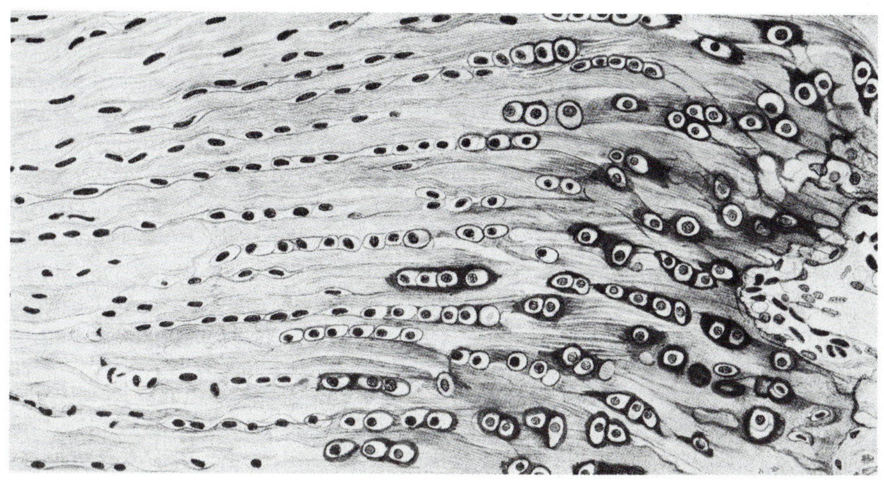

FIGURE 3-22
Diagrammatic representation of fibrocartilage at the insertion of tendon into bone. [From D. W. Fawcett, *A Textbook of Histology*, 9th ed. W. B. Saunders Company, Philadelphia, p. 219, 1968.]

cartilage is found on the surfaces of bones at joints, the ventral ends of the ribs (costal cartilages), the dorsal tips of the thoracic vertebrae (buttons), and as the cartilaginous tissue associated with the lumbar and sacral vertebrae. It is elastic, semitransparent, and has a bluish white tint. Elastic cartilage has a yellow tint, is less transparent and more flexible (or elastic) than hyaline cartilage. Its matrix has a number of branched elastin fibers. Elastic cartilage is a component of the epiglottis and internal and external portions of the ear. Fibrocartilage is characterized by the presence of numerous collagenous fibers, and is associated with the attachment of tendons to bone, and with the capsules and ligaments of joints.

Bone

Bone, like the other connective tissues, contains cells, fibrous elements, and the extracellular matrix (ground substance). Unlike the other connective tissues, the extracellular matrix is calcified. This provides the rigidity and protective properties characteristic of the skeleton. Bones are storage sites for calcium, magnesium, sodium, and other ions. Bone is a dynamic tissue, as it is continually being renewed and reconstructed.

In long bones, such as in the femur, the long central shaft, called the *diaphysis*, is a hollow cylinder of compact bone. On both ends of the diaphysis are enlargements of the bone, called the *epiphyses*. The entire bone is covered with a thin membrane of specialized connective tissue called the *periosteum*. The periphery of the epiphysis is covered with a thin layer of compact bone, but the inside is composed of spongy bone. These features of bone are illustrated in Figure 4-6. Articulating surfaces of the epiphyses are covered with a thin layer of hyaline cartilage, called the *articular cartilage*. The joint formed between the articular cartilage of adjacent bones contains the specialized connective tissue substance known as synovial fluid. The cartilaginous region separating the diaphysis and epiphysis of growing animals is called the *epiphyseal plate*. The cartilage of the epiphyseal plate lies adjacent to a transitional region of spongy bone in the diaphysis called the *metaphysis* (Figure 4-6). The spongy bone of the metaphysis and the adjacent epiphyseal cartilage constitute the region of the bone where growth in length occurs. The central interior of the diaphysis contains marrow which, in the young animal, is primarily red marrow. As the animal grows older, the red marrow gradually is converted into yellow marrow.

In compact bone, the calcified bone matrix is deposited in layers. Cavities, each of which contains a bone cell, are distributed throughout the matrix. Small tubular elements called *canaliculi* radiate in all directions forming networks of canals between the cell cavities. It is believed that the network of canaliculi is important in the bone cell nutrition. Three different kinds of bone cells are recognized, namely osteoblasts, osteocytes, and osteoclasts. The *osteoblasts* are involved in the formation of bone tissue, and they are present in the advancing regions of developing and growing bone. *Osteocytes* are the principal cells of fully formed bone, and they are located in the cell cavities within the calcified matrix. *Osteoclasts* are primarily connected with the process of bone resorption.

The two principal components of the bone matrix are the organic matrix and inorganic salts. The organic matrix consists of a ground substance that contains protein–polysaccharide molecular complexes. Embedded in the ground substance matrix are numerous collagenous fibers. The inorganic component of bone matrix consists primarily of calcium phosphate salts. Crystals of these calcium salts are deposited in the organic matrix between the collagen fibers.

Blood and Lymph

Blood and lymph, and their respective vessels, are derived from connective tissues. It has already been noted that connective tissues are present in the walls of the blood and lymph vessels. Blood provides a medium, both for nutrient (including oxygen) transport to the cell for metabolism and, for the removal of waste by-products. Blood also serves as a defense mechanism against infectious agents and as a regulator of pH and fluid balance.

Blood consists of a fluid medium, the plasma, in which the various cellular components are suspended. The blood cells are generally classified as *platelets*, *red cells* (erythrocytes), and *white cells* (leukocytes). These cells constitute approximately 4 percent of the total blood volume in meat animals. The primary function of the erythrocytes is that of carrying oxygen. The function of the leukocytes within the blood stream is largely unknown. Outside the vascular system, some leukocytes are *phagocytic* (digest foreign matter) and others function in immune reactions. Platelets participate in the blood clotting mechanism. The erythrocytes contain the pigment *hemoglobin*, which gives blood its red color, and is directly responsible for the oxygen transport to the cells. Blood normally constitutes about 7 percent of the body weight of mammals. The color of meat is largely due to the presence of the sarcoplasmic protein *myoglobin* (muscle pigment), but hemoglobin also contributes to the color because, even with ideal slaughtering techniques, a small amount of blood remains in the muscles.

Lymph is the fluid found within the lymphatic vessels. The cells in lymph are predominately lymphocytes, although a few erythrocytes and leukocytes may be present. Lymph circulates continually, and passes through all the tissues, organs, and lymph nodes of the body into the lymph of the thoracic duct, and ultimately into the blood stream.

MUSCLE ORGANIZATION AND CONSTRUCTION

Thus far, the discussion of skeletal muscle has dealt with the muscle fiber, its organelles, and their relationship to overall functional properties. In the skeletal muscles, groups of muscle fibers are bound together by connective tissues (Figure 3-3). A small amount of epithelial and nervous tissue is also associated with skeletal muscles. The organization, role, and function of each of these tissues in muscle is the topic of the following section.

Muscle Bundles and Associated Connective Tissues

In structural terms, muscle is composed of many individual fibers that are grouped together into bundles. The number of fibers varies from one bundle to another; consequently, bundle size is variable. A number of bundles, in turn, are grouped in various patterns to form a muscle. In cross-section the bundles appear as irregularly shaped polygons (Figure 3-3). They assume these shapes largely because of mutual deformation, but the pressure of the *connective tissue septa* (partitions) between adjacent bundles also affects their shape. The sizes of the bundles, and the thickness of their connective tissue septa determine the texture of muscle; those with small bundles and thin septa have a fine texture, and muscles with large bundles and thicker connective tissue septa have a coarse texture. The finer the texture, the greater is the precision of movement in the living muscle. In postmortem muscle, fine texture is generally associated with more tender meat.

The sarcolemma (outer cell membrane) of individual muscle fibers is surrounded by the *endomysium*, a delicate connective tissue covering that is shown in Figures 3-3 and 3-23. It should be emphasized that the sarcolemma and endomysium are two separate and distinct structures, even though both encase the muscle fiber. The sarcolemma is a char-

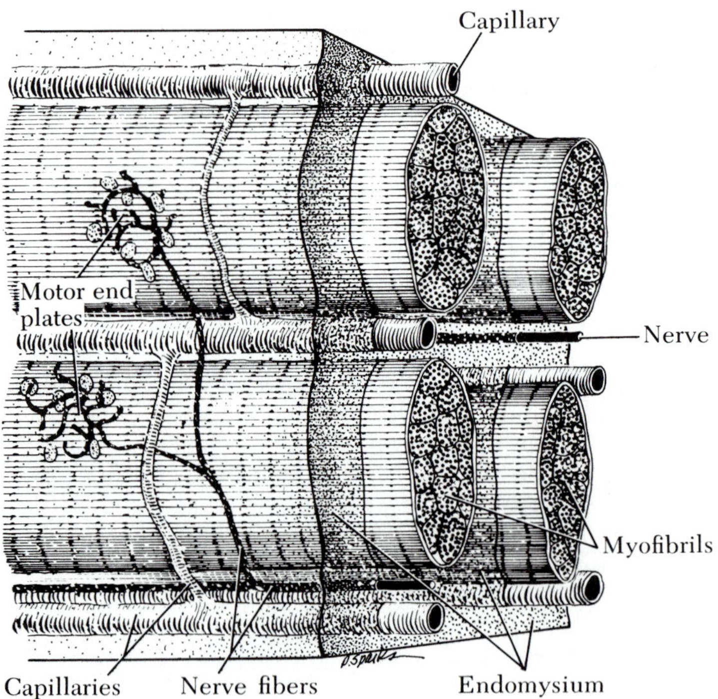

FIGURE 3-23
Diagrammatic illustration of the structure of a primary muscle bundle, showing the relationship of fibers, endomysium, blood vessels, and nerves. [Modified from J. E. Crouch, *Functional Human Anatomy*, 2nd ed. Lea & Febiger, Philadelphia, p. 183, 1972.]

FIGURE 3-24
Photomicrograph of a bovine *semitendinosus* muscle showing muscle fibers, primary and secondary bundles, and the associated endomysial and perimysial connective tissues (× 50). [From American Meat Institute, *The Science of Meat and Meat Products*. W. H. Freeman and Company. Copyright © 1971.]

acteristic lipid-protein unit-membrane structure, and the endomysium is composed of a protein-polysaccharide coating. It includes a delicate network of connective tissue fibers composed of a few fine collagen fibers and numerous reticular fibers.

Approximately 20–40 muscle fibers, and the associated endomysium, are grouped into structures called the *primary bundles*. A variable number of primary bundles are grouped together to form larger bundles known as the *secondary muscle bundles*. Both the primary and secondary bundles are ensheathed in the *perimysial connective tissue* (Figures 3-3 and 3-24). The *perimysium* consists of a sheath of collagenous connective tissue. Finally, a variable number of secondary bundles are grouped together to form a *muscle*, which is, itself, contained in a connective tissue sheath called the *epimysium*. The collagen fibers of the epimysium are continuous with those of the perimysial septa surrounding the individual bundles which, in turn, are continuous with the collagen and reticular fibers of the endomysial septa that surround the

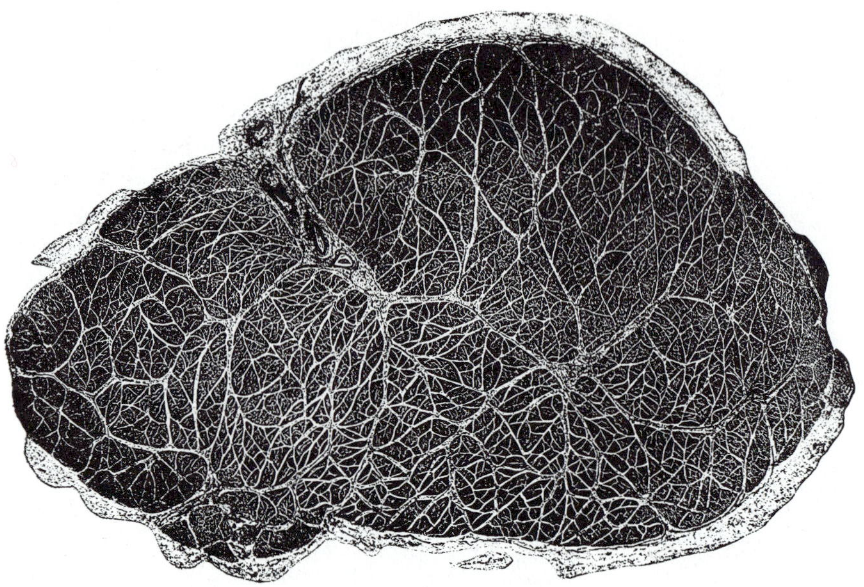

FIGURE 3-25
A photomicrograph of a human *sartorius* muscle in cross section, showing its subdivision into bundles of various sizes, and the endomysial, perimysial, and epimysial connective tissues (× 3). [From D. W. Fawcett, *A Textbook of Histology*, 9th ed. W. B. Saunders Company, Philadelphia, p. 271, 1968.]

individual muscle fibers. These connective tissue septa bind the fibers and bundles together and they support the associated nerves and blood vessels. A few fine elastin fibers are frequently found associated with the collagenous and reticular fibers in these connective tissue sheaths. These features of muscle are illustrated in Figures 3-3 and 3-25. The individual muscles in a group, such as in limbs, may have interconnections in the form of loose connective tissue septa between the epimysium or bundles of adjacent muscles.

NERVE AND VASCULAR SUPPLY. Each muscle is supplied with at least one artery and vein, as well as with nerve fibers from a nerve trunk (Figure 3-2). The latter contains both sensory and motor fibers. The nerve fibers and blood vessels supplying the muscle are associated with the epimysium and they enter its interior via the perimysium. In turn, the branches that arise from these nerves and blood vessels supply the individual muscle fibers, and are supported by the endomysium (Figure 3-23). The finer arterioles and venules are oriented transversely to the muscle fibers, and the bulk of the capillaries are arranged parallel to the long axis of the fibers. Each fiber is supplied with a rather ex-

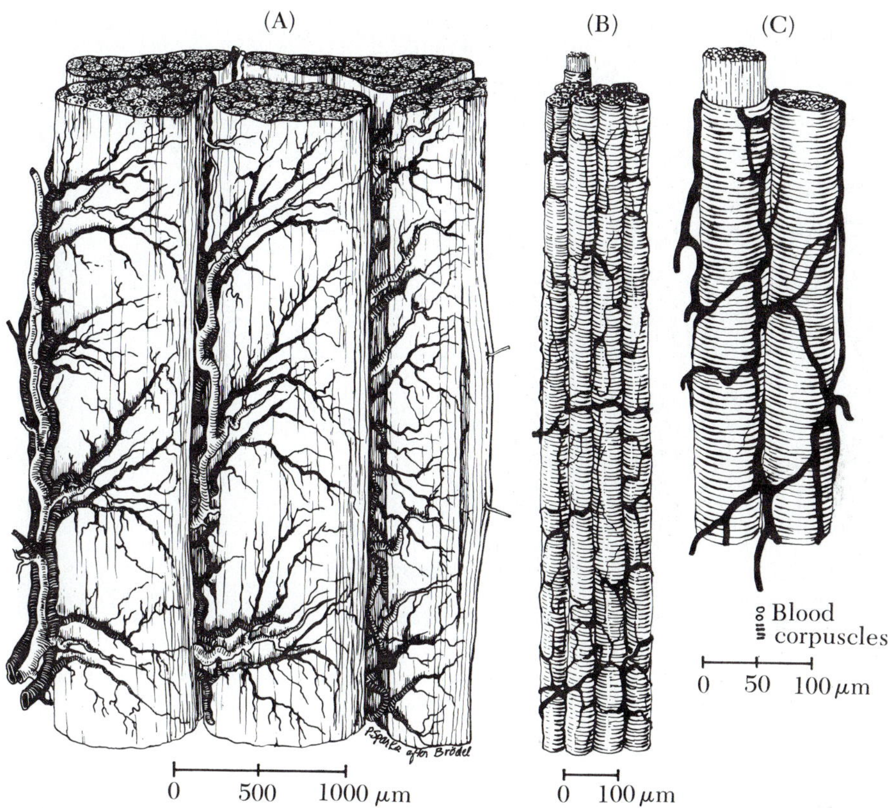

FIGURE 3-26
A diagrammatic illustration of the vascular supply to, and network associated with, skeletal muscle fibers. (A) The vascular network that surrounds the muscle fiber bundles. (B) The capillary network supplying muscle fibers. (C) A close up view of the capillary network and two muscle fibers. [After M. Brödel, Johns Hopkins Hosp. Bull. 61:295, 1937. © The Johns Hopkins University Press.]

tensive vascular bed composed principally of these longitudinally oriented capillaries, lying in the endomysium, with occasional transverse branches encircling them, as shown in Figure 3-26. This arrangement of blood vessels facilitates exchange of the nutrients and waste products of muscle fiber metabolism. Like blood vessels, the lymph vessels enter and exit the muscle via the connective tissue septa. Similarly, the nerve fibers and their branches enter and leave the muscle along the avenues of the connective tissue septa. As a nerve fiber approaches the interior of a muscle, its myelin sheath is lost. The nerve axon, its associated Schwann sheath, and the connective tissue sheath are found in, and become continuous with, the endomysium surrounding the muscle fibers. The axon endings make contact with the muscle

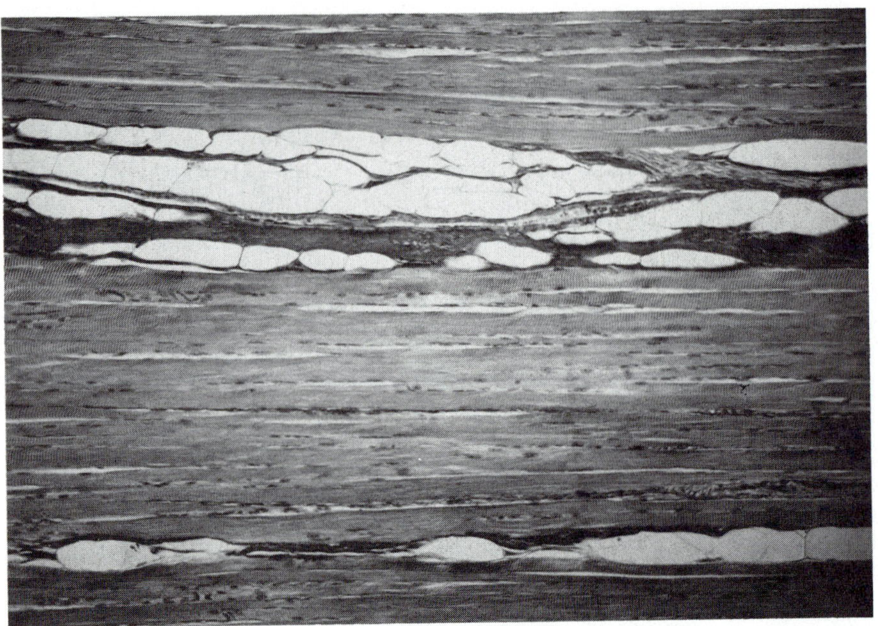

FIGURE 3-27
A photomicrograph (× 175) of porcine *semitendinosus* muscle, showing the interfascicular (bundle) fat depots or marbling fat (white vacular areas).

fiber at the myoneural junction (Figure 3-23). Each nerve fiber may branch and innervate numerous muscle fibers.

Skeletal muscle also contains *muscle spindles*, unique sensory organs that function as strain gauges, that provide the central nervous system with information about the position, movements, and tension within a muscle. These spindles are composed of a group of muscle fibers and their nerve endings. The specialized muscle fibers in the interior of the spindle, usually numbering about six, are called *intrafusal fibers.* The entire spindle complex is enclosed in a common connective tissue sheath.

Intramuscular fat, called marbling in meat, is deposited within the muscle in the loose networks of the perimysial connective tissue septa in close proximity to blood vessels, as shown in Figure 3-27. Likewise, the connective tissue septa between individual muscles contains variable quantities of adipose tissue. This latter type of fat deposit is referred to as *intermuscular fat* or *seam fat.*

MUSCLE–TENDON JUNCTION. The structural details of how the force of the contracting myofibrils is transmitted to the muscle tendon are unresolved. However, it is believed that the ends of the myofibrils are connected to tendon fibrils. As the myofibrils approach the tapered ends of the muscle fiber the characteristic banding pattern is lost, and the myofibrils become continuous with strands of noncontractile fibers (Figure 3-28). Tendon fibrils are inserted in the sarcolemmal membrane, opposite the point at which the noncontractile connecting fibers of each myofibril are attached. This entire *myotendinal* structure is continuous with the tendon. The epi-, peri-, and endomysial network of connective tissue sheaths encases the muscle; in effect, its bundles and fibers form a harness that transmits the contraction force to the tendon. The tendons, in turn, are attached to bones, and these tendinal attachments, as well as the myotendinal junctions, are referred to as *aponeuroses*. However, some muscles are attached by a connective tissue *fasciae* to other structures, such as skin, ligaments, and other muscles. This fasciae is composed primarily of white collagenous fibers with a few elastin fibers, and forms a sheet of connective tissue that replaces the function of the tendon in aponeuroses. Some muscles are actually attached to bones by fasciae.

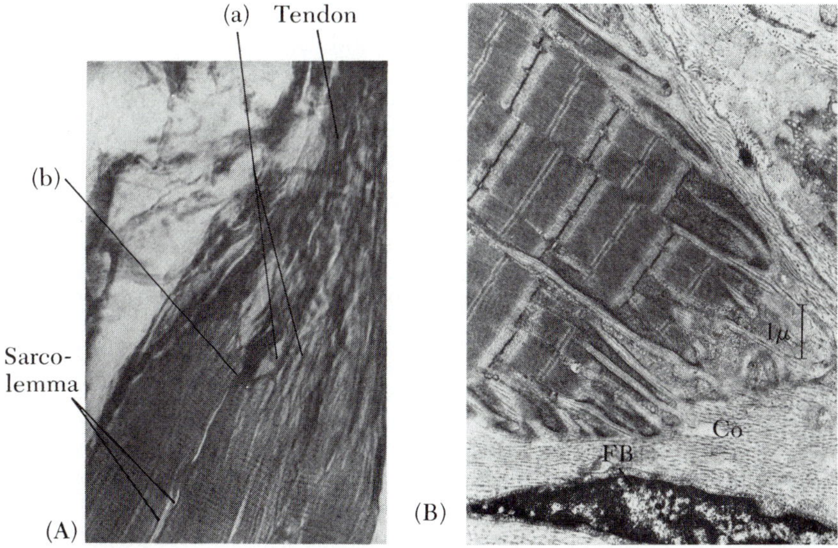

FIGURE 3-28
Photomicrographs of the muscle–tendon junction. (A) Visible light photomicrograph showing a muscle–tendon connection (× 500). (B) An electron photomicrograph of the myotendinal junction, showing the relation of the sarcomere to tendon collagen fibers. [(A) From Maximow and Bloom, *Textbook of Histology*, 6th ed. W. B. Saunders Company, Philadelphia, p. 149, 1952. (B) From D. S. Smith, Muscle. Academic Press, Inc., New York, p. 56, 1972; Courtesy of Dr. D. S. Smith.]

Muscle and Fiber Types

Muscles are usually classified as red or white, based principally on their color intensity, which is attributable to the proportion of *red* and *white fibers* they contain. It should be emphasized that few muscles are composed of all red or all white fibers, but most are mixtures of red and white fibers. Additionally, most muscles of meat animals contain a higher proportion of white fibers than red fibers, even in muscles that are visibly red. Thus, red muscles are those with a higher proportion of red fibers than found in white muscles; or, alternatively, white muscles have fewer red fibers than do red muscles. A few muscles, such as the *semitendinosus* in pigs, exhibit regions that are predominately red, while other portions of the same muscle are distinctly white. Fibers with characteristics that are intermediate between those of the red and white types also exist in muscle, and are called, naturally enough, the intermediate fiber type.

The structural, functional, and metabolic characteristics of red, intermediate, and white muscle fibers are different, and some of these features are presented in Table 3-1. We want to emphasize that these differences are relative, and that considerable variation exists for each characteristic within each fiber type; thus, some overlap between types occurs. Photomicrographs of red, white, and intermediate fiber types are shown in Figure 3-29. In most mammalian species, red and white fibers

TABLE 3-1
Characteristics of red, intermediate, and white muscle fibers in domestic meat animals and birds*

Characteristic	Red Fiber	Intermediate Fiber	White Fiber
Color	red	red	white
Myoglobin content	high	high	low
Fiber diameter	small	small – intermediate	large
Contraction speed	slow	fast	fast
Contractile action	tonic	tonic	phasic
Number of Mitochondria	high	intermediate	low
Mitochondrial size	large	intermediate	small
Capillary density	high	intermediate	low
Oxidative metabolism	high	intermediate	low
Glycolytic metabolism	low	intermediate	high
Lipid content	high	intermediate	low
Glycogen content	low	high	high

*The characteristics are relative to the other fiber types.

(A)

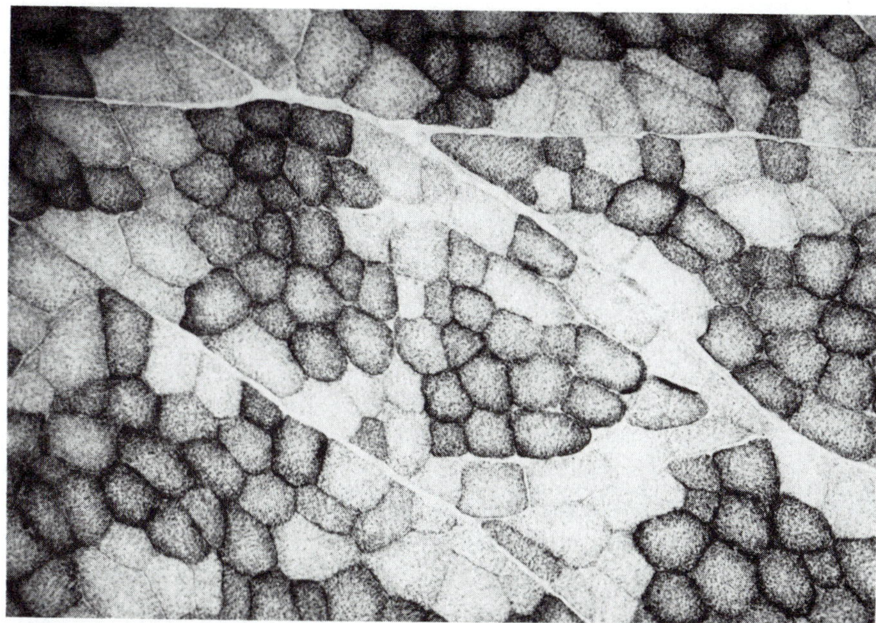

(B)

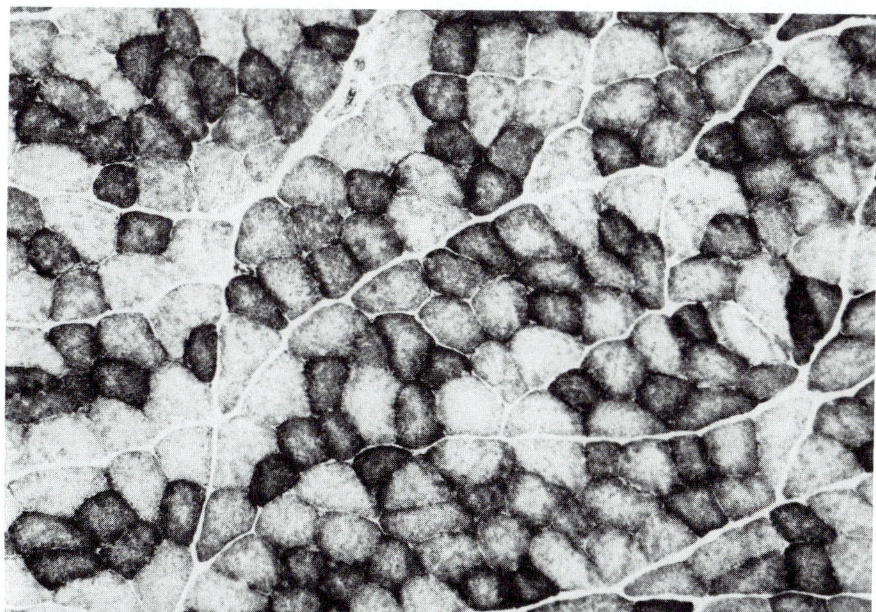

FIGURE 3-29
Photomicrographs of cross sections of (A) bovine and (B) porcine *longissimus* muscle, showing their red and white fibers (dark and light, respectively). (A) The bovine muscle shows random intermingling of red and white fibers, as well as evidence of intermediate fibers (grey). (B) The porcine muscle shows the grouping and central position of red fibers more or less surrounded by white fibers in the fasciculus (× 175).

are more or less randomly intermingled within the muscle bundles. However, in pig muscles the red fibers are located near the center of the fascicles in discrete groups, and they are surrounded by white fibers that form the periphery of the bundles, as shown in Figure 3-29.

The higher myoglobin content of the red fibers, as compared to the white fibers, accounts for their red color. Red and intermediate fibers generally have smaller diameters than white fibers. Red fibers contain a high proportion of the enzymes involved in *oxidative metabolism* (requires oxygen for energy production) and low levels of the glycolytic enzymes. White fibers, on the other hand, have a high content of glycolytic enzymes and a low oxidative enzyme activity. Consistent with metabolic activity, the mitochondria of red fibers are more numerous and larger in size than in white muscle fibers. These features are a reflection of the greater oxidative metabolic activity in red fibers, since these enzymes are associated with the mitochondria. Red fibers also have a greater capillary density, which facilitates the transfer of "metabolic wastes" and nutrients (particularly oxygen) to and from the vascular bed. They also have a greater lipid content, some of which presumably serves as a source of metabolic fuel, and a lower content of glycogen than white fibers. Glycolytic metabolism, which predominates in white fibers, can occur in either the presence or absence of oxygen (i.e., it is *anaerobic*). Thus, white fibers have a lower capillary density than red fibers. White fibers have a more extensively developed sarcoplasmic reticulum (SR) and T tubule system. This is consistent with their more rapid contraction speed, as compared to that of red fibers.

White fibers have a *phasic* mode of action; they contract rapidly in short bursts and they are relatively easily fatigued. Red fibers contract more slowly, but for a longer duration of time. This slower but sustained mode of action is generally referred to as *tonic* (continual tension) contraction. Red fibers are functionally important in posture and, because of their metabolism, they are less easily fatigued, so long as an oxygen supply is available. The intermediate fibers contract faster than red fibers and are less easily fatigued than white fibers.

CHEMICAL COMPOSITION OF THE ANIMAL BODY

The animal body normally contains about one-third of the approximately one hundred chemical elements. About 20 of these are essential to life. An additional 30–35 elements have been detected in mammalian and avian tissues, but their presence is probably coincidental with their presence in the animal's environment during its lifetime. The most abundant elements (by weight) in the animal body are those present in water and in the organic compounds, such as proteins, lipids, and carbohydrates: oxygen, carbon, hydrogen, and nitrogen. These four elements account for approximately 96 percent of the total body chemical composition, as shown in Table 3-2. Since water constitutes 60 to 90 percent of the soft tissues in the animal body, the reason for the high

TABLE 3-2
Elemental composition of the animal body

Major elements	*% of body weight*	
Oxygen	65.0	96.0%
Carbon	18.0	
Hydrogen	10.0	
Nitrogen	3.0	
Calcium	1.5	3.4%
Phosphorus	1.0	
Potassium	0.35	
Sulfur	0.25	
Sodium	0.15	
Chlorine	0.15	
Magnesium	0.05	

Trace elements	
Aluminum	Lithium
Arsenic	Manganese*
Barium**	Molybdenum*
Boron	Nickel
Bromine**	Rubidium
Cadmium**	Selenium*
Chromium**	Silicon
Cobalt*	Silver
Copper*	Strontium**
Fluorine**	Titanium
Iodine*	Vanadium
Iron*	Zinc*
Lead	

*Essential microelements.

**These microelements are normally present in the animal body but evidence of their essentiality is lacking. (The microelements without asterisks are present in measurable quantities, but they have no known metabolic function.)

percentage of oxygen and hydrogen is obvious. The lipid and carbohydrate materials are primarily combinations of carbon, hydrogen, and oxygen atoms. Protein molecules contain these three elements plus nitrogen, sulfur, phosphorus, and small quantities of several other elements. Numerous other elements are present in the body as inorganic constituents (Table 3-2). The abundance of these inorganic elements varies from a high of 1.5 percent of the total body weight, as in the case of calcium, to barely detectable quantities. However, the total amount of an element in the body is not necessarily an accurate indication of its functional importance.

Water

Water is the fluid medium of the body, and some is also associated with cellular structures; in particular, with the colloidal protein molecules. Water serves as a medium for transporting nutrients, metabolites, hormones, and waste products throughout the body. It is also the medium in which most of the chemical reactions and metabolic processes of the body occur.

Proteins

The proteins constitute a very important class of chemical compounds in the body. Some are necessary for its structure, and others function in vital metabolic reactions. Except in fat animals, protein ranks second only to water in abundance (by weight) in the animal body. In the body, most of the protein is present in muscle and connective tissues. The proteins of the body vary in size and shape; some are globular while others are fibrous. The structural differences of the protein molecules contribute to their functional properties. For instance, the fibrous proteins form structural units, and the globular proteins include the numerous enzymes that catalyze metabolic reactions.

Lipids

The animal body contains several classes of lipids, but the *neutral lipids* (fatty acids and glycerides) predominate. Of the various lipids in the body, some serve as sources of energy for the cell; others contribute to cell membrane structure and function; still others, such as some of the hormones and vitamins, are involved in metabolic functions. Most of the lipids in the body are found in the various fat depots in the form of *triglycerides*, glycerol esters of long chain fatty acids.

$$\begin{array}{ll} H_3C{-}(CH_2)_n{-}\overset{O}{\overset{\|}{C}} & {-}O{-}CH_2 \\ & \qquad\;\; | \\ H_3C{-}(CH_2)_n{-}\overset{O}{\overset{\|}{C}} & {-}O{-}CH \\ & \qquad\;\; | \\ H_3C{-}(CH_2)_n{-}\overset{O}{\overset{\|}{C}} & {-}O{-}CH_2 \\ \text{(Fatty Acids)} & \text{(Glycerol)} \end{array}$$

TRIGLYCERIDE

The fatty acid and triglyceride composition of some fat depots in several species of meat animals is presented in Table 3-3. Except for milk fat, fatty acids having chains of 10 carbon atoms or less are rarely found in animal fats. On the other hand, the C_{16} and C_{18} (molecules with chains of 16 and 18 carbon atoms, respectively) fatty acids predominate, with C_{12}, C_{14}, and C_{20} acids present only in small quantities. Of the saturated fatty acids, palmitic and stearic (C_{16} and C_{18}, respectively) predominate in animal fats.

$$H_3C{-}(CH_2)_{14}{-}\overset{O}{\overset{\|}{C}}{-}OH$$

PALMITIC ACID

$$H_3C{-}(CH_2)_{16}{-}\overset{O}{\overset{\|}{C}}{-}OH$$

STEARIC ACID

TABLE 3-3
Fatty acid and triglyceride composition of some animal fat depots (percent by weight)

Component	Chicken*	Pig			Cattle		Sheep	
		Subcutaneous** Outer	Subcutaneous** Inner	Perirenal***	Subcutaneous**	Perirenal***	Subcutaneous**	Perirenal***
Fatty acids								
Lauric	–	trace	trace	trace	0.1	0.2	0.1	0.1
Myristic	0.1	1.3	0.1	4.0	4.5	2.7	3.2	2.6
Palmitic	25.6	28.3	30.1	28.0	27.4	27.8	28.0	28.0
Stearic	7.0	11.9	16.2	17.0	21.1	23.8	24.8	26.8
Arachidic	–	trace	trace	trace	0.6	0.6	1.6	2.6
Total Saturated	32.7	41.5	47.3	49.0	53.7	55.1	57.7	59.5
Palmitoleic	7.0	2.7	2.7	2.0	2.0	2.2	1.3	1.9
Oleic	20.4	47.5	40.9	36.0	41.6	40.1	36.4	34.2
Linoleic	21.3	6.0	7.1	11.8	1.8	1.8	3.5	4.0
Linolenic	–	0.2	0.3	0.2	0.5	0.6	0.5	0.6
Archidonic plus chipandonic	0.6	2.1	1.7	1.0	0.4	0.2	0.6	0.8
Total Unsaturated	67.3	58.5	52.7	51.0	46.3	44.9	62.3	41.5

TABLE 3-3 (*continued*)

Component	Chicken*	Pig: Subcutaneous** Outer	Pig: Subcutaneous** Inner	Pig: Perirenal***	Cattle: Subcutaneous**	Cattle: Perirenal***	Sheep: Subcutaneous**	Sheep: Perirenal***
Approximate triglyceride composition								
Fully saturated								
Tripalmitin		1			3		trace	
Dipalmitostearin		2			8		3	
Palmitodistearin		2			6		2	
Tristearin		–			–		–	
Mono-oleo-disaturated								
Oleodipalmitin		5			15		13	
Oleopalmitostearin		27			32		28	
Oleodistearin		–			2		1	
Dioleo-monosaturated								
Palmitodiolein		53			23		46	
Stearodiolein		7			11		7	
Triolein		3			0		0	

*Fat associated with the skin.
**Fat from the dorsal, thoracic, and lumbar regions.
***Fat around the kidneys.

The predominant unsaturated fatty acids in animal fats are palmitoleic, oleic (C_{16} and C_{18}, respectively, with 1 double bond), linoleic (C_{18}, 2 double bonds), and linolenic (C_{18}, 3 double bonds).

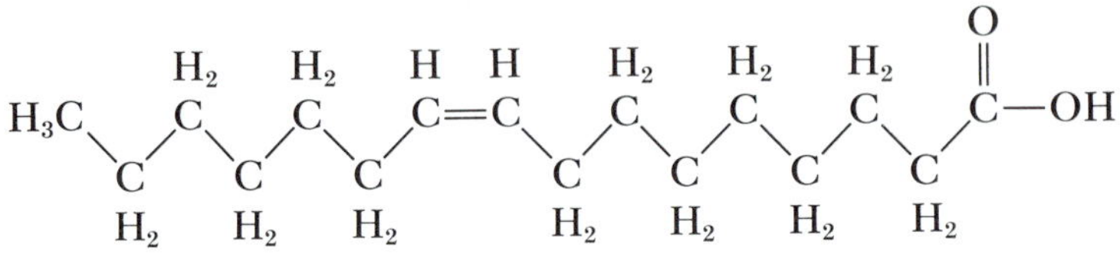

```
                                                         O
       H2   H2   H  H   H2   H2   H2   ||
H3C    C    C    C=C    C    C    C    C—OH
   \  / \  / \  /   \  / \  / \  / \  /
    C    C    C      C    C    C    C
    H2   H2   H2     H2   H2   H2   H2
```

PALMITOLEIC ACID

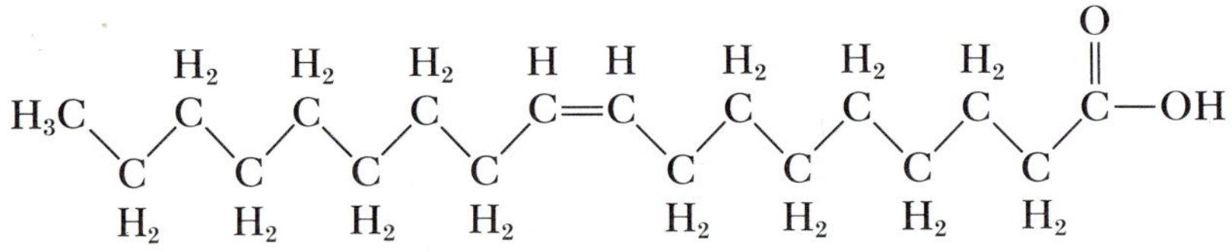

```
                                                              O
       H2   H2   H2   H  H   H2   H2   H2   ||
H3C    C    C    C    C=C    C    C    C    C—OH
   \  / \  / \  / \  /   \  / \  / \  / \  /
    C    C    C    C      C    C    C    C
    H2   H2   H2   H2     H2   H2   H2   H2
```

OLEIC ACID

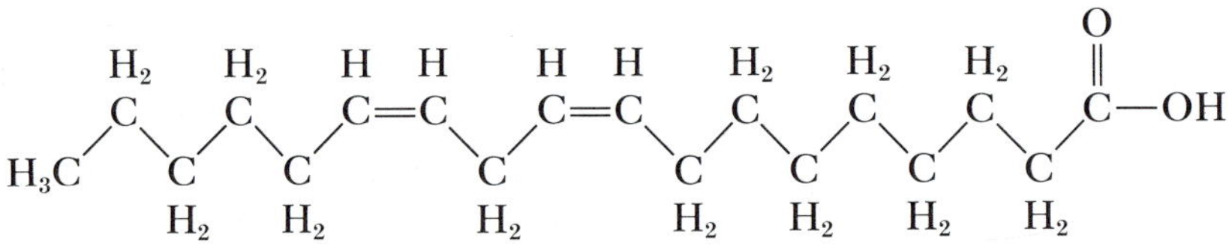

```
                                                                   O
    H2   H2   H  H   H  H   H2   H2   H2   ||
    C    C    C=C    C=C    C    C    C    C—OH
   / \  / \  /   \  /   \  / \  / \  / \  /
H3C   C    C      C      C    C    C    C
      H2   H2     H2     H2   H2   H2   H2
```

LINOLEIC ACID

```
                                                               O
       H  H   H  H   H  H   H2   H2   H2   ||
H3C    C=C    C=C    C=C    C    C    C    C—OH
   \  /   \  /   \  /   \  / \  / \  / \  /
    C      C      C      C    C    C    C
    H2     H2     H2     H2   H2   H2   H2
```

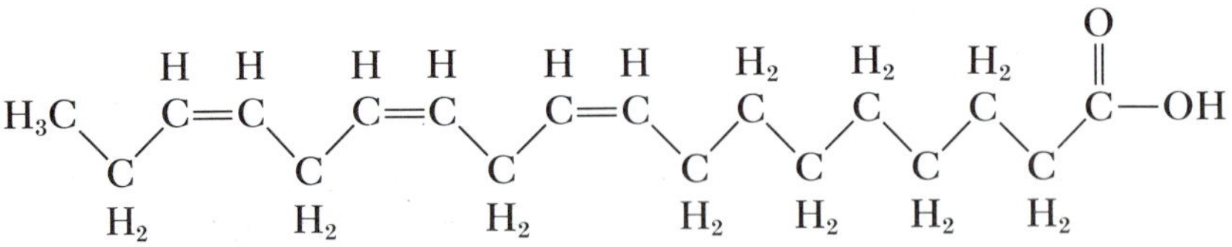

LINOLENIC ACID

Oleic acid is the most abundant fatty acid in the animal body. Of the triglycerides in animal body fats, those that contain one palmitic fatty acid and two oleic fatty acid molecules in each triglyceride molecule are the most abundant, followed by triglycerides containing one molecule each of oleic, palmitic, and stearic acid.

$$
\begin{array}{r}
\text{Oleic acid—O—}CH_2 \\
| \\
\text{Palmitic acid—O—}CH \\
| \\
\text{Oleic acid—O—}CH_2
\end{array}
$$

PALMITODIOLEIN

$$
\begin{array}{r}
\text{Oleic acid—O—}CH_2 \\
| \\
\text{Palmitic acid—O—}CH \\
| \\
\text{Stearic acid—O—}CH_2
\end{array}
$$

OLEOPALMITOSTEARIN

Carbohydrates

The animal body is a poor source of carbohydrates, but most of those present in the animal body are found in muscles and liver. *Glycogen*, the most abundant carbohydrate, is present in the liver (2–8 percent of the fresh liver weight), but muscle generally contains only very small amounts. The other carbohydrate compounds found in the animal body include the intermediates of carbohydrate metabolism and the mucopolysaccharides present in the connective tissues. Although carbohydrates make up a small proportion of the body weight they have extremely important functions in energy metabolism and in the structural tissues.

CHEMICAL COMPOSITION OF SKELETAL MUSCLE

Since muscle is the principal component of meat, a brief discussion of its composition is necessary. Like the animal body, muscle contains water, protein, fat, carbohydrate, and inorganic constituents (Table 3-4). Muscle contains approximately 75 percent water (range: 65–80 percent), by weight. Water is the principal constituent of the extracellular fluid and numerous chemical constituents are dissolved or suspended in it. Because of this it serves as the medium for the transport of substances between the vascular bed and muscle fibers.

TABLE 3-4
Approximate composition of mammalian skeletal muscle (percent fresh weight basis)

	Percent		Percent
WATER (range 65 to 80)	**75.0**	NON-PROTEIN NITROGENOUS SUBSTANCES	**1.5**
PROTEIN (range 16 to 22)	**18.5**		
Myofibrillar	9.5		
myosin	5.0	Creatine and Creatine phosphate	0.5
actin	2.0		
tropomyosin	0.8	Nucleotides (Adenosine triphosphate (ATP), adenosine diphosphate (ADP), etc.)	0.3
troponin	0.8		
M protein	0.4		
C protein	0.2	Free amino acids	0.3
α-actinin	0.2		
β-actinin	0.1	Peptides (anserine, carnosine, etc.)	0.3
Sarcoplasmic	6.0		
soluble sacroplasmic and mitochondrial enzymes	5.5	Other nonprotein substances (creatinine, urea, inosine monophosphate (IMP), nicotinamide adenine dinucleotide (NAD), nicotinamide adenine dinucleotide phosphate (NADP))	0.1
myoglobin	0.3		
hemoglobin	0.1		
cytochromes and flavo-proteins	0.1	CARBOHYDRATES AND NON-NITROGENOUS SUBSTANCES (range 0.5 to 1.5)	**1.0**
Stroma	3.0		
collagen and recticulin	1.5	Glycogen (variable range 0.5 to 1.3)	0.8
elastin	0.1	Glucose	0.1
other insoluble proteins	1.4		
LIPIDS (variable range: 1.5 to 13.0)	**3.0**	Intermediates and products of cell metabolism (hexose and triose phosphates, lactic acid, citric acid, fumaric acid, succinic acid, acetoacetic acid, etc.)	0.1
Neutral lipids (range: 0.5 to 1.5)	1.0		
Phospholipids	1.0	INORGANIC CONSTITUENTS	**1.0**
Cerebrosides	0.5	Potassium	0.3
Cholesterol	0.5	Total phosphorus (phosphates and inorganic phosphorus)	0.2
		Sulfur (including sulfate)	0.2
		Chlorine	0.1
		Sodium	0.1
		Others (including magnesium, calcium, iron, cobalt, copper, zinc, nickel, manganese, etc.)	0.1

Proteins constitute 16–22 percent of the muscle mass (Table 3-4) and they are the principal component of the solid matter. Muscle proteins are generally categorized as *sarcoplasmic*, *myofibrillar*, or *stromal*, based primarily upon their solubility. The sarcoplasmic proteins are readily extractable in water or low ionic strength buffers (0.15 or less). However, the more fibrous of the myofibrillar proteins require intermediate to high ionic strength buffers for their extraction. The *stroma proteins* that constitute the connective tissue and associated fibrous proteins are comparatively insoluble. The sarcoplasmic proteins include myoglobin, hemoglobin, and the enzymes associated with glycolysis, the citric acid cycle, and the electron transport chain. Although the enzymes of the citric acid cycle and electron transport chain are contained within the mitochondria, they are readily extractable, along with those found directly in sarcoplasm. The myofibrillar proteins discussed previously constitute the proteins associated with the thick and thin filaments. They include actin, myosin, tropomyosin, troponin, α- and β-actinin, C protein, and M proteins. These salt soluble proteins are required for emulsion stabilization in the manufacture of emulsion type sausage products. In addition to proteins, other nitrogenous compounds are present in muscle. They are categorized as nonprotein nitrogen (NPN) and include a host of chemical compounds. Notable among these are amino acids, simple peptides, creatine, creatine phosphate, creatinine, some vitamins, nucleosides, and nucleotides including adenosine triphosphate (ATP), as shown in Table 3-4.

The lipid content of muscle is extremely variable, ranging from approximately 1.5 to 13 percent (Table 3-4). It consists primarily of the neutral lipids (triglycerides) and phospholipids. While some lipid is found intracellularly in muscle fibers, the bulk of it is present in the adipose tissue depots associated with the loose connective tissue septa between the bundles (Figure 3-27). As noted earlier the latter type of fat deposit is called *marbling* or *intramuscular fat*.

The carbohydrate content of muscle tissue is generally quite small. Glycogen, the most abundant carbohydrate in the muscle, has an abundance that varies from approximately 0.5–1.3 percent of the muscle's weight. The bulk of the remainder of the carbohydrate is comprised of the mucopolysaccharides associated with the connective tissues, glucose, and other mono- or disaccharides, and the intermediates of glycolytic metabolism. Finally, muscle contains numerous inorganic constituents notable among which are cations and anions of physiological significance, calcium, magnesium, potassium, sodium, iron, phosphorus, sulfur, and chlorine. Many of the other inorganic constituents found in the animal body (Table 3-2) are also present in muscle.

CARCASS COMPOSITION

Carcass composition is generally of greater concern for animal- and meat-scientists than is the animal's total body composition. Of the carcass compositional components, the proportions of muscle, fat, and bone are of greatest interest with regard to the evaluation of livestock production practices. Table 3-5 gives the approximate percentages of muscle, and bone plus tendon, for beef and lamb carcasses having various degrees of fat. As can be seen, when the percentage of fat increases, both the percentage of muscle, and bone plus tendon decrease. These compositional characteristics affect carcass value, and are influenced by genetic as well as environmental factors during the growth and development of the animal. The growth, development and fattening processes are discussed in the next chapter in order to provide an understanding of compositional data such as these.

TABLE 3-5
Approximate percentages of muscle, and bone plus tendon, in beef cattle and sheep carcasses having varying percentages of fat

Fat	*Muscle*	*Bone plus tendon*
8	66	26
12	62	26
16	62	22
21	61	18
26	59	15
32	54	14
37	49	14
42	46	12

REFERENCES

Bendall, J. R., *Muscles, Molecules and Movement* (American Elsevier Publishing Company, Inc., New York), 1971, pp. 3–57.

Bloom, W. and D. W. Fawcett, *A Textbook of Histology* (W. B. Saunders Company, Philadelphia), 9th ed., 1969, pp. 131–357.

Bourne, G. H., *The Structure and Function of Muscle* (Academic Press, New York), 1st ed., 1960, 3 vols.

Cassens, R. G., "Microscopic Structure of Animal Tissues," in *The Science of Meat and Meat Products*, J. F. Price and B. S. Schweigert, eds. (W. H. Freeman and Company, San Francisco), 1971, pp. 11–77.

Cassens, R. G. and C. C. Cooper, "Red and White Muscle," in *Advances in Food Research*, C. O. Chichester, E. M. Mrak and G. F. Stewart, eds. (Academic Press, New York), 1971, Vol. 10, pp. 1–74.

Ham, A. W., "Muscle (Muscular) Tissue," in *Histology* (J. B. Lippincott Company, Philadelphia), 6th ed., 1969, pp. 546–577.

Lawrie, R. A., "Chemical and Biochemical Constitution of Muscle," in *Meat Science* (Pergamon Press, New York), 1st ed., 1966, pp. 66–115.

Murray, J. M. and A. Weber, "The Cooperative Action of Muscle Proteins," Scientific American, Vol. 230, No. 2, pp. 58–71 (1974).

Peachey, L. D., "Form of the Sarcoplasmic Reticulum and T system of Striated Muscle," in *The Physiology and Biochemistry of Muscle as a Food*, 2, E. J. Briskey, R. G. Cassens and B. B. Marsh, eds. (University of Wisconsin Press, Madison), 1970, pp. 273–310.

Ross, R. and P. Borstein, "Elastic Fibers in the Body," Scientific American, Vol. 224, No. 6, pp. 44–52 (1971).

Smith, D. S., *Muscle* (Academic Press, New York), 1972, pp. 5–59.

West, E. S. and W. R. Tood, "Composition and Metabolism of Specialized Tissues," in *Textbook of Biochemistry* (The MacMillan Company, New York), 3rd ed., 1961, pp. 1161–1203.

Wilke, D. R., *Muscle* (Edward Arnold, London), 1968, pp. 1–22.

CHAPTER 4

Growth and Development

The production of meat and all other food is almost entirely dependent upon the growth process, even in an era which has produced tremendous advances in meat production and processing technology. The growth phenomenon is the essential core of the livestock and meat industry, both from the standpoint of animal growth and in the production of most of the raw materials that the animal converts to meat. A basic understanding of animal growth has potential for direct application to problems of efficiency in meat production. Not all aspects of growth are completely understood, particularly the mechanisms involved in initiation of growth, regulation of growth rate, and termination of growth at maturity. However, much that is known about animal growth still remains to be applied at the production level.

What is growth? This is an interesting and personal question since we have all experienced physical growth, and growth has been observed in every form of living matter. Many would answer by saying that when an animal grows it gets larger, but that is not all that is implied by the term growth. The adult is not merely an enlarged version of the newborn animal. The body's structure, its functions, proportions, and composition all change as the individual grows. Closely associated with animal growth is the development of new structures and changes in functional capabilities.

Many definitions exist for the term growth. For the purpose of this text we will define *growth* as: a normal process of increase in size, produced by accretion of tissues similar in constitution to that of the original tissue or organ. This increase in size may be accomplished by any of the following processes: (1) *hypertrophy,* which is the enlargement of existing cells; (2) *hyperplasia,* which is a multiplication or production of new cells; or (3) *accretionary growth,* due to an increase in non-cellular structural material.

From the standpoint of animal production, it is important to distinguish between *true growth,* which involves an increase in the structural tissues such as muscle, bone, and the vital organs, and *fattening,* which is essentially an increase in adipose tissue (a storage depot). Both phenomena fit our definition of growth but must be considered separately when applied to practical animal production.

Development, which must be considered and defined along with growth, is a gradual progression from a lower to a higher stage of complexity, as well as a gradual expansion in size. Other terms that are used in our discussion of growth and development are defined as follows. *Differentiation* is the process by which cells and organs acquire completely individual characteristics. This occurs in the progressive diversification of the cells of the embryo into muscle cells, brain cells, liver cells, etc. Differentiation results in morphological or chemical heterogeneity. *Morphogenesis* is the organization of various dividing cells into specific organs, each with a characteristic makeup. *Maturation* is the process of becoming fully developed. *Maturity* is used in meat grading to indicate the stage of development of the animal producing the carcass. Fully *mature* tissues have attained their highest stage of complexity. It should be recognized that different tissues and body parts grow, develop, and mature at different rates. It is therefore difficult to say when an animal becomes mature, since some tissues are still developing long after others have attained a high stage of complexity. *Senescence* begins in individual tissues and organs when they are no longer maintained in their mature form, but degenerate without complete replacement or repair.

Growth begins with the fertilized ovum or zygote and proceeds through several distinct phases. Prenatal growth (conception to birth) may be divided into three phases, *ovum, embryonic,* and *fetal.* The ovum phase lasts from fertilization until implantation (attachment to the uterine wall) takes place. This period may last 11 days in some meat animal species. During this period the shape of the embryo remains approximately spherical. During the embryonic phase the tissues, organs, and the various systems are differentiated. The embryonic phase may be 25–45 days long, depending upon the species of meat animal. During this phase the embryo undergoes a series of complex changes with little gain in weight. The fetal phase lasts from the embryonic phase until birth, and is characterized by different growth rates in the various organs and parts of the fetus. Organs and systems, such as the central nervous system, heart, liver, and kidneys, which have functional importance during fetal growth, undergo a greater proportion of their growth in the early stages. The digestive tract, which is not vitally important until birth, undergoes a greater proportion of its growth in the later stages. Each of the prenatal growth periods is shorter in animals having short gestation periods, as compared to animals with longer gestation periods. Continuous changes both in the chemical composition and in the conformation of the fetus, occur during the fetal phase.

The phases of postnatal growth and development are not quite as distinct as the prenatal phases. As an animal grows and develops from birth to maturity, continuous changes are occurring in its body conformation and composition. A fundamental principle of postnatal growth is that the shape of the growth curve is similar in all species and is represented by a sigmoid curve, such as that shown in Figure 4-1. For a period of time after birth an animal goes through a phase of very slow growth. This is followed by a phase of rapid growth, during which the rate of increase in size may be nearly constant and the slope of the curve remains almost unchanged. During the later stages of this phase, the growth of muscles, bones, and vital organs begins tapering off, and fattening begins to accelerate in meat animals. Finally, as the animal reaches mature size, a retardation of growth occurs.

The postnatal growth of all animals, including such extreme examples as a mouse and an elephant, fits the sigmoid curve, if appropriate time

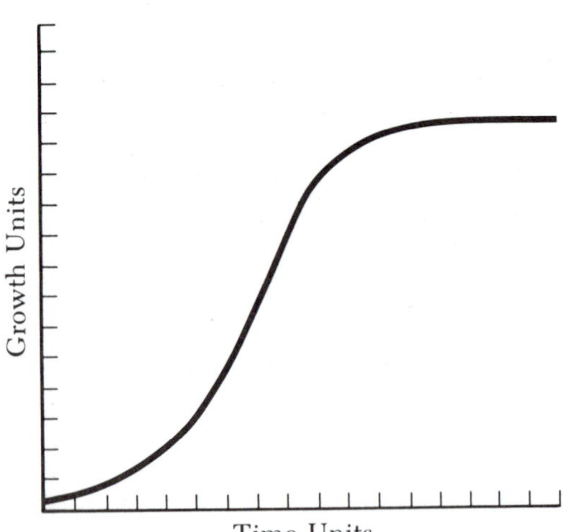

FIGURE 4-1
A typical growth curve.

and growth units are used. Within a species, animals having a larger mature size require a longer time for each growth phase than do animals with a smaller mature size.

GENERAL EMBRYOLOGICAL DEVELOPMENT

An animal's ultimate potential for growth and development is determined at the time of conception. After conception, any modification or failure of an animal to attain its full potential is due to numerous environmental factors. Seldom, if ever, does an animal attain its full potential. In other words, the environment determines the extent to which an animal achieves its genetic potential, and its influences are very important from conception to maturity. Future research on embryonic growth and intra-uterine environment could have far reaching effects in increasing the potential for animal food production. It is important to know how the tissues that are used for food develop, and to keep in mind that these early stages of development are critical in determining the total amount of any one tissue that an animal is capable of producing.

Consideration is given first to the early differentiation that occurs in the ovum and embryonic phases. Later, the complete development of the economically important tissues of the meat animal, muscle, bone, and fat is described; and then the interrelationships of growth among these various tissues are discussed.

Initial Embryological Development of Muscle and Associated Tissues

After the ovum is fertilized, rapid cleavage of the zygote soon results in the formation of a spherical cluster of cells called the *morula*. These cells appear to be anatomically similar, except for slight differences in size. The cells of the morula form an outer capsule, the *trophoblast*, and a central core. The outermost layer of trophoblast cells is involved in the initial nutrition of the embryo rather than in its formation. Next, a cavity filled with a gelatinous matrix appears within the trophoblast. As the cavity is formed a central core of cells, attached at one point to the inner surface of the capsule, projects into the cavity, forming the *inner cell mass*. The entire structure is termed the *blastocyst* (Figure 4-2).

Two secondary cavities soon appear within the inner cell mass. The cavity nearest to the uterine wall becomes the *amniotic* cavity and the other cavity becomes the *yolk sac*. The layer of cells lining the amniotic

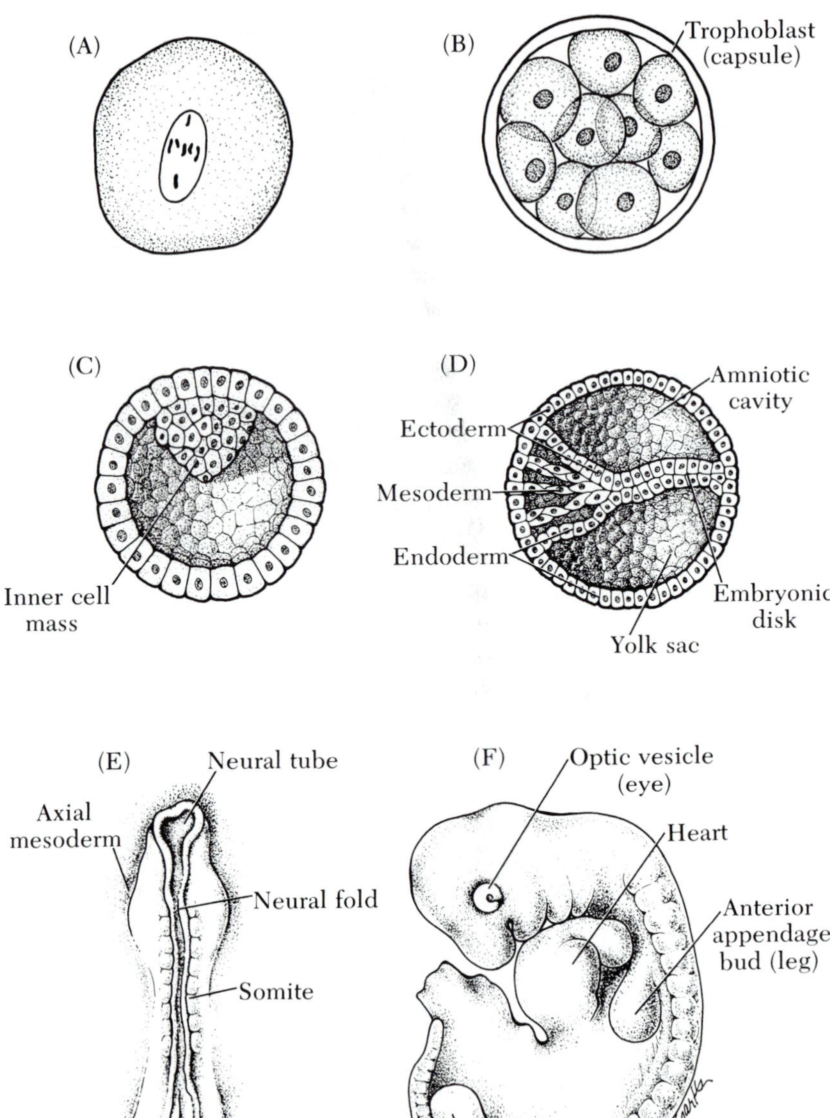

FIGURE 4-2
The morphogenesis of a single egg cell (A) into a morula (B), and a blastocyst (C). Diagram (D) shows the stage at which two cavities have formed in the inner cell mass; an upper (amniotic) cavity and a lower cavity yolk sac. The embryonic disc, containing the three germ layers, the ectoderm, mesoderm, and endoderm, is located between these cavities. Diagram (E) is of a pig embryo, showing the formation of the neural tube and the somites. Diagram (F) illustrates the well developed somites of an 18–19 day pig embryo.

cavity (the *ectoderm*), gives rise to the nervous system and other epithelial tissues, such as the skin, hair, hoofs, and the nasal passage lining. The layer of cells lining the yolk sac (the *endoderm*), will produce glands, and the linings of internal organs, such as those of the stomach and lungs. Between the ectoderm and endoderm is a third layer of cells, the *mesoderm*, which will form the muscles, bones, cartilage, heart, veins, and arteries. The three layers of primary germ cells, the ectoderm, endoderm, and mesoderm, form a flattened plate of cells, the *embryonic disc.*

The first active differentiation of tissues occurs with the appearance of a linear group of proliferating ectodermal cells, known as the *primitive streak*, from which the mesoderm develops. The primitive streak later forms a rod-like structure, the *notochord,* which gives rise to the vertebral column. Ectodermal cells above the notochord multiply to form a thickened strip that later folds into a tube, the *neural tube*, which is the precursor of the brain and spinal cord.

Multiplication of cells in the primitive streak forms a longitudinal mass of tissue, the *axial mesoderm*, on both sides of the midline. Simultaneously, cells of the axial mesoderm undergo segmentation forming a series of paired blocks, called *somites*, on either side of the neural tube. Certain portions of the somites later form muscle plates or *myotomes.* These plates thicken and their cells differentiate into *myoblasts*, which, in turn, give rise to elements of the skeletal musculature. Other portions of the somites form the axial skeleton and connective tissues present throughout the musculature. This later embryonic mesodermal tissue is termed *mesenchyme.*

The digestive system begins as a tubular arrangement of mesodermal cells, lined with cells from the endoderm. As the tube forms, specialized groups of cells along its length give rise to the esophagus, lungs, stomach, liver, pancreas, and intestines. The progressive differentiation of tissues during the prenatal period of growth and development is illustrated in Figure 4-3. Note that the origin of the various organs and tissues can be traced to the three primary germ layers, the ectoderm, endoderm, and mesoderm. Of special significance for us is the fact that the mesoderm is the origin of the majority of the tissues used as meat.

The process of prenatal growth and development is governed not only by the genetic materials in the nucleus of the fertilized egg, but also by structures in the developing embryo. Hans Spemann, the German recipient of the Nobel Prize in medicine and physiology for 1935, had discovered an "organizer effect" in embryonic development. He observed that, at every stage of embryonic development, those structures that are already present act as organizers, inducing the emergence of subsequent structures on the embryological timetable. The primitive streak is apparently the primary organizer in higher vertebrates.

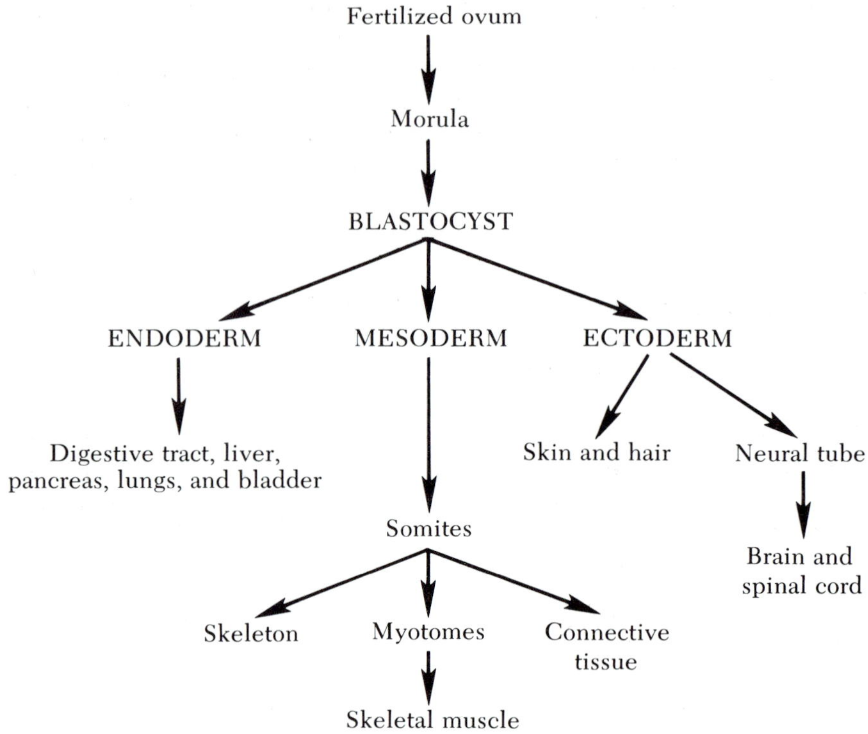

FIGURE 4-3
The progressive differentiation of tissues during prenatal growth and development.

Significance of Protein Synthesis in Growth and Development

All prenatal or postnatal growth, development, and differentiation depends upon the phenomenon of protein synthesis. Proteins form the structure and functional basis of every cell, tissue, and organ. The enzymes that catalyze the synthesis of cell, tissue and body constituents are proteins. Proteins serve as a means of transport and communication within the cell, between cells, and between tissues. The instructions for synthesizing the numerous proteins, and the information necessary to control protein synthesis, are contained in the deoxyribonucleic acid (DNA) chains of the chromosomes present in nuclei. The mechanisms by which information is extracted from DNA, processed, and translated into complete proteins, is protein synthesis. If one realizes by how much the solids content of the body increases during growth, and that the bulk of this increase is protein, then the importance of protein synthesis is obvious.

It is beyond the scope of this text to give a thorough discussion of protein synthesis mechanisms. Students who are interested in the subject of animal growth and development are encouraged to seek out information on this subject because of its vital importance in animal production.

DIFFERENTIATION, DEVELOPMENT, GROWTH AND MATURITY OF STRIATED MUSCLES

Muscle tissue is the prime reason that meat is highly valued as a food by the meat eating people of the world. An understanding of the mechanism of its origin and development is essential for those who wish to play a major role in the improvement and supply of this prized food.

Embryological Development of Striated Muscle

In the early stages of growth and development, *mitosis* is the primary mechanism involved in the formation of new tissues. When tissues are differentiating, there are two types of mitosis occurring. In the first type, *proliferative mitosis*, each daughter cell is identical to the parent cell. This type of mitosis increases cell number without changing the character of the tissue. In the second type, *quantal mitosis*, one or both of the daughter cells of a single mitotic division may be different from the parent cell.

The embryonic connective tissue found in the mesoderm is called the *mesenchyme*. The mesenchyme differentiates into two different cell types—myoblasts and fibroblasts. The connective tissue proper develops from fibroblasts, and myoblasts are primitive muscle cells.

MYOGENESIS. *Myogenesis* is the embryonic formation of muscle tissues. Mesenchyme cells in the mesoderm form myogenic and fibrogenic cells through quantal mitosis. The fibrogenic cells differentiate further into the various connective tissues, and the myogenic cells differentiate into muscle tissues. Once the myogenic cells appear they undergo a period of proliferative mitosis until differentiation occurs.

The next distinct stage in muscle formation is the *myoblast*. Myoblasts are cells which have the ability to fuse with one another to form *myo-*

tubes. Young myotubes synthesize myosin and actin, and begin forming myofibrils. The myotubes are elongated, multinucleated structures formed by myoblast fusion. Structurally, the myotube has a central core of cytoplasm containing the nuclei, while the periphery is a complete cylinder of myofibrils. Initially, a few non-striated myofibrils appear in the myotube. As myogenesis progresses the myofibrils just beneath the sarcolemma are the first to become striated. The myofibrils increase in number within each myotube by longitudinal fission and, concurrently, the nuclei move from their central position to the periphery, and increase in number by further fusion of myoblasts. As the myofilaments become aligned within the myofibrils they develop the typical cross striations of skeletal and cardiac muscle. At this stage of myogenesis the myotube has developed the characteristics that are typical of a muscle fiber.

During embryonic and fetal development, the fibers in a muscle increase in number and become grouped into distinct bundles. The individual fibers increase only slightly in diameter during the first two-thirds of the prenatal period. Most of the increase in muscle weight during this period is due to *hyperplasia* (increase in the number of fibers). During the last one-third of the prenatal period, *hypertrophy* (increase in the size of existing fibers) contributes comparatively more to muscle growth than it does during the initial two-thirds. Developing muscle fibers grow in length by the addition of complete sarcomere units to the ends of existing myofibrils.

Postnatal Muscle Growth and Development

The greatest increase in muscle size occurs after birth. The rate of increase in size declines as an animal approaches maturity. Muscle fibers grow by increasing both their diameter and length. However, the exact mechanism of this growth has not been completely elucidated. After birth, the number of muscle fibers does not appear to increase, to any significant extent. This indicates that postnatal muscle growth is accomplished primarily through hypertrophy. The diameter of the individual muscle fibers is increased by proliferation of the myofibrils. The number of myofibrils within a single muscle fiber may increase by 10–15 times during the lifetime of the animal. This proliferation occurs through longitudinal splitting of large myofibrils into two small daughter myofibrils. A single myofibril may split 4 or more times. Maximum muscle fiber diameter of the pig is attained after about 150 days of postnatal growth (Figure 4-4). Obviously the age at which maximum fiber diameter is attained is not constant among animals. Animals that mature

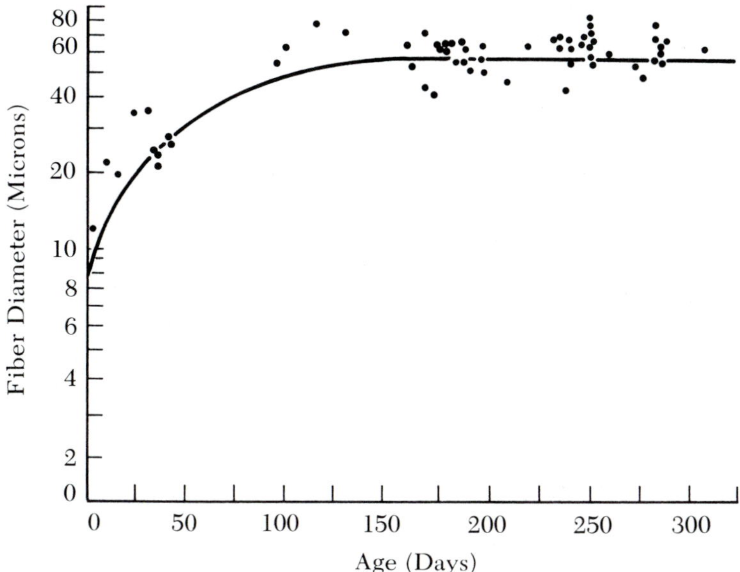

FIGURE 4-4
Change in swine *longissimus* muscle fiber diameter with increasing age. [Data from B. B. Crystall, University of Missouri.]

at an early age would be expected to attain maximum fiber size sooner than animals that mature at a later age.

Increases in muscle length can be accomplished either by an increase in the length of existing sarcomeres or by the addition of new sarcomere units. In the laboratory mouse, some of the postnatal increase in length has been atttibuted to an increase in sarcomere length. However, the major portion is due to the formation of new sarcomeres. There is good evidence that after the addition of new sarcomeres is complete, the muscle cannot increase its length by more than 20 percent by changes in sarcomere lengths.

Individual muscles vary in their rates of growth. The larger muscles, such as those of the legs and back, have the greatest rate of postnatal growth. The variation in maximum adult size among animals of a given species is due to differences in muscle fiber number and not necessarily to fiber size. Fiber size in a given animal is therefore not directly proportional to body size.

During postnatal growth, the number of nuclei in skeletal muscle increases. However, the hypertrophy of muscle fibers that is due to the proliferation of myofibrils results in a progressive increase in muscle weight per nucleus. Hypertrophy also progressively increases the amount of muscle per blood capillary. This can prevent the muscles of very muscular animals from receiving an adequate blood supply and could lead to serious consequences during stress. However, in the case of work-induced muscle hypertrophy, the capillary density increases in approximately equal proportion to the degree of hypertrophy.

Muscle fiber diameter is influenced by many factors—the species, breed, sex, age, nutritional level, and physical activity of the animal. The muscles of sheep usually have smaller fiber diameters than those of pigs and cattle. Also, the muscle fibers of sheep tend to be more uniform in their diameter than are those of pigs and cattle. The muscle fibers of males are usually larger than those of females or castrates, with those of male castrates being intermediate in fiber size. In general, muscle fiber diameter increases with age, nutritional adequacy, and physical activity.

FIBER TYPE DIFFERENTIATION. All muscle fibers appear to be red at birth, but shortly thereafter some of them differentiate into white and intermediate fiber types. Present evidence suggests that the nerve which innervates a given muscle fiber determines, in part, whether it differentiates into a white or intermediate fiber, or remains a red fiber.

MATURITY. After the animal is fully developed, muscle and fiber size may increase or decrease due to work-induced hypertrophy or atrophy (decrease in size) due to inactivity. As the animal grows older and enters senescence the total number of muscle fibers will decrease and the remaining fibers will become larger. The hypertrophy probably compensates for some of the function performed by the atrophied fibers. Thus, skeletal muscles in old age contain fewer fibers which are larger than those at maturity.

GROWTH AND DEVELOPMENT OF ADIPOSE TISSUE

Embryological Development

Adipose or fat tissue cells are derived from the embryonic mesenchyme of the mesoderm, and the primitive fat cell is indistinguishable from a fibroblast. However, it is apparently a specific cell type from the standpoint of future development and function. The general structure of adipose tissue is first outlined by elements of the mesenchyme that, in the future, will provide its structural support. Adipose tissue gradually develops into lobes and lobules that are enclosed in a delicate sheath of collagenous fibers, and supplied with a network of blood capillaries.

In the embryo, this supporting tissue arrangement develops in areas where lipid will later be deposited, such as beneath the skin, between muscles, and in the mesentery.

When the primitive adipose cell begins to accumulate lipids it is known as an *adipoblast*: and, when it is filled with lipid, as a mature *adipocyte*. The primitive cell contains cytoplasm, mitochondria, free ribosomes, and an endoplasmic reticulum. As the adipoblast begins to accumulate lipid in the form of droplets near the center of the cell, among the free ribosomes and mitochondria, the number and size of the mitochondria increase. The lipid droplets fuse as the adipoblast continues to accumulate lipid and forms one large lipid globule. The cytoplasm, nucleus, mitochondria, and other organelles are pushed to one side of the adipocyte as it becomes filled with lipid. The mature adipocyte is large; it can have a diameter of up to 120 μm, which is huge, compared with the 1–2 μm diameter of the primitive cell.

When nutrition of the fetus is adequate, some fat cells may become filled with lipid. Under conditions of inadequate nutrition, lipid will be mobilized and removed from cells that then appear to revert to embryonic mesenchyme. When the nutrition of the fetus is improved, the same cells may again fill with fat.

As mentioned in the previous chapter, two types of adipose tissue may be present in the body of meat animals at birth, namely white and brown fat. The histogenesis of brown fat is comparable to that already described for white fat. The brown adipose tissue provides the newborn animal with an available energy source, such as it might need in order to cope with low environmental temperatures. Most brown fat either disappears within a few weeks after birth or, possibly, is converted to white fat.

Postnatal Adipose Tissue Growth and Development

The adipose tissues continue to grow and develop in growing and adult animals, as a smaller proportion of the nutritional intake is required for the growth of other tissues, and a greater proportion is available for storage. Therefore, fat cells are numerous and widespread in the connective tissues of the normal healthy adult animal. These cells develop at widely varying rates in different parts of the body, which results in marked variations in the proportions of fat present in different parts of the body. In young animals, deposits of fat usually appear around the viscera and kidneys. Then, if nutrient intake is adequate as the animal increases in age, fat will be deposited between the muscles, beneath the skin, and finally, as marbling between the muscle fibers. Intramuscular fat deposits occur in the fat cells associated with the perimysial connective tissues. These fat cells accumulate and grow in the

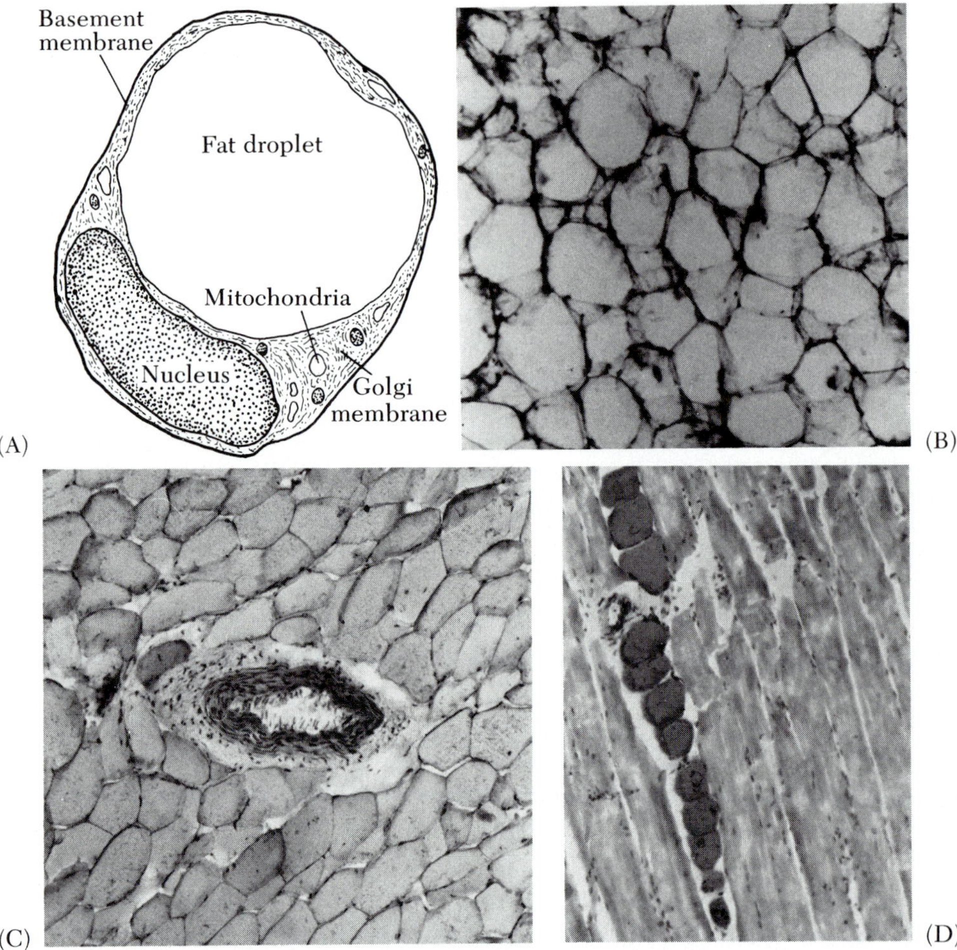

FIGURE 4-5
Diagram and photomicrographs of fat cells. (A) Diagram of a typical fat cell. (B) Photomicrograph of bovine subcutaneous fat cells. (C) Intermuscular bovine fat cells surrounding an artery. (D) Intramuscular bovine fat cells arranged longitudinally between muscle fibers. [Photomicrographs courtesy W. G. Moody, University of Kentucky.]

extravasicular spaces in close proximity to the arteries of the circulatory system that permeate the muscle (Figure 4-5).

During the fattening phase of growth, the percentage of lipid in adipose tissue increases and the percentages of water, protein and other constituents decrease. However, the lipid content of all adipose tissues does not increase at the same rate. For example, the pig has three distinct layers of backfat, the middle of which increases in lipid content most rapidly, while the inner layer shows the slowest rate of increase.

Adipose tissue is a dynamic tissue, which means that it is in a constant state of flux. It is the site of considerable metabolic activity, and as such it is well supplied with blood vessels and rich in enzyme systems. Most adipose tissues represent a storage of nutritive material that can be drawn upon, in response to the needs of other tissues. There is a constant turnover of the stored lipids in adipose tissue; they are not merely deposited and left until needed. As was pointed out earlier, the amount of lipid stored at any adipose tissue site at a given time will vary with the nutritional status of the animal. During periods of poor nutrition, and consequently of limited energy intake, the adipocytes release their stored lipid into the vascular system. If energy intake were to be restricted to a low level over a prolonged period of time, the adipocytes would revert to aggregations of irregularly shaped cells within the supporting connective tissues of the depot. These cells retain their ability to again accumulate lipid, if and when the animal's energy intake is restored to a level above that needed for maintenance. This series of events may occur many times throughout the life of the animal.

GROWTH AND DEVELOPMENT OF BONE

Bone is formed, both in prenatal and postnatal life, by a transformation of connective tissue. During the embryonic phase of growth, connective tissue cells are derived from embryonic mesenchyme. Many of these cells form *osteoblasts*, *fibroblasts*, and *chondroblasts*. The osteoblasts are directly active in the production of bone. The fibroblasts are primarily responsible for the formation of tendons, ligaments, reticular tissue, and *areolar* (fibrous) tissue. The chondroblasts form cartilage. Both prenatal and postnatal formation of bone may be preceded by the formation of a cartilage "model," or it may occur by the direct transformation of connective tissue. When a bone is formed by the replacement of cartilage the process is known as *endochondral ossification* or *cartilaginous ossification*. Growth in both the length and thickness of bones occurs by this process. When bone formation occurs directly, without the presence of cartilage, the process is known as *intramembranous ossification*. In intramembranous bone formation, ossification occurs in and replaces connective tissue. An example is bone growth within or beneath the periosteum. Bone *remodeling* is the process that accomplishes structural change or a continuous turnover of bone constituents. This process occurs in every bone of the skeleton during the continuous growth and internal reconstruction of bone throughout the life of the animal. This process continues, even after the length and thickness of the long bones has ceased changing.

Bone varies considerably in density, but consists essentially of a complex calcareous material deposited in a fine matrix of collagenous fibers, and it is supplied with blood vessels and nerves. The internal architecture and external form of a living bone can change in response to stresses to which it is subjected. The primary function of bone is that of a supporting framework for the soft tissues of the body. The skeleton forms a series of mechanical levers to which muscles are attached by ligaments, allowing mobility while providing rigidity for the body. Certain parts of the skeleton, such as the thorax and skull, provide protective functions. Bone also serves as a reservoir for calcium and other minerals that can be drawn upon when necessary to sustain various metabolic activities.

The structure of a typical long bone is illustrated in Figure 4-6. Each bone is ensheathed by a tough membrane called the *periosteum*, except at articular surfaces. The strength of the adherence of periosteum to bone varies in different phases and at different ages. In bone of young animals, the periosteum is easily stripped off. In adult animals, it is more firmly attached. The periosteum is composed of coarse fibrous connective tissue that contains few cells, but it does have numerous blood vessels and nerves. It serves as an anchorage for muscles, tendons, and ligaments. The ends of bone are covered with a thin layer of hyaline (articular) cartilage. This cartilage is resilient and serves as a cushion between bones.

At birth, a bone such as the femur or tibia has a *diaphysis* (bony shaft) with a medullary cavity, and cartilaginous extremities. Growth in

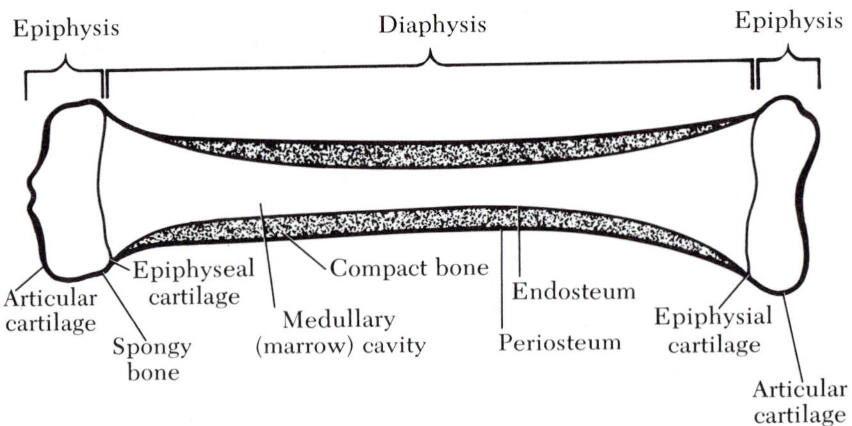

FIGURE 4-6
Diagram of a long bone showing the longitudinal structure.

thickness occurs as the result of the continuous deposition of layers of bone on the surface of the shaft directly beneath the periosteum. Osteoblasts are directly active in the growth of bone. Concurrently, the marrow cavity becomes enlarged, as the bone grows in size, due to the continuous resorption of the medullary wall of the shaft by the activity of *osteoclasts*, cells which are associated with the resorption and removal of bone. The above series of changes in growth of bone are illustrated in Figure 4-7. It is by the control of these two processes, deposition and resorption, that the size and shape of the skeleton is changed during

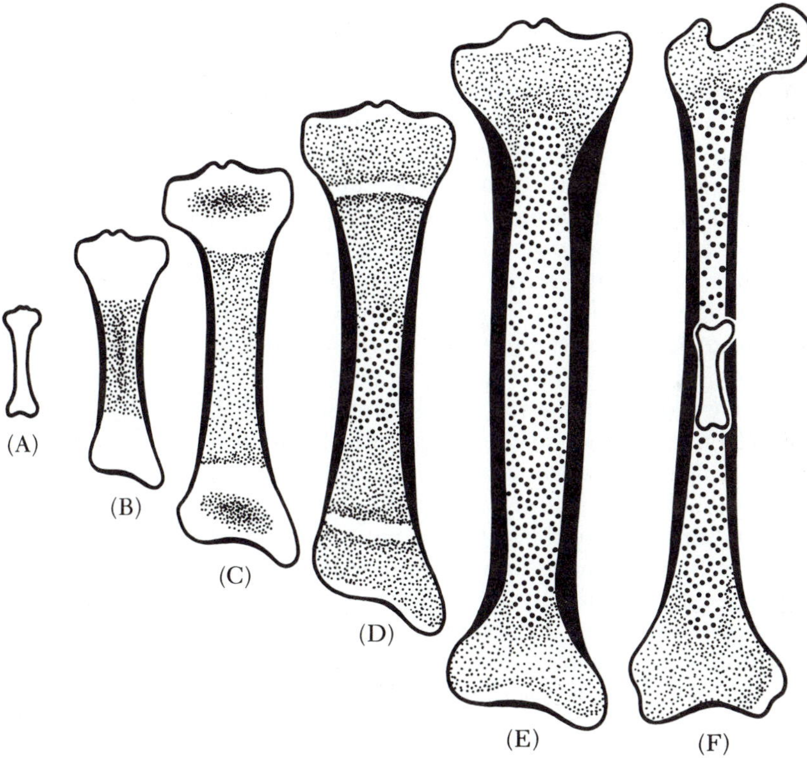

FIGURE 4-7
Progressive ossification and growth of a long bone. (A) The cartilaginous stage. (B) Deposition of spongy, endochondral bone (stipple), and the compact, perichondral bone (black). (C) An epiphysis appears at each end. (D) The marrow cavity (sparse stipple) appears as endochondral bone is resorbed. (E) Each epiphysis ossifies, leaving articular cartilage at both ends. Notice that the enlargement of the marrow cavity continues by the resorption of bone centrally as deposition continues on the periphery. (F) A bone at birth superimposed over the same bone of an adult to show their relative sizes, and the amount of deposition and internal resorption that occurs during growth.

growth. These same processes continue during adult life, and are responsible for a continuous remodeling of bone tissue. The two opposed forces are normally balanced in any particular part of the skeleton. In adult animals, the main part of the shaft has been derived from the intramembranous growth of bone beneath the periosteum. The site of the original bony diaphysis has been replaced by the *medullary cavity.* This cavity is occupied by *marrow,* a soft tissue, composed of meshes of delicate reticular connective tissue filled with various cells. Two types of marrow are recognized: red and yellow. The red is the chief blood forming organ of the adult body, being the main source of red blood cells and granular leucocytes. Red marrow is found in spongy bone of the epiphysis of long bones, and in the main bodies of the sternum, ribs, and vertebrae. Red marrow also functions in the formation of the thrombocytes (platelets) that aid in clotting of blood. Phagocytosis, the destruction of old, worn-out red blood cells also occurs in the red bone marrow. The yellow marrow consists mainly of adipose tissue, and is mainly found in the medullary cavity of bones.

Growth in the length of a typical long bone results from growing epiphyseal cartilage being gradually replaced by bone. As long as the bone of an animal is growing cartilage, complete ossification does not occur. When the bone has attained its full size and growth has ceased, then the epiphyseal cartilage does not exist. The full extent of bone growth can be seen by comparing the fetal skeleton to that of an adult animal, as is done for one bone in Figure 4-7f.

At about the time of birth, ossification centers appear in each end of the long bones, between the epiphysis and diaphysis. The epiphysial cartilage in these centers passes through the same changes as does the cartilage on the outermost ends of the bone. It is at these ossification centers where most of the actual growth in the length of a bone occurs. Growth in length is maintained by the extension of ossification from the diaphysis into the diaphysial surface of the epiphysial cartilage. This growth process occurs at a similar rate on both ends of the diaphysis. The epiphysial cartilage continues to separate the bony epiphysis from the diaphysis until the bone reaches its maximum length, at which time the proliferation of the cartilage ceases. Then the epiphysis becomes fused with the diaphysis (ossified) and no further growth in length of the bone takes place.

The extent of ossification at the epiphysial plate (break joint) of the lamb foreshank, and the restructuring and calcification beneath the periosteum of rib bones are used by meat graders in determining the maturity designation of lamb carcasses. In the case of the lamb foreshank the forefeet are removed at the epiphysial cartilage junction, between the epiphysis and diaphysis. Complete ossification at this junction necessitates removal of the feet at the spool joints, and consequently these carcasses are classified as mutton carcasses. Likewise, in beef carcasses, the degree of ossification between the sacral vertebrae, at the ends of spinous processes, in the breast bone, and in the ribs is used to determine maturity. Termination of growth does not occur simultaneously in all bones—the sacral vertebrae reach maturity before the lumbar, and the lumbar before the thoracic vertebrae.

GROWTH AND DEVELOPMENT OF FIBROUS CONNECTIVE TISSUE

Fibrous connective tissues have the same embryonic origin as do bone and adipose tissues. All of these tissues, constituting the connective tissue group, arise from a differentiation of the mesenchymal cells of the mesoderm, which ultimately forms adult tissues with distinctly different structural details and mechanical properties. For example, the fibroblasts and osteoblasts extract appropriate raw materials from the circulation, elaborate upon them in their cytoplasm, and deposit their characteristic secretions as fibrous connective tissue and bone, respectively. Since the tissues that constitute the connective tissue group have closely related functions, and are derived from a common type of ancestral cell, one type of connective tissue may be converted into or be replaced by another.

Early in the embryonic phase, mesenchymal cells rapidly occupy most of the free space between the deeper lying structures and the superficial ectoderm. Although supported in a common gelatinous matrix, the cells at this stage still tend to remain fairly independent of one another. Later, adjacent cells begin to unite and form a fibrous network. At the same time, delicate fibrils make their appearance along the peripheral areas of the cytoplasm. Delicate fibrils that are attached to the exterior surface of the cells also make their appearance at this time. Toward the end of the embryonic phase, and early in the fetal phase, the fibrils are quite abundant. Since the cells are continually shifting their position through amoeboid activity, some of the fibrils become detached and appear free in the intercellular spaces. When these fibers begin to become a conspicuous part of the young connective tissue, the term *fibroblast* is used in referring to those cells which are producing the connective tissue.

After these early stages, the young connective tissue soon takes on a distinctly characteristic appearance. The fibers become arranged in definite patterns, depending upon the mechanical conditions under which the tissue is beginning to function. There is apparently a chemical change by which the primitive embryonic fibers are converted into true collagenous (white) fibers. Elastin (yellow) fibers are secreted by fibroblasts in the same manner as the white fibers, but do not make as early an appearance. The chemical composition and physical properties of the elastin fibers are quite different. They also tend to appear as single fibers, rather than in bundles, as do the white fibers. The reticulin fibers are probably formed in a similar manner to collagen and elastin fibers.

Connective tissue is of great physiological importance to the living animal, and also influences the qualitative characteristics of meat. Fibrous connective tissue, constituted of varied amounts of collagen, elastin, and reticulin, is an integral part of organs and muscles. The total amount of connective tissue in muscle varies throughout the body. Muscles used in locomotion (leg muscles) contain more connective tissue than do supporting muscles, such as those in the lumbar and thoracic regions. The less tender cuts of meat are obtained from muscles that contain more connective tissue. During postnatal growth, the total amount of connective tissue increases, as does muscle, fat, and bone. But more important, from a meat quality viewpoint, are the molecular changes that occur in connective tissue during growth that adversely affect meat tenderness.

The fibroblast synthesizes a collagen precursor, tropocollagen, into the extracellular space. These submicroscopic fibrils can grow in length and thickness by the accumulation of monomers by aggregation (Figure 3-20). The collagen fibers of newborn animals usually are small and imbedded in the viscous mucopolysaccharide ground substance. Newly synthesized fibers are held together by cross-links, which increase in number as the fiber matures. As the number of cross links increases, the collagen fibers become less soluble in salt solutions. Finally, as the whole fiber matures, it increases in strength due to the increased numbers of chemical cross-linkages within its collagen macromolecules.

In muscle, reticulin and elastin fibers are interspersed among the more numerous collagen fibers. The greatest concentration of elastin fibers occurs in specialized structures, such as ligaments and arterial walls. More attention has been paid to collagen and its relationship to meat tenderness than has been given to that of the other two fibers.

CHANGES IN PRENATAL BODY FORM AND COMPOSITION

Due to different gestation periods, the times at which similar changes in form and composition take place vary among cattle, swine, and sheep. The fetuses of cattle take a longer time to undergo comparable morpho-

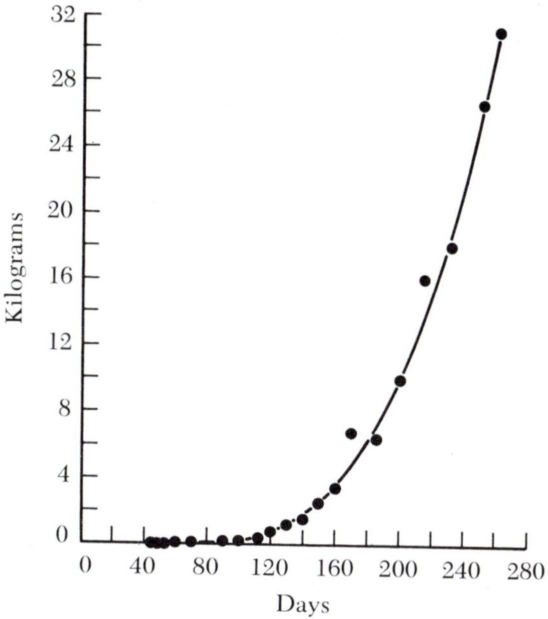

FIGURE 4-8
Prenatal weight growth of the bovine fetus. [From Winters, L. M., W. W. Green and R. E. Comstock, "Prenatal Development of the Bovine," Minn. Agr. Exp. Sta. Tech. Bul. 151 (1942).]

logical changes than do the fetuses of sheep, and sheep take longer than swine. However, at birth, calves and lambs are more mature physiologically than pigs.

The fetus undergoes a tremendous change in body form and shape during prenatal growth and development. The head is comparatively larger than the body early in the prenatal period. Later, the body and limbs grow more rapidly. Figure 4-8 illustrates a typical curve of weight increase in the fetus during prenatal growth and development. Note that the actual gain in weight is slow at first, then increases rapidly during the later part of the prenatal period.

During growth, changes occur in the composition of various parts of the fetus, and of the fetus as a whole. Composition of fetal muscle during growth is of particular interest. The water content of fetal muscle declines with age. Concurrently the contents of solids (ash, protein, and ether extractable constituents) increases. The decline in water content continues during postnatal growth, and many research workers have suggested that this decline is a universal accompaniment of growth.

The order in which an animal's anatomical parts and their composite tissues develop is the same in all species. Their relative importance for the survival of the animal dictates that those which are more vital for growth and development be formed first. Thus, the order of tissue growth follows a sequential trend determined by physiological importance, starting with the central nervous system and progressing to bones, tendons, muscles, intermuscular fat, and subcutaneous fat.

Effect of Prenatal Nutrition on Fetus Condition at Birth

The nutritional level of the dam has been shown to affect the weight of young at birth. Severely underfed dams are likely to produce young that are below average in weight and vigor. The proportion of separable muscle and fat at birth has been shown to be lower, and that of skeletal and internal organs higher, in calves whose dams were poorly fed during gestation. Further, the proportion of chemically determined fat and nitrogen were less, and moisture, ash, and phosphorus were higher, in calves whose dams were poorly fed.

In addition to the effect that the amount of food has, the qualitative nutrition of the dam also affects the size and vigor of the young. For example, protein or calcium deficiency in the dam's diet has been shown to decrease the weight and vigor of young pigs. Severe fetal undernutrition will increase the incidence of death loss in the newborn, and may have permanent effects on growth and behavior. However, there apparently is no conclusive evidence that birth weight is related to mature size, especially in view of the large individual variations that exist among healthy newborn animals.

POSTNATAL GROWTH AND DEVELOPMENT

Body Measurements

During postnatal growth, so many changes occur in body measurements that the structural form of an animal at any particular time is only temporary. An animal continues to change in structural form, due to the differential growth rates among the various parts of the body. The lengths of body and legs are proportionately greater in an immature animal than they are in a mature animal. As an animal grows there is a decline in growth rate in terms of body depth and thickness. Data presented in Table 4-1 illustrate that carcass length does not increase in direct proportion to the increase in liveweight. For example, the carcass length of pigs weighing 22 kilograms was 62 percent of the length of

TABLE 4-1
Comparison of linear measurements, lean cuts, and fat trim of pigs at various liveweight intervals

Live weight (kg)	Carcass length (cm)	Backfat thickness (cm)	*Longissimus* muscle area 10th rib (cm^2)	Yield of four lean cuts of carcass (percent)	Fat trim of carcass (percent)
23	51.3	1.3	11.6	57.5	10.7
45	62.5	2.3	18.1	55.1	15.3
57	67.1	3.0	18.1	49.5	19.5
68	69.6	3.0	20.6	51.2	19.3
80	71.1	3.0	21.9	48.1	21.3
91	72.6	3.6	27.1	48.6	21.5
102	73.9	3.8	26.5	47.0	22.6
114	72.2	4.3	31.0	45.0	25.3
125	79.2	4.3	31.6	46.0	23.8
137	81.8	4.6	32.9	45.4	24.7

SOURCE: Zobrisky, et al., (1958).

pigs weighing 137 kilograms. Note also that this is true for the change in backfat thickness, *longissimus* muscle area, yield of lean cuts, and fat trim.

Dressing Percent

The ratio of dressed carcass weight to the weight of the live animal, expressed as a percentage, is known as *dressing percent*. This may be calculated either on a hot or a chilled carcass weight basis. When calculated on the latter basis, it will be about two percent less than if done on a hot carcass weight basis. Intestinal contents and individual animal variation have considerable influence on dressing percent. As shown in Table 4-2, there is a progressive increase in dressing percent of beef and lamb from the lower to higher grades, and in pork from U.S.D.A. No. 1 to U.S.D.A. No. 4 grade carcasses.

Body Components

GROSS PHYSICAL CHANGES IN COMPOSITION. Body components, such as the hide, organs, stomach, and intestines, change during growth. The percentage of carcass to liveweight usually increases with an increase in liveweight (Table 4-3). Concurrent with this increase, the percentages

TABLE 4-2
Dressing percentages of various kinds of livestock by grades*

Grade	Range	Average
CATTLE		
Prime	62–67	64
Choice	59–65	62
Good	58–62	60
Standard	55–60	57
Commercial	54–62	57
Utility	49–57	53
Cutter	45–54	49
Canner	40–48	45
*CALVES AND VEALERS***		
Prime	62–67	64
Choice	58–64	60
Good	56–60	58
Standard	52–57	55
Utility	47–54	51
Cull	40–48	46
LAMBS (Wooled)		
Prime	49–55	52
Choice	47–52	50
Good	45–49	47
Utility	43–47	45
Cull	40–45	42
*BARROWS AND GILTS****		
U.S. No. 1	68–72	70
U.S. No. 2	69–73	71
U.S. No. 3	70–74	72
U.S. No. 4	71–75	73
Utility	67–71	69

SOURCE: United States Department of Agriculture, Consumer and Marketing Service, Livestock Division.
*All percentages are based on hot weights.
**Based on hide-off carcass weights.
***Based on packer style dressing (ham facings, leaf fat, kidneys, and head removed).

of hide, blood, stomach and intestines, and liver decline. Thus, a higher proportion of the body of young animals is composed of these parts than is the case in older and larger animals. On an empty body basis (liveweight minus the gastro-intestinal tract contents) the percentage of skeleton and lean muscle declines, and that of the total fatty tissues increases, during growth.

TABLE 4-3
Distribution of body components of cattle during growth and development*

	Liveweight (kg)					
	111	204	313	517	853	882
	Age (months)					
Component	3	5	10	19	44	47
Percentage of empty weight** to live weight	88.0	84.3	87.6	88.8	90.4	92.2
Percentage of carcass weight to live weight	54.2	53.7	56.5	60.5	65.3	69.0
Percentage of carcass weight to empty weight	61.6	63.7	64.5	68.1	72.6	74.8
Percentage of hide and hair weight to empty weight	10.5	8.2	8.4	7.4	5.9	6.2
Percentage of blood weight to empty weight	6.2	5.2	5.0	4.1	3.3	3.5
Percentage of stomach and intestines weight to empty weight	4.8	4.6	5.0	4.7	2.5	2.3
Percentage of liver weight to empty weight	1.8	1.8	1.5	1.2	0.8	0.8
Percentage of offal fat weight to empty weight	1.4	3.9	4.7	5.2	7.0	4.7
Percentage of skeleton weight to empty body weight	21.2	17.0	13.8	13.6	10.0	9.9
Percentage of lean weight to empty body weight	41.9	39.1	41.2	41.8	34.3	31.9
Percentage of total fatty tissue weight to empty body weight	5.6	17.6	18.3	21.3	37.1	35.4
Percentage of offal and kidney fat weight to total fatty tissue weight	33.2	32.9	36.9	36.3	23.9	17.8

SOURCE: Moulton, C. R., P. F. Trowbridge and L. D. Haigh, "Studies in Animal Nutrition. II. Changes in Proportion of Carcass and Offal on Different Planes of Nutrition," Mo. Agr. Exp. Sta. Res. Bul. 54 (1922).

*Cattle were fed *ad libitum*.

**Empty weight is the liveweight minus the intestinal tract contents.

PROPORTION OF MUSCLE, FAT, AND BONE. In the appraisal of carcass composition, three variables need to be considered—bone, muscle, and fat. If there is a greater proportion of one, there must be less of one or both of the remaining variables. It is the relative proportions of these three tissues in carcasses of similar weight that, to a large extent, determines the carcass value. An increase in proportion of fat results in a concurrent decline in both muscle and bone. These general principles are illustrated by the data from beef animals, full fed from birth to slaughter, presented in Figure 4-9. Note that the decline in the percentage of bone was greater than that of muscle, up to 10 months of age. After 10 months, the decline in the percentage of muscle was greater than that of bone. Again, these changes in the proportions of bone, muscle, and fat tend to parallel the order of the inherent stimulus for growth of these tissues: first bone, then muscle, and finally, fat.

The proportions of bone, muscle, and fat in the body change continually during growth. The rate and extent of change is variable among

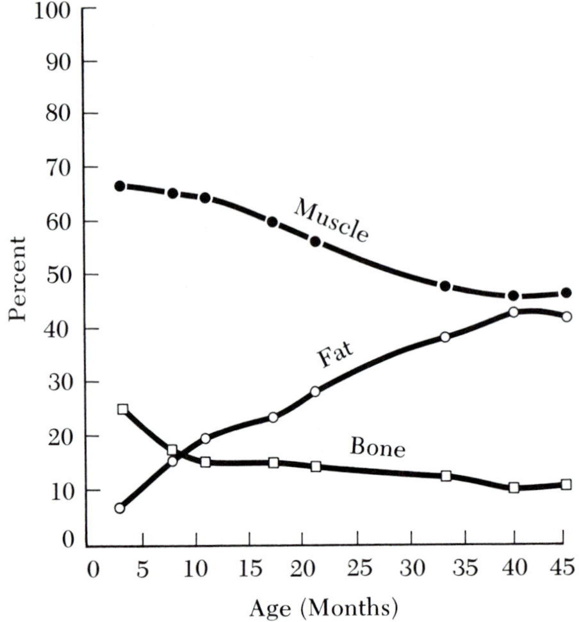

FIGURE 4-9
Changes in the percentage of bone, muscle, and fat in beef carcasses during growth. [Data from Moulton, C. R., P. F. Trowbridge and L. D. Haigh, "Studies in Animal Nutrition; III. Changes in Chemical Composition on Different Planes of Nutrition," Mo. Agr. Exp. Sta. Res. Bul. 55 (1922).]

animals of the same species, and is influenced by factors such as level of nutrition, breed, sex condition, environmental conditions, and health. The effects that these factors have upon body composition are discussed later in this chapter.

PROTEIN DISTRIBUTION. Data on the quantitative changes that occur during the growth of pigs, in the total amount of protein in the empty body are presented in Figure 4-10. This semilogarithmic plot of total protein in the empty-body of pigs is similar in shape to the curve (Figure 4-4) representing changes in the muscle fiber diameter of pigs during growth and development. Data presented in Table 4-4 illustrate the distribution of protein in various parts of the empty body of beef steers during growth. The total protein, expressed as a percentage of empty body weight, declines about 7.5 percent during growth, when animals are full fed, but changes very little when animals receive restricted diets. Blood protein constitutes about 5.5 percent of the total

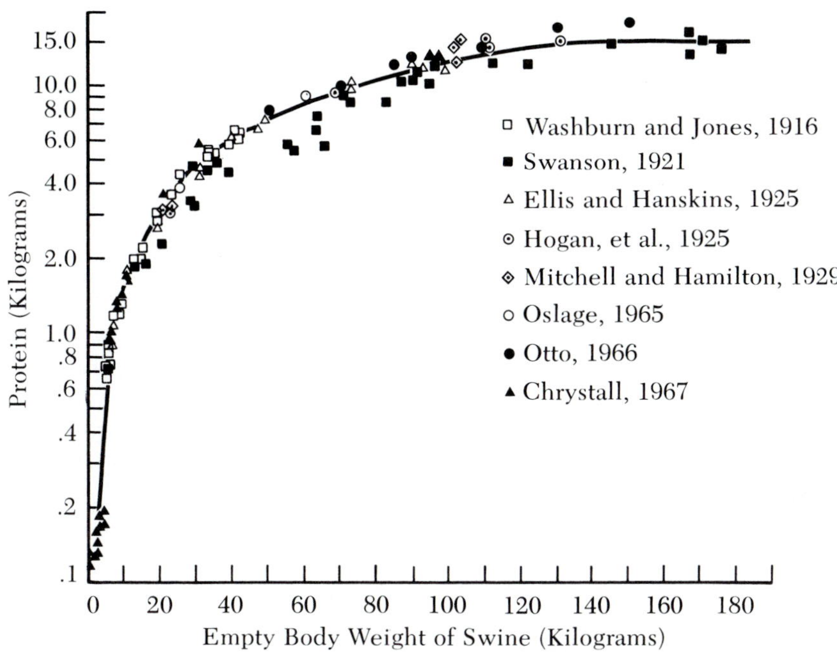

FIGURE 4-10
Graph of the change in kilograms of protein *vs* increases in the empty-body weight of swine. [Data were obtained from various sources and presented by Milton E. Bailey and S. E. Zobrisky, "Changes in Proteins During Growth and Development of Animals," in *Body Composition in Animals and Man* (National Academy of Sciences, Washington, D.C.), 1968.]

TABLE 4-4
Distribution of Protein in the Empty Body of Cattle

Age (months)	Empty body weight (kg.)	Total protein (in percent)	Distribution of protein (in percent)				
			Blood	Organs	Hair and hide	Skeleton	Fat and lean
3	98	19.7	5.7	8.7	17.1	23.2	45.5
8	171	18.3	5.0	9.0	15.6	18.7	51.8
11	281	17.3	4.6	8.2	17.4	16.6	53.3
18	454	16.6	4.5	6.7	14.4	17.2	57.8
21	475	16.1	5.7	7.7	18.5	16.5	51.6
48	813	12.3	5.9	6.6	17.2	19.8	50.5

SOURCE: Data from Moulton, C. R., P. F. Trowbridge and L. D. Haigh. Studies in animal nutrition. III. Changes in chemical composition on different planes of nutrition. Mo. Agr. Exp. Sta. Res. Bul. 55, 1922.

protein in the body. The portion of protein in the internal organs is about 7.7 percent, regardless of nutritional treatment. Liberally fed cattle will have a smaller portion of total body protein present in the hide and skeleton, and a greater percentage in the skeletal muscle and fat, than animals on a restricted diet.

With the exception of some small changes with age and nutritional level, the proportions of the total body protein present in each of the body parts remains remarkably stable. Age and nutritional level have a considerably greater influence on the absolute amount of protein that is present than on its percentages of distribution throughout the body.

During growth, and in the entire life of an animal, the various cellular constituents are involved in continuous building-up (*anabolism*) and breaking-down (*catabolism*) processes. The gross changes in body composition actually occur at the cellular level. The rates at which these processes occur are variable for different tissues, and for the constituents of various tissues. For example, muscle proteins can be divided into those that are rapidly metabolized, slowly metabolized, and inert. Sarcoplasmic proteins are classed as rapidly metabolized, myofibrillar proteins as slowly metabolized, and elastin, collagen, and reticulin as rather inert. The rates of catabolism and anabolism vary from tissue to tissue, depending on their activity. The turnover rate of proteins in the very active tissue of the gastro-intestinal tract is a few days, a longer turnover period for skeletal muscle proteins, and much longer in the case of the more inert connective tissues. The overall turnover rate of protein in cell tissues of the body exceeds the daily protein intake in the diet. The catabolic and anabolic processes are influenced by many factors, such as the animal's age, stress conditions, nutritional level, and certain hormones.

Muscle Components

The chemical composition of a muscle changes during its growth. Changes that occur in total amount of muscle in the body, as well as the total chemical composition of muscle are influenced by age, species, nutritional intake, and sex condition.

During growth, there is an increase in the concentration of intracellular proteins in skeletal muscle. Data presented in Table 4-5 show that there is a consistent increase in the concentration of sarcoplasmic and myofibrillar proteins in mammalian skeletal muscle during growth. The concentration of myofibrillar proteins is consistently greater than that of sarcoplasmic proteins, in both immature and mature animals. Myoglobin, a sarcoplasmic protein, increases in concentration, and is partly responsible for the darker color of muscle from mature animals compared to that of immature animals.

TABLE 4-5
Changes in concentration of proteins in mammalian skeletal muscle during growth

Species	Immature*		Adult*	
	Sarcoplasmic	Myofibrillar	Sarcoplasmic	Myofibrillar
Pig	3.5	10.6	5.2	12.8
Cattle	5.0	8.4	5.3	13.0
Human	3.2	11.0	4.3	12.8
Rat	4.1	10.0	5.2	11.2

SOURCE: Bailey and Zobrisky, 1968.
*Grams/100 grams of tissue.

The time at which increase in concentration of proteins occurs varies among species, as well as among muscles within the body. Some species reach maturity earlier and attain a maximum concentration of intracellular protein sooner than animals that mature later.

The amount of connective tissue protein in muscle increases during prenatal growth, when muscle cells are increasing in number, but becomes proportionately less while cells are increasing in size (postnatal growth) due to dilution.

After an animal reaches maturity, very few changes occur in the quantitative protein composition of skeletal muscle, provided that the animal's nutritional level and muscular activity remain reasonably constant. The amount of extracellular protein increases proportionately in senescence, due to a decrease in myofibrillar and sarcoplasmic proteins.

Nucleoprotein constitutes a large percentage of the total protein in muscle during early prenatal growth. The large nuclei in prenatal muscle cells may constitute as much as 66 percent of all the intracellular protein. During postnatal growth, the percentage of nucleoprotein in muscle fibers declines as the fibers increase in size.

During growth, the percentage of water in muscle decreases as the concentration of proteins and lipids increases. The decrease of water content (in percent) is more rapid during early stages of growth. The percentage decrease becomes progressively less rapid with maturity and senescence.

In general, during growth, there is an overall increase in the total amount of ash (minerals) present in muscle. Even though the total amount of mineral increases, changes do occur in the proportions of one mineral with respect to another. For example, at birth, the concentration of sodium declines, while that of potassium increases.

Quantitatively speaking, fat is the most variable chemical constituent in muscle, and in the animal body. Muscle growth does not necessarily depend upon an increase in fat (lipid) content, but it is generally accompanied by an increase in intramuscular and intracellular lipids. The rate at which intramuscular fat is deposited in muscle increases during periods of high caloric or nutrient intake, and decreases during periods of reduced nutrient intake. The animal's breed is a prime determinant of the age at which it enters the fattening phase of growth. The early maturing breeds and strains are noted for their early fattening ability. An animal's sex influences the rate of fat deposition in muscle. Generally, the muscles of male animals contain less fat than do those of females or castrates.

FACTORS THAT INFLUENCE GROWTH AND DEVELOPMENT

Genetics

Prior to the work of Robert Bakewell (1725–1795) and his contemporaries in England, most European meat animals were nondescript in appearance. The various breeds of livestock as we know them today were developed by mating animals of similar hereditary characteristics over a period of several generations. Even though the offspring of such matings appear uniform in coat color patterns, considerable variation exists in other characteristics, due to the great number of possible combinations of hereditary material. There is little chance that any two animals will be exactly alike, except in the case of identical twins. The female's egg and the male's sperm each contain a specific number of chromosomes. When these come together in the zygote, the chromosomes form pairs. The zygote of swine contains 20 chromosome pairs; that of sheep, 27 pairs, and that of cattle, 30 pairs. There are over one million possible gene combinations in the chromosomes of swine, and over one billion possible gene combinations in cattle.

Animals within a given breed may vary in growth rate and body composition. However, there is a tendency for animals of a given breed to grow and develop in a characteristic manner, and to produce carcasses with distinctive characteristics that are peculiar to the breed. For example, Duroc pigs and Angus cattle are known for their characteristic tendency to deposit intramuscular fat. A major difference between the dairy and beef breeds of cattle is the manner in which fat is distributed among the various fat depots. Carcasses from dairy type animals tend to have a higher proportion of kidney and pelvic fat, and a smaller proportion of subcutaneous fat, than do beef types. Mature size is also a breed characteristic. For example, the mature size of Holstein cattle is usually much greater than that of Jersey cattle. Likewise, Southdown sheep are smaller than Hampshire or Suffolk breeds.

The *phenotypic variations* (outward visible expression of the hereditary constitution of an individual) of meat animals are due to heredity, environment, or to an interaction of both. Both heredity and environment are of great importance in determining the characteristics of any animal. Heredity provides the necessary potential for growth and development, and the environment will tend to maximize or minimize the realization of this potential. The interaction of heredity and environment means that animals with a certain *genotype* (fundamental hereditary constitution of an individual), might perform better in one environment than in another.

Many investigations have been conducted to determine the amount of phenotypic variation due to heredity, and the amount due to environment. The hereditary portion of the phenotypic variance, expressed as a percentage, is called the *heritability estimate.* The percentage of heritability subtracted from 100 then gives an estimate of the variation due to environment.

The heritability estimates presented in Table 4-6 show the extent to which important production and carcass traits are heritable. Furthermore, these data indicate that the livestock producer can make improvements in animals by selecting breeding animals which possess these desirable traits.

TABLE 4-6
Heritability Estimates of Growth and Carcass Characteristics of Cattle, Sheep, and Pigs

Characteristic	Species	Heritability estimate* approximate average
Weaning weight	Cattle	25
	Sheep	33
	Pigs	17
Post weaning rate of weight gain	Cattle	57
	Sheep	71
	Pigs	29
Feed efficiency in feed lot	Cattle	36
	Sheep	15
	Pigs	31
Carcass grade	Cattle	48
Tenderness**	Cattle	61
	Sheep	33
Fat thickness	Cattle	38
	Sheep	20
	Pigs	49
Longissimus muscle cross sectional area	Cattle	70
	Sheep	48
	Pigs	48

*Estimates were compiled from various sources.
**Further information on tenderness is found in Chapter 7.

Nutrition

Although heredity dictates the maximum amount of growth and development that is possible, nutrition governs the actual rate of growth and the extent to which development will be attained. The utilization of ingested nutrients is partitioned among the various tissues and organs according to their metabolic rate and physiological importance. Maintenance and function of vital physiological systems such as the nervous, circulatory, digestive, and excretory systems, take precedence over muscle growth and fat deposition. The order of precedence is as follows: (1) tissues that constitute vital organs and physiological processes, (2) bone, (3) muscle, and (4) fat deposition. During pregnancy, the developing fetus holds a priority similar to that of the vital tissues and organs of the dam herself.

When food is plentiful, all tissues of the body receive sufficient nutrients for maintenance, normal growth, and fattening. However, if the food supply is limited, the tissues are affected in reverse order of physiological importance. Vital tissues are maintained at the expense of others, so that a severe restriction of nutrients will result in body tissues of less importance being utilized to maintain those of more vital importance.

PLANE OF NUTRITION. It is possible to control the rate at which different tissues and parts of the body grow and develop by altering the nutritional level of the animal and selecting the time at which nutritional level is altered. For example, data presented in Figure 4-11 show that when pigs were maintained on a high level of nutrition during the first 16 weeks of postnatal growth, the growth rate of bone, muscle, and fat in the carcass was higher than that of pigs on a low level of nutrition. At 16 weeks, when animals were changed from a low to a high level of nutrition, there was a marked recovery in the growth rate, known as *compensatory* growth. This recovery was greater for fat than for muscle and bone. If the pigs were started on a high level of nutrition and switched to a low level they produced carcasses with more muscle and less fat. However, if the nutritional levels were reversed, low level to high level, the carcasses had more fat and less muscle.

The stage of postnatal growth over which the nutritional treatment is imposed will affect the nature of the response. In the case of the response noted in the low nutritional level that was changed to a high level (Figure 4-11), the greater deposition of fat to the higher nutritional level may be explained on the basis that it occurred during a period when growth intensity of bone and muscle was declining and that of fat was increasing. Thus, if undernutrition occurs early in the postnatal growth period, the long lasting effect is greatest on the earliest maturing tissue (bone) and least on the latest maturing tissue (fat).

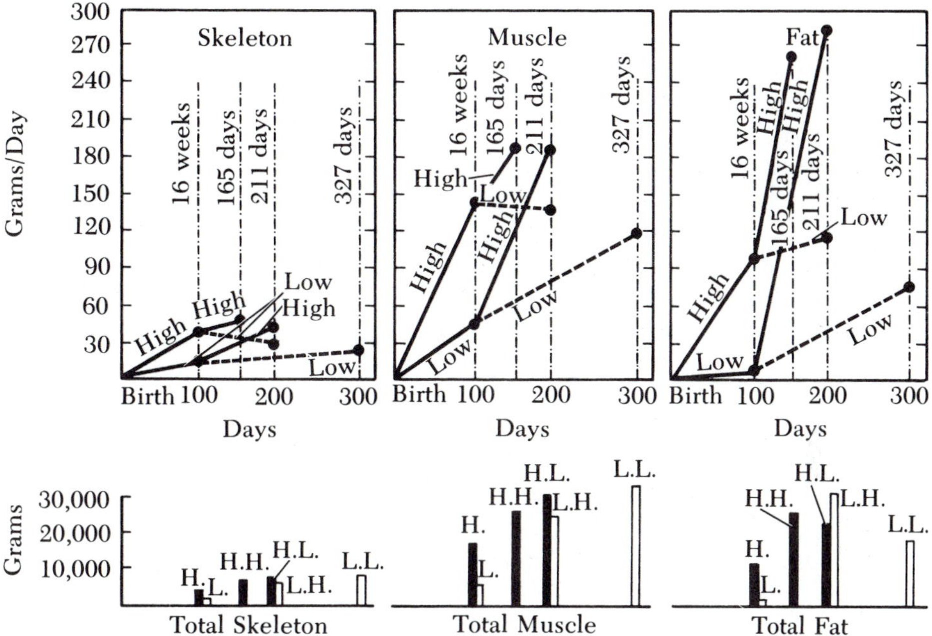

FIGURE 4-11
Effect of plane of nutrition on rate of skeleton, muscle, and fat growth in the carcasses of pigs with liveweights of up to 91 kilograms. (Key: H = high plane of nutrition; L = low plane of nutrition; HL = high plane, then low plane (to 91 kg); HH = high plane (to 91 kg); LH = low plane, followed by high plane (to 91 kg); LL = low plane of nutrition (to 91 kg).) [Data from McMeekan, C. P., "Growth and Development in the Pig, with Special Reference to Carcass Quality Characters; III. Effect of the Plane of Nutrition on the Form and Composition of the Bacon Pig," J. Agr. Sci. 30, 511 (1940).]

The various tissues and organs of the body of a growing animal which have been retarded in development by restricted nutrition may exhibit remarkable compensatory growth when changed to a high level of nutrition. If the animal has not been subjected to severe malnutrition for a long period, and if sufficient time is allowed, the underdeveloped organs or tissues may completely recover from the retarding effects sustained earlier. However, if the undernutrition is severe enough and of long enough duration, irreversible damage may occur.

The efficiency of meat animals, in converting the feed that they consume into meat, is generally related to the level of feed intake. Animals that are full-fed are usually more efficient in converting feed to edible meat than are limited fed animals. Fast growing animals are more efficient than slow growing animals. Also, the efficiency of conversion declines as animals increase in size. As body weight increases more food is required to maintain existing tissues and physiological processes.

The efficiency of animals in converting feed to meat varies among species. Efficiencies range from that of the beef animal, at 10–15 percent, to those of the pig (25–30 percent), chicken (40–45 percent), and fish (65–70 percent).

PROTEIN. Protein is a principal constituent of muscle, organs, and other soft tissues in the animal body. Therefore, an adequate and continuous supply of protein is needed in the animal's food, throughout its life, for the growth and maintenance of its tissues. An important difference between proteins and other nutrients, such as fats and carbohydrates, is that proteins contain nitrogen in addition to carbon, hydrogen, and oxygen. Proteins are comprised of varied amounts and kinds of amino acids, some of which cannot be synthesized in the animals' own body. These are called *essential amino acids*, and must be present in the diet of the animal.

Every animal has a daily need for nitrogen in the form of protein. Although an animal cannot be forced to synthesize tissue proteins beyond its genetic potential by being fed excess protein, the rate of tissue synthesis (growth) is readily reducible by not providing adequate protein. Growth rates in the monogastric will be reduced by an inadequate total amount of protein, a deficiency of any one of the essential amino acids, or an imbalance of amino acids. Likewise, an inadequate total amount of protein, or poor quality protein, will reduce growth rate in the ruminant. If the animal consumes a surplus of protein, the excess is broken down and used as energy, or stored as fat. Any nitrogen that is left over is excreted in the urine and wasted.

FAT. Fats represent a more concentrated form of energy than either proteins or carbohydrates. For this reason fats require more food for their production than does muscle. Fats are used by the animal for energy, and certain fatty acids are essential for growth.

The composition of deposited fat varies among species. Although all meat animals are able to synthesize fat from carbohydrates and proteins, the fat which is synthesized and deposited is characteristic of the species. Fats in the diet of the monogastric animal are assimilated and deposited in a relatively unchanged form; whereas, to a large extent, dietary fats consumed by ruminants undergo changes by rumen bacteria before assimilation. Thus, the fat deposits derived from dietary fat by the ruminant are similar to the fat synthesized in its body from carbohydrates and proteins.

Carcasses from pigs that have been fed a diet containing a specific type of fat will have fat deposits of similar chemical composition to that dietary fat. For instance, when pigs have received a diet high in unsaturated fat, such as peanuts, the fat deposits in the carcass will be very soft and oily. The quality characteristics of such carcasses are objectionable.

Furthermore, the characteristics of the fat in the animal's diet during the last few weeks before slaughter have a greater influence on carcass fat characteristics than do the fats in the animal's diet 1 to 2 months prior to slaughter. This is due to the turnover of depot fats in the adipose tissue, which are continually mobilized into the bloodstream, while fats in the blood are continually being deposited. The newly synthesized fatty acids are formed largely in the liver, passed into the bloodstream, and are deposited either in the place of mobilized fat, or as additional depot fat. Therefore, if pigs on a diet high in unsaturated fat are changed to a diet containing saturated fat, or one that is low in fat content, their fat depots will change from an unsaturated, soft and oily nature to a more saturated, firmer nature.

WATER. Water is absolutely essential for life and growth. An animal can survive long periods without food, but will soon die if deprived of water.

MINERALS. A few examples will illustrate the relationship between growth rate and adequate mineral intake. An early symptom of several mineral element deficiency syndromes is *anorexia* (loss of appetite). Low dietary levels of sodium, phosphorus, zinc, and cobalt will reduce appetite, and, as food intake is lowered, the growth rate is slowed.

The elements, copper, iron, manganese, magnesium, potassium, phosphorus, and zinc are essential components in the activators (*cofactors*) of enzyme systems involved in metabolic processes. Iron is part of the structure of myoglobin and hemoglobin and an iron deficiency causes anemia and weight loss. Iodine is part of the structure of thyroxine, a thyroid hormone that influences growth and metabolic rate.

VITAMINS. Vitamins are organic compounds that are required in minute amounts by an animal, and function as accessory nutrients in facilitating maintenance and growth. Vitamins do not perform a structural or energy-yielding role like amino acids, fats, minerals, and carbohydrates, but function at the cellular level by catalyzing enzymatic processes that are involved in regulation of metabolic processes and growth and maintenance of various tissues. Most vitamins are stored in the liver, but members of the vitamin B complex are stored both in the liver and in muscle.

The dietary requirements for vitamins are not the same for all meat animals. Although various B-complex vitamins are essential for normal ruminant metabolism, their presence in the diet is not essential, since they are synthesized by bacteria in the rumen. On the other hand, swine and poultry require B-complex vitamins in their diets since they are unable to synthesize or otherwise obtain these compounds.

The lack of vitamin A (or of its precursor carotenoid pigments) depresses the growth rate, and a severe vitamin A deficiency may result in death. Vitamin D is required for normal bone growth, and is also related to calcium and phosphorus absorption and utilization by other tissues.

The B-complex vitamins are cofactors of enzymes that are necessary in many physiological processes associated with energy utilization, protein synthesis, and fatty acid synthesis and oxidation. The body's regulatory processes are accomplished by enzymatic processes in which vitamins function as essential components of the enzyme systems. Pyridoxine (vitamin B_6) influences the metabolism of amino acids and essential fatty acids. Niacin aids in the synthesis and metabolism of tryptophan, an essential amino acid. Vitamin A functions in the photoreception process in the retina of the eye, and vitamin K aids in the clotting of blood.

Hormones

Hormones are substances that are secreted into the body fluids by ductless endocrine glands or other tissues. Although ductless glands are commonly designated as the principal source of hormones, other tissues (such as the intestinal tract) are associated with hormone production. The hormones are considered to act as regulators in helping to control various chemical reactions concerned with the maintenance of tissues, growth of tissues, and other physiological processes. Combinations of hormones are involved in the growth process.

The locations of the various endocrine glands and tissues which produce hormones in the female pig are presented diagrammatically in Figure 4-12. The testes of the male (not shown in this figure) also produce hormones that are associated with growth and development.

The ultimate control of all endocrine glands associated with growth rests with the *hypothalamus*, a portion of the brain. Next in the chain of command is the *pituitary gland* which produces *growth hormone*, the hormone that is most directly responsible for growth. It also produces *tropic hormones* that stimulate the activity of other endocrine glands associated with growth and various other physiological processes. Removal of the anterior part of an animal's pituitary, the part which

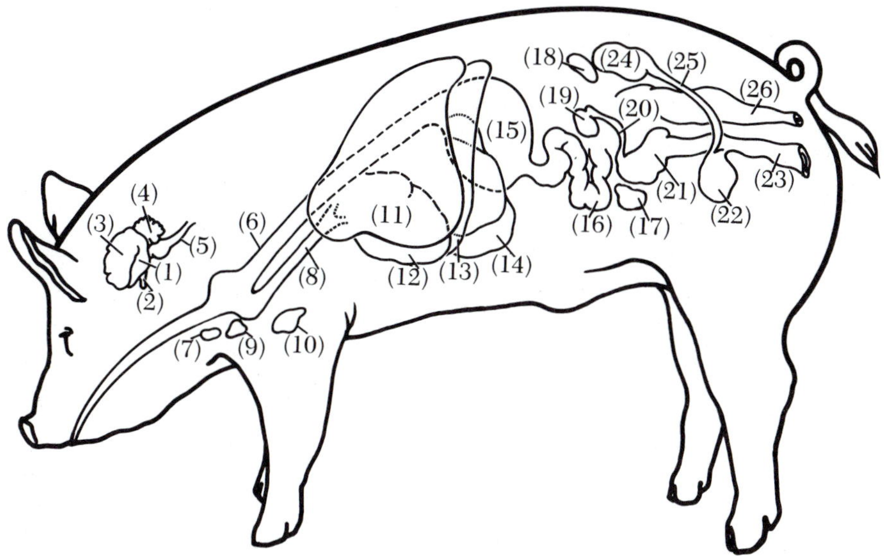

FIGURE 4-12
Location of the endocrine glands in relation to other organs in a female pig. Hormones are secreted by the gastric and intestinal mucosa. Thus, these tissues are also endocrine organs. Endocrine glands are marked with an asterisk: (1) hypothalamus*; (2) pituitary gland*; (3) cerebral cortex; (4) cerebellum; (5) medulla oblongata; (6) esophagus; (7) parathyroid gland*; (8) trachea; (9) lobe of thyroid gland*; (10) thymus gland; (11) lung; (12) heart; (13) diaphragm; (14) liver; (15) stomach; (16) small intestine; (17) pancreas*; (18) adrenal gland*; (19) ovary*; (20) uterine horn; (21) cervix; (22) urinary bladder; (23) vagina; (24) kidney; (25) ureter; (26) rectum.

produces growth hormone, will markedly retard its growth. In such an animal, the administration of extracts made from other anterior pituitaries will restore or accelerate the growth of various tissues, particularly that of bone and muscle. The growth rate of body tissues is believed to decline, as the amount of growth hormone that is available per unit of body tissue declines.

Each of an animal's two *adrenal glands* are composed of two sections, called the *medulla* and the *cortex*. The medulla produces the hormones *epinephrine* and *norepinephrine*, which both profoundly affect the cardiovascular system. In addition, epinephrine mobilizes glycogen from the liver and muscle. The cortex produces three types of *steroid hormones*; the *glucocorticoids, mineralocorticoids,* and *sex hormones.*

The glucocorticoids (sometimes called sugar hormones) are associated with the preservation of the carbohydrate reserves of the body. They *catabolize* (break down) proteins by promoting the mobilization of amino acids from the cells. These amino acids are then available for conversion into glucose (*gluconeogenesis*). In addition to facilitating gluconeogenesis, the glucocorticoids promote both the deposition of glycogen and the oxidation of fat in the liver, and the mobilization of depot fat. The secretion of these hormones is also increased when the animal is subjected to stressful situations, such as heavy muscular exercise, or exposure to cold, trauma, or infection.

The mineralocorticoids are of vital importance in the maintenance of fluid and electrolyte balance. They are often called the salt and water hormones, since they regulate the absorption and the excretion of sodium, potassium, and chloride ions, and water.

The sex hormones are the third group of hormones produced by the adrenal cortex. They have functions in the body similar to those of the hormones produced by the testes and ovaries.

A number of characteristic symptoms occur when there is a deficiency of adrenocortical hormones. Young animals cease to grow, and mature animals lose body weight. Thus, the animal's two adrenal glands play an important role in the metabolic activities of its body, and influences its growth and body composition.

The hormones of the testes and ovaries play an important role in the growth and development of the body. The rate of growth and the development of an animal is associated with its sex condition. It is a commonly accepted practice to castrate male animals, and a less common practice to castrate females. Castrates deposit more fat but muscle growth is less. In boars, castration (in most cases) eliminates the sex odor associated with the mature intact male.

The principal hormones produced by the ovaries are the *estrogens* and *progesterone*. These steroid hormones exert pronounced effects upon tissues of the female reproductive tract.

The principal hormones produced by the testes are *androgens*, steroid hormones that cause pronounced changes in body metabolism. Androgens stimulate the growth of muscles. They increase the rate of protein synthesis, resulting in increased body weight. The increased protein synthesis in skeletal muscles is accompanied by a decrease in fat deposition. Muscles located in the forequarter of the male, especially those in the neck and crest region, show greater development than in females or castrates. The effects of androgens explains this characteristic growth pattern.

When the carcass characteristics of intact male and female pigs are compared at similar weights, the carcasses of boars will yield more lean cuts and less fat than those of gilts. Carcasses from both castrate male (barrows) and female pigs have a lower yield of lean cuts and more fat than carcasses from intact males and females. In numerous studies where the boar and barrow have been compared, the boar had better feed conversion, greater carcass length, less backfat, more lean cuts, but a lower dressing percent than the barrow.

The ram lamb, compared to the wether lamb, will gain faster and produce a carcass that has more muscle and less fat. Likewise, the bull, compared to the steer, makes more rapid and efficient gains and produces a carcass with a higher yield of lean meat and less fat. However, the bull carcass has less intramuscular fat than either the steer or heifer, and consequently has a lower carcass quality grade at the same age or weight. In terms of fattening rates, the heifer normally fattens faster than the steer, and the barrow fattens faster than the gilt.

Synthetic organic chemical compounds with estrogen- and progesterone-like activity may be added to the growing and fattening diets of beef cattle, or administered as subcutaneous implants. The most commonly used of these compounds is *diethylstilbestrol* (DES). The effects of DES on growth in the ruminant is somewhat similar to those of androgens and growth hormone. In steers and heifers DES generally increases appetite, increases the rate of gain, improves feed conversion efficiency, increases muscle growth, and decreases fat deposition.

The effect of DES on growth is variable among species and difficult to explain. In chickens, it increases the rate of fattening. In barrows and gilts, it tends to depress body weight gain. In bulls, it tends to increase the rate of fattening.

In the human female, the administration of large doses of DES during pregnancy has resulted in the development of cases of a rare form of cancer in the adolescent child. Because it is a possible cancer causing agent, DES has been banned in some countries (Australia, Canada, and New Zealand) for use on animals being produced for food. This has been done because, at the time of retail sale, small residues of the compound were detected in the liver of some treated animals. As of early 1975, subcutaneous implants were allowable in the United States, but the future of this compound in animal production is uncertain.

Environment and Management

The environmental conditions under which animals are reared can have a marked influence on growth rate, and even on body composition. Probably the most important environmental condition from the standpoint of livestock production is temperature, which varies from below −18°C to over 38°C in livestock producing areas of the United States.

A warm-blooded animal (*homeotherm*) must maintain a relatively constant body temperature. Heat loss must equal heat production, or else normal physiological processes will be affected. When the environment is hot the animal must dissipate heat; when the environment is cold the animal must conserve body heat, and produce additional heat to keep the body temperature constant. The loss or gain of heat from the body is accomplished by the processes of radiation, conduction, convection, and evaporation. Most heat is lost through evaporation. The homeotherm has the unique ability to conserve body heat by several means. For example, in cold weather, animals grow a heavier hair coat and increase heat production by muscular exercise or by increased production of thyroxine and epinephrine. They also reduce heat lost through evaporation by decreasing their respiration rate. Blood flow to the surface may be reduced, thus reducing heat exchange between the air and the warm skin. In addition, meat animals consume large quantities of feed to "fuel" the high rate of heat production that keeps them warm. During hot weather, the above processes are reversed in the homeotherm and more heat is dissipated, or the production of heat in the body is reduced. The amount of hair in the coat is reduced, more blood is moved to the surface for cooling, respiration rate is increased, feed intake is reduced, and the rate of metabolic processes is slowed down.

The optimum environmental temperature for meat animals is about 15°C to 25°C. For the most part, temperatures below this range have less effect on the overall productivity and efficiency of meat animals than do temperatures that are above this range. The growth rate and efficiency of meat animals is little affected by environmental temperatures down to as low as −10°C. When environmental temperatures approach that of body temperature, then feed consumption and growth rate are depressed.

Cattle native to hot climates thrive in hot environments better than those that are native to countries of temperate climates. In general, animals native to cold or temperate climates are large, compact, and smooth skinned. Animals that have become adapted to the tropics are more rangy in type and have a larger surface area. The Zebu cattle, for example, (*Bos indicus*) can stand heat better than the European breeds (*Bos typicus*). The Zebu cattle have more surface area, large dew lap, pendulous skin with large folds and large ears, and more sweat glands. Therefore, they can dissipate more heat than the European breeds. Crossbreeding of the Zebu and European breeds has resulted in cattle that are better adapted to hot environments than are the original European breeds. Thus, one way to improve (or maximize) growth rate is to select those animals that are best adapted to the particular environment, as well as those having the best genetic potential for growth. In addition, the environmental conditions for animals can be improved by the use of artificial cooling methods. It is a common practice in hot climates to supply water sprays and shade for animals.

Animals are subject to disease, and infestation by insects and parasites, all of which may affect growth rate and body composition. The livestock producer can exert a marked degree of control over the growth and performance of meat animals by controlling these three problems.

ABNORMAL GROWTH

Several factors can cause abnormal growth and development of meat animals. Some of these factors are of a pathological nature. When this occurs the carcass is generally condemned and not used for human food. Pathological changes in skeletal muscle and associated tissues will not be discussed in this chapter. Some examples of abnormal growth and development that are of a nonpathological nature will be presented in the following discussion.

Dwarfism

This condition is characterized in cattle by small size and stocky form. At birth the calf is often weak, and mortality rate is higher than in normal calves. If the calf survives, it soon develops a large stomach and sometimes exhibits labored breathing. The rate of growth is slow compared to that of normal cattle. The head, leg, and body lengths fail to grow in proportion to the width and depth of the body. This results in a very small mature individual that is abnormally compact in form.

The dwarf condition is of genetic origin. Most cattle producers dispose of dwarf calves, and seldom make any attempt to raise the calf to maturity. However, of major concern to the cattle industry is the fact that an individual may be a carrier of the recessive dwarf gene and, if mated to another carrier, the result may be the production of a dwarf. The mating of two dwarfs will always produce dwarf calves. The mating of two drawf carriers will produce, on the average, 50 percent dwarf carriers, 25 percent nondwarf carriers, and 25 percent dwarfs. Mating of a dwarf carrier to a nondwarf carrier will produce, on the average, 50 percent dwarf carriers and 50 percent nondwarf carriers. The dwarf carrier grows and develops similar to an otherwise normal individual.

The frequency of dwarfism in some blood lines, coupled with the fact that dwarfs represent an almost complete economic loss, because of poor growth rate, is a problem of considerable importance. The most successful method of reducing the incidence of dwarfism has been the use of "clean pedigree" animals (animals with no known dwarf-gene carrier in their pedigrees). This has required the culling of many otherwise desirable animals with a known carrier in their pedigrees.

(A)

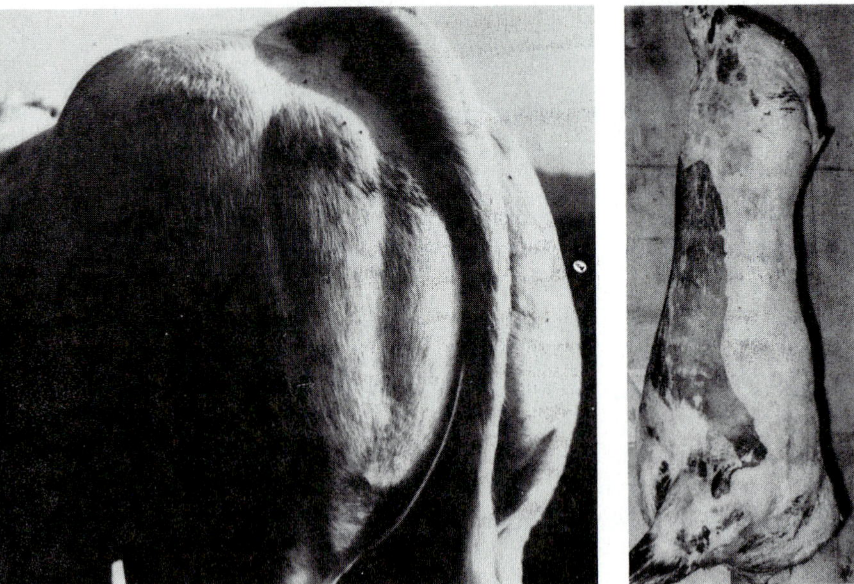

(B) (C)

FIGURE 4-13
Bull and carcass that exhibit "double muscling". Note the two depressions in the muscling of the round (B), and in the forequarter of the live animal (A), and the bulging round and depression of the carcass (C).

Double Muscled Cattle

This condition is characterized by individual animals with unusually thick bulging muscles, especially in the round, as illustrated in Figure 4-13. The term *double muscling*, which is commonly applied to this condition may be a misleading term, in that such individuals have the same number of muscles as normal individuals. Animals exhibiting double muscling have been observed to have a lower proportion of red fibers and a higher proportion of white fibers than normal animals. These white fibers are larger than fibers from normal animals which, along with the increased proportion of white fibers, would account for the muscle hypertrophy. There is also an indication that double muscled animals may have a greater total number of muscle fibers than normal.

This condition is of genetic origin and is quite varied in phenotypic expression. There is not complete agreement as to the mode of inheritance. When double muscled cattle are compared to normal cattle, the growth and development of muscle and rate of gain are greater, but fat deposition is less. The muscle to bone ratio is greater, meaning a greater proportion of muscle in relation to bone. Some interest has developed in the production of such animals, with the objective of increasing the muscle content of the carcass. However, there is evidence that selection of animals for extremes in body form may result in a simultaneous selection for undesirable traits. In the case of double muscled cattle, their fertility is often lower than normal, and calving difficulties are encountered.

Skeletal Muscle Degeneration

Visible symptoms associated with skeletal muscle degeneration include retarded growth, weight loss, lack of vigor (unthriftiness), stiffness with reluctance to move, abnormal gait, and recumbency. When examined histologically, skeletal muscle from animals exhibiting the above symptoms may show degenerative changes. The extent of the histological abnormalities will depend upon the etiology and duration of the condition. These abnormalities of muscle fibers may include an atrophy of the fibers, a swelling of scattered fibers, hyalinization (transparency), granular degeneration, changes in normal staining characteristics, calcification fibrosis, nuclei proliferation, and the transposition of nuclei. In general, skeletal muscle degeneration may result from many artifactual, congenital, traumatic, infectious, toxic, or nutritional factors. Two common examples of muscle degeneration; steatosis, and white muscle disease are discussed in this section.

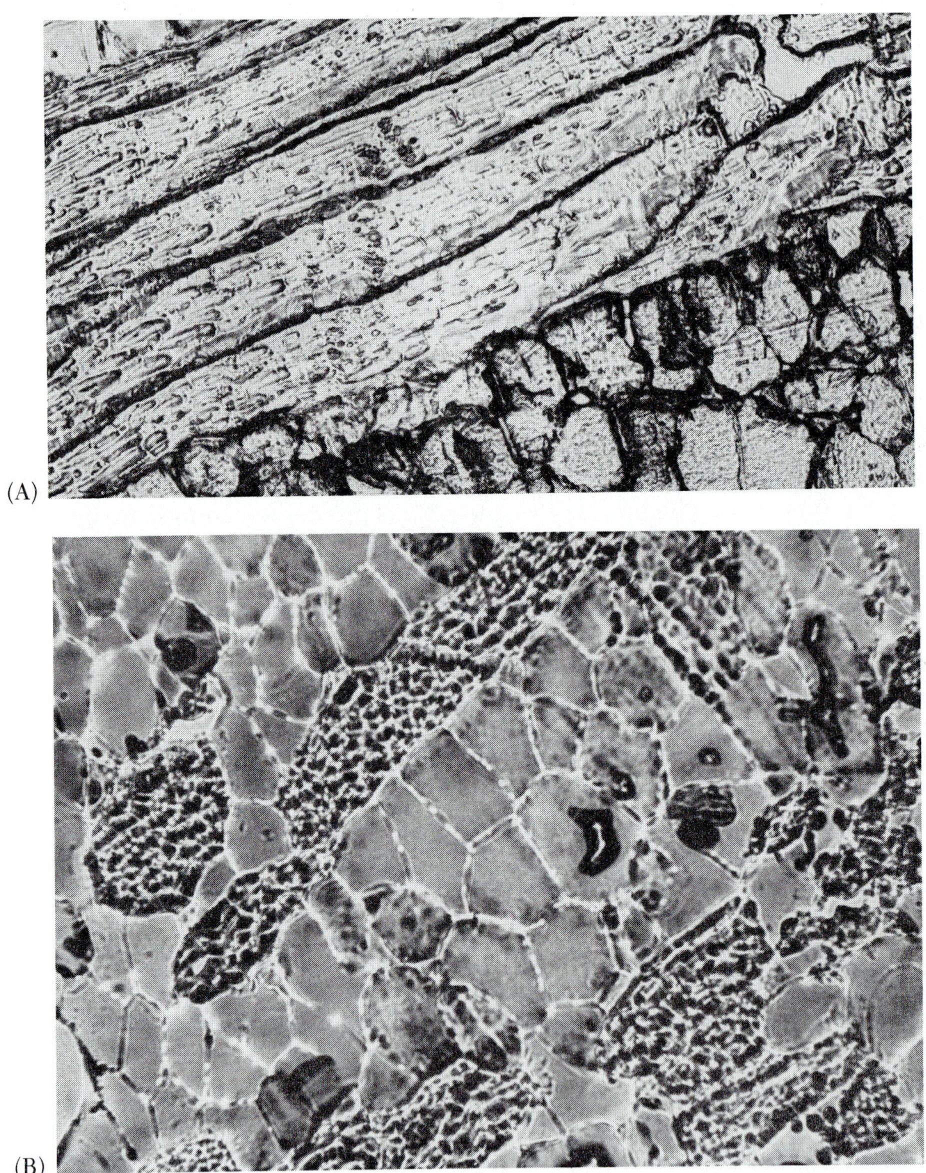

FIGURE 4-14
Longitudinal section (A) and cross section (B) of beef muscle affected with steatosis (fatty degeneration). The granular appearing fibers have been infiltrated with fat. [Photomicrographs courtesy B. B. Chrystall, University of Missouri.]

Steatosis is characterized by extensive fat infiltration into muscle fibers, as illustrated in Figure 4-14. Fat has apparently replaced the contents of muscle fibers. The fresh muscle appeared to be almost completely comprised of fat. The condition may occur in any muscle in the carcass. It is a noninflammatory, nonpathological condition, where no evidence of the condition is visible until the carcass is cut into various wholesale or retail cuts. The condition may occur in beef, lamb, or pork. The presence of the condition detracts from the acceptability of the meat because of the excess fat.

Experimental evidence has shown that nerve stimuli influence fat mobilization and fat deposition. Excessive nerve stimulation will cause a fat loss from adipose tissue, and paralysis or denervation will stimulate deposition of fat. Steatosis could possibly be caused by denervation of the affected muscle.

White muscle disease is a form of dystrophy that may occur in the muscles of young lambs, calves, or pigs. The condition is characterized by white or grayish areas of degeneration, often localized in extent and involving a large group of fibers. The affected muscle is usually edematous and will tear easily. Histologically, single muscle fibers appear transparent and show coagulation necrosis. The cause of the condition is generally recognized as being of nutritional origin. Vitamin E and selenium deficient diets will cause the condition. A dietary excess of unsaturated fatty acids can produce a vitamin E deficiency, and thus can also cause the condition.

REFERENCES

Bailey, Milton E. and S. E. Zobrisky, "Changes in Proteins During Growth and Development of Animals." Body Composition in Animals and Man. National Academy of Sciences, Washington, D. C. (1968).

Bodemer, Charles W., *Modern Embryology* (Holt, Rinehart and Winston Inc., New York), 1968.

Bourne, G. H., *The Biochemistry and Physiology of Bone* (Academic Press Inc., New York), 1956.

Bourne, G. H., *The Structure and Function of Muscle* (Academic Press Inc., New York), 1960, Vol. 1, 2, 3.

Brody, S., *Bioenergetics and Growth* (Reinhold Publishing Corp., New York), 1945.

Hedrick, H. B., "Bovine Growth and Composition," Mo. Agr. Exp. Sta. Res. Bul. 928; N.C. Regional Res. Publication No. 181, (1968).

Joubert, D. M., "A Study of Prenatal Growth and Development in the Sheep," J. Agr. Sci., 47, 382 (1956).

Lodge, G. A. and G. E. Lamming, *Growth and Development of Mammals* (Plenum Press, New York), 1968.

Palsson, H., "Conformation and Body Composition" and "Progress in the Physiology of Farm Animals," Butterworths Scientific Publications, London, 2, 340 (1955).

Winters, L. M., W. W. Green and R. E. Comstock, "Prenatal Development of the Bovine," Minn. Agr. Exp. Sta. Tech. Bul. 151 (1942).

Zobrisky, S. E., H. D. Naumann, J. F. Lasley, D. E. Brady and A. M. Mullins, "Physical Composition of Swine During Growth and Fattening," Mo. Agr. Exp. Sta. Res. Bul. 672 (1958).

CHAPTER 5

The Mechanism of Muscle Contraction

Living muscle is a highly specialized tissue that is capable of converting chemical energy into mechanical energy through its contractions. It becomes a highly nutritious food when converted to meat following the slaughter of an animal. Muscles are positioned and attached to the skeleton in such a way that their contraction and relaxation lead to movement and locomotion. The ability to contract and relax is lost when muscle is converted to meat. Yet, certain aspects of contraction and relaxation in the living muscle relate directly to the shortening and decreased tenderness that occurs in meat during this postmortem conversion. The biochemical processes that provide energy for muscle function in the living animal are the same processes that cause lactic acid production and loss of water holding capacity during the postmortem period. Thus, an understanding of how living muscles function aids in understanding many postmortem properties of muscles in their role as food.

In this chapter, muscle function is discussed from the standpoint of the stimulus that initiates contraction, the mechanical events occurring during contraction and relaxation, and the chemical reactions and processes that furnish energy to the muscle. Before studying the mechanism whereby muscle contracts, generates force, and performs work, a clear understanding of the fine structure of the muscle fiber and myofibril (see Chapter 3) is necessary. Particular emphasis should be given to visualizing the transverse tubules, the sarcoplasmic reticulum, the three-dimensional positions of the sarcoplasmic reticulum and myofibrils, as well as to the structure, banding pattern, and proteins of the myofilaments. It cannot be overemphasized that muscles can contract *only* because of their unique structure.

NERVES AND THE NATURE OF STIMULI

A muscle contraction is initiated by a stimulus that arrives at the surface of the muscle fiber (the sarcolemma). In skeletal muscle, contraction is usually initiated by a nervous stimulus that starts in the brain or spinal cord, and is transmitted to the muscle via a nerve. The structure of neurons, their cell body, dendrites, and axon was described in Chapter 3. Nerve fibers that transmit contractile stimuli to skeletal muscles are called *motor nerves*. They are among the larger fibers in the peripheral nervous system, and are enveloped in a layer of Schwann cells and a myelin sheath. Thus, they are often referred to as *myelinated nerves*. At intervals along the motor nerve fiber, the Schwann cell layer and its associated myelin sheath are interrupted by nodes of Ranvier, at which locations the axon's plasma membrane is partly exposed. The myelin sheath acts as an insulator around the nerve fiber and, in concert with the nodes, allows for a much greater rate of stimulus conduction along the fiber. Myelinated nerve fibers have 30–40-fold greater conduction velocities than do unmyelinated fibers.

Membrane Potentials In Nerve and Muscle

In living cells, under normal resting conditions, an electrical potential always exists between the inside and outside of the cell. These potentials may vary from 10–100 millivolts, depending upon the type of cell, but in resting nerve and muscle fibers the potential is 80–85 millivolts. The fluids that are inside and outside of these fibers contain approximately equal concentrations of positive and negative ions. Generally, a slight excess of negative ions accumulates in the intracellular fluid along the inner surface of the membrane, and a slight excess of positive ions is present along the extracellular surface of the membrane. This separation of opposite charges causes an electrical potential to appear across the cell membrane. This *membrane potential* is positive on the outer surface and negative inside the cell, as is illustrated in Figure 5-1.

The membrane potential in nerve and muscle is the result of (1) the active transport of ions through the membrane, (2) the selective permeability characteristics of the membrane to the diffusion of ions and small

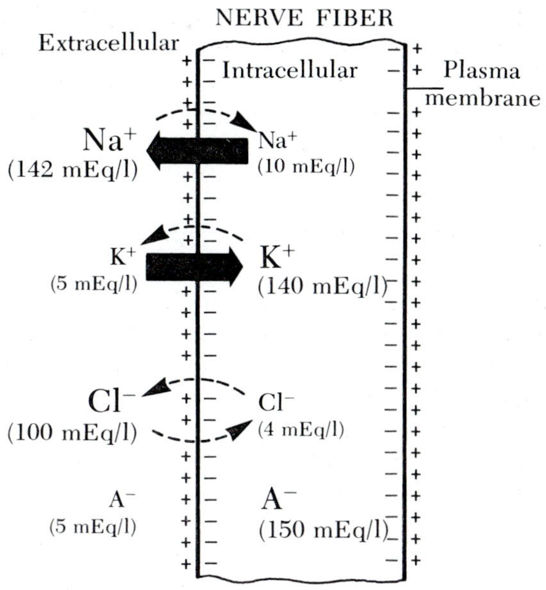

FIGURE 5-1
The establishment of a membrane potential in the normal resting nerve fiber, and the development of concentration differences of sodium (Na^+), potassium (K^+), chloride (Cl^-), and nondiffusable negative ions (A^-), between the two sides of the membrane. Dashed arrows represent diffusion and solid arrows represent active transport ("pumps"). The concentration of each ion is indicated in parentheses in units of milliequivalents/liter. [From Guyton, A. C., *Textbook of Medical Physiology*, 4th ed. W. B. Saunders Co. Philadelphia, 1971.]

molecules, and, (3) the unique ionic composition of the intracellular and extracellular fluids. Extracellular fluid contains high concentrations of Na^+ and Cl^- ions, but very low concentrations of K^+ and nondiffusible negative ions (Figure 5-1). In contrast, concentrations of K^+ and non-diffusible negative ions are very high in the intracellular fluid, and Na^+ and Cl^- ions are quite low. The concentration gradients of Na^+ and K^+ across the plasma membrane are maintained by the active transport of Na^+ out of the cell, and K^+ into the cell. The system which accomplishes the active transport of Na^+ and K^+ is located in the plasma membrane, and is commonly referred to as the *Na^+–K^+ pump*. The energy that is required to pump both types of ions across the membrane against a concentration gradient is furnished by the hydrolysis of adenosine triphosphate (ATP). The permeability of a plasma membrane to the diffusion of K^+ is 50–100 times greater than is its permeability to Na^+ diffusion. Thus K^+, compared to Na^+, passes through the membrane with relative ease. If the ionic concentration gradients across the membrane are also considered, along with the membrane's permeability characteristics, it is evident that K^+ will diffuse out of the cell along its concentration gradient much more rapidly than Na^+ will diffuse in. Of course, the nondiffusible negative ions in the intracellular fluid pass through the membrane only with extreme difficulty. Therefore, the net flow of electrical charges across the membrane is due to the diffusion of positively charged potassium ions, K^+, into the extracellular fluid leaving fewer positive charges inside the cell. As a result, oppositely charged ions will line up on the membrane, with the positive ions on the outer surface attracting negative ions onto the inner surface, thus establishing the membrane potential.

However, the net diffusion of positive charges out of the cell does not continue indefinitely. Once the membrane potential is established, it impedes the further flow of K^+ ions out of the cell. As the potential increases, the greater will be the force opposing K^+ diffusion. Thus, an equilibrium is established between the outward diffusion of K^+ (along its concentration gradient) and the force that opposes this outward diffusion; the positive potential established by the presence of K^+ ions on the outer surface of the plasma membrane.

Action Potentials: the Stimuli

Nerve and muscle fibers exhibit membrane potentials, as do other cells, but they have a unique capability not shared by any other cell type. They are able to transmit an electrical impulse, called an *action potential*, along their membrane surfaces. When an action potential is transferred from a motor nerve to muscle fibers, it initiates muscle contractions. An action potential travels along the membrane surface of the nerve fiber, and is actually a wave of reversing electrical polarization that results from chemical changes in the membrane. Thus, it is often referred to as an *electrochemical process*. As described in the previous section, in the resting state, the membrane is positive on its outer surface. The action potential is initiated by a several hundred to several thousandfold increase in the sodium ion permeability of the membrane. If Na^+ permeability increases to a higher value than that which exists for K^+ ions, the high concentration of Na^+ in the extracellular fluid will cause a more rapid rate of diffusion of Na^+ into the cell, than of K^+ out of the cell. This results in an excess of positive charges inside the cell membrane and negative charges on its outer surface, thus reversing the membrane po-

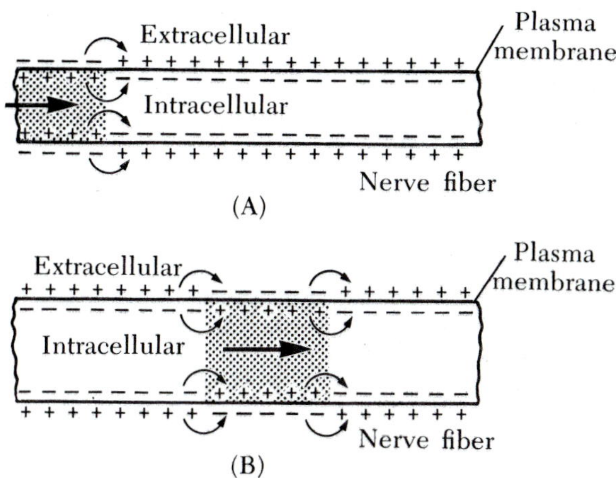

FIGURE 5-2
(A) The beginning of an action potential in a nerve fiber showing the reversed membrane potential. (B) The action potential, as it progresses along a nerve fiber, showing the depolarization at the front and repolarization at the rear. The large solid arrows indicate the direction of the action potential's movement. The curved arrows indicate points of reversing potential in the membrane.

tential (Figure 5-2a). However, the increased permeability to Na^+ ions lasts only for a small fraction of a millisecond. The newly reversed membrane potential reduces the Na^+ permeability to its former low level. The outward flow of the potassium ions can now continue and reestablish the resting membrane potential (Figure 5-2b). These electrochemical events occur on the cell surface during the transmission of the stimulus and can be recorded as an electrical depolarization wave (Figure 5-3). The entire sequence of events in the transmission of an action potential past any point on the nerve fiber requires about 0.5–1 millisecond.

MYONEURAL JUNCTION. The stimulus (action potential) that initiates muscle contraction is transferred from the nerve fiber to the muscle fiber at the *myoneural* (neuromuscular) *junction.* At this junction, the motor nerve branches into several terminal endings that are implanted in small invaginations of the sarcolemma (Figure 5-4). These terminal endings adhere tightly to, but do not penetrate, the sarcolemma. The combined structures of the myoneural junction form a small mound on the surface of the muscle fiber, called the *motor end plate.*

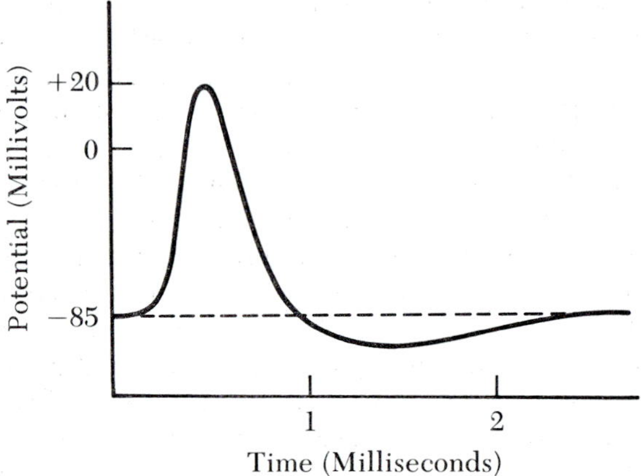

FIGURE 5-3
An action potential recorded at a point on the cell surface as the depolarization-repolarization wave passes it.

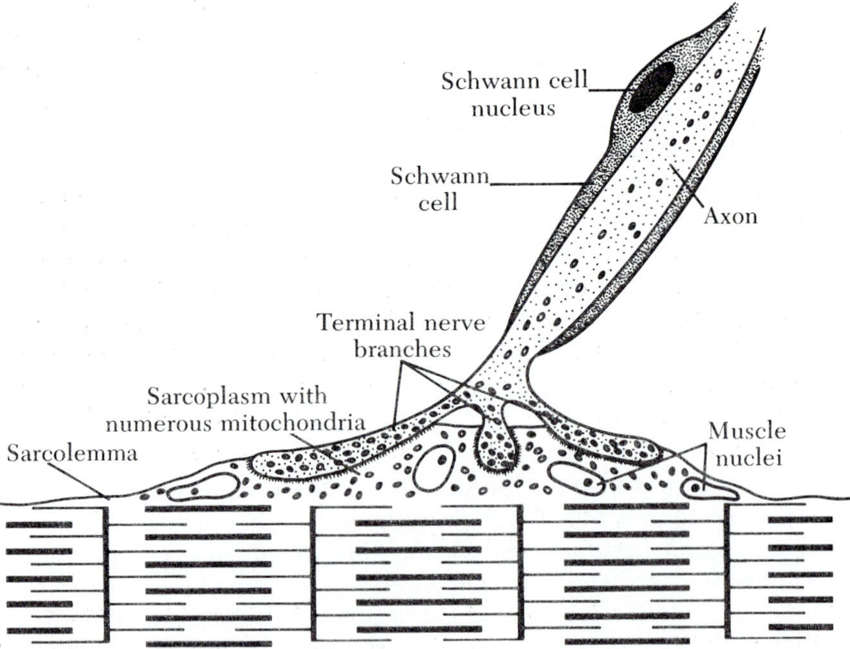

FIGURE 5-4
A schematic representation of the motor end plate, as seen in a histological section along the longitudinal axis of a muscle fiber. [Modified from R. Couteaux, *Structure and Function of Muscle*, G. H. Bourne, ed. Academic Press, New York, 1960.]

Although muscle fibers can be stimulated into contraction by a strong man-made electrical impulse, the electrical impulse of the action potential in the nerve is not strong enough to accomplish this task by itself. A special mechanism "amplifies" the electrical signal, and transfers it to the muscle fiber. When the action potential arrives at the end plate, it causes a chemical "transmitter," *acetylcholine*, to be released. Acetylcholine is stored in small vesicles found in the terminal nerve branches, and the contents of a few vesicles are released with each action potential. When it is in contact with acetylcholine, the sarcolemma becomes more permeable to Na^+ ions, its polarity reverses, and an action potential (similar to that in a nerve fiber) is propagated along its length. Acetylcholine acts on the sarcolemma for only a few milliseconds, since it is rapidly destroyed after its release by *cholinesterase*, an enzyme that is present in high concentrations in myoneural junctions. Some extremely potent poisons act at the myoneural junction. The toxin produced by the bacterium *Clostridium botulinum* (responsible for botulism food poisoning), and curare (of poison arrow fame) both prevent the transmission of impulses from nerves to muscles by interfering with the release or action of acetylcholine.

ACTION POTENTIALS IN MUSCLE. The action potential that occurs in a muscle fiber when acetylcholine contacts the sarcolemma is nearly identical to that which occurs in a nerve fiber. The major difference is in the duration of the action potential; 5–10 milliseconds in skeletal muscle fibers, as compared to 0.5–1 millisecond in nerve fibers.

Most muscle fibers have only one myoneural junction from which the stimulus is transmitted to all parts of the fiber. The action potential begins at the myoneural junction and progresses longitudinally, in both directions, along the sarcolemma and stimulates the entire length of the fiber. It is transmitted to each myofibril in the interior of the fiber by the transverse tubule (T tubule) system. The T tubules originate as invaginations of the sarcolemma and course inward from the sarcolemma across the muscle fiber (Chapter 3). The action potential enters the interior of the fiber along the T tubules and, at the triad, is transferred to the sarcoplasmic reticulum that surrounds each myofibril.

CONTRACTION OF SKELETAL MUSCLE

The contraction of skeletal muscle directly involves four of the myofibrillar proteins: actin, myosin, tropomyosin, and troponin. Actin and myosin are the contractile proteins, and form the actin and myosin filaments of the myofibril. Crossbridges formed between the filaments by myosin generate the contractile force during contraction. In the relaxed state, a muscle generates very little tension, and can easily be stretched by a force pulling on it. In terms of an explanation at the level of the myofilaments, this means that there are no crossbridges between the actin and myosin filaments, and the filaments of each sarcomere slide passively over one another. During the development of rigor in postmortem muscle, permanent crossbridges form and prevent the sliding of these filaments, so that the muscle becomes inextensible. Tenderness changes associated with rigor are discussed in Chapters 6 and 12. In contrast to actin and myosin, tropomyosin and troponin play the role of regulatory proteins; they assist in turning the contractile process "on" and "off".

Relaxed muscle has a very low level of ionic calcium (Ca^{2+}) in the sarcoplasmic fluid that bathes the myofibrils. Expressed in terms of molarity, there is a concentration of less than 10^{-7} moles/liter of free Ca^{2+}. The total concentration of calcium in skeletal muscle is more than 1000 times this level (greater than 10^{-4} moles/liter). However, nearly all of the calcium that is present in the muscle fiber is bound in the sarcoplasmic reticulum, and is not found free in the sarcoplasm. In order for muscle to remain in the relaxed state, it must also contain a relatively high concentration of adenosine triphosphate (ATP). Most of the ATP is found in the form of a magnesium ion (Mg^{2+}) complex. The Mg–ATP complex must be present in order to prevent the interaction of actin and myosin (crossbridge formation). It may be helpful to make the analogy that Mg–ATP is the "lubricant" that permits actin and myosin filaments to slide easily past one another. When the sarcoplasmic Ca^{2+} concentration is low (less than 10^{-7} moles/liter), and the Mg–ATP concentration is high, troponin and tropomyosin inhibit crossbridge formation between the actin and myosin filaments. Following the death of an animal the Mg–ATP supply is depleted, allowing the formation of actin–myosin crossbridges, and the development of rigor, as is more fully explained in chapter 6.

When the action potential is transmitted from the sarcolemma to the interior of the fiber along the T tubules, it causes bound calcium to be released from the sarcoplasmic reticulum into the sarcoplasmic fluid. The increased free Ca^{2+} concentration in the sarcoplasm is the trigger that initiates the contractile mechanism. It is necessary to increase the concentration of free Ca^{2+} ions to about 10^{-6} or 10^{-5} moles/liter (a 10- to 100-fold increase) to initiate a contraction.

When calcium ions are released into the sarcoplasm, they are bound by troponin. This relieves the inhibition that troponin and tropomyosin exert on crossbridge formation between actin and myosin in the relaxed state. Myosin is now free to form crossbridges between the filaments. These crossbridges develop a contractile force, and the actin filaments in each half of the sarcomere are pulled toward the center of the sarcomere (Figure 5-5). The protein complex formed when actin and myosin interact at the crossbridge is called *actomyosin*. During contraction, the

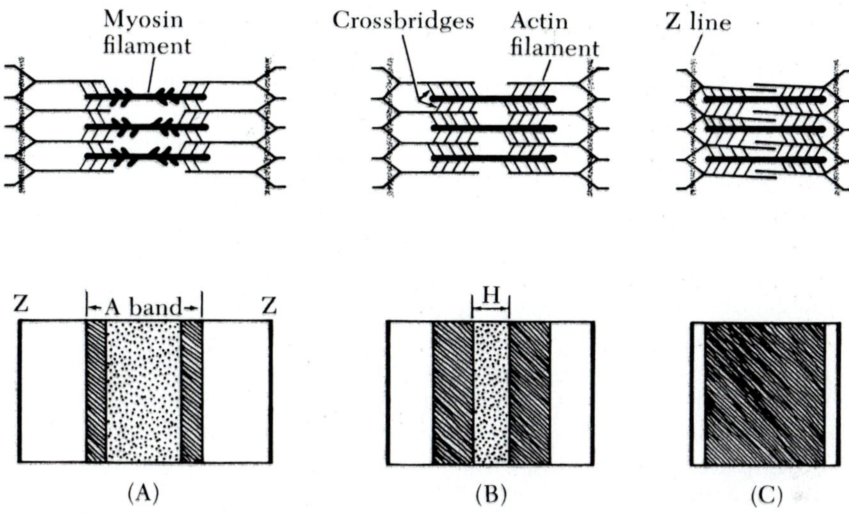

FIGURE 5-5
One sarcomere is shown at various stages of shortening during contraction (top). The banding pattern in a muscle fiber at corresponding degrees of contraction (bottom). Contraction is shown for the following stages: (A) in a muscle that is extended, (B) at rest length, and (C) severely shortened. [Modified from W. Bloom and D. W. Fawcett, *A Textbook of Histology* (W. B. Saunders Co., Philadelphia), 1968.]

length of individual actin and myosin filaments does not change. Rather, the filaments slide along each other, pulling the Z lines closer to the myosin filament, thereby decreasing the sarcomere length. Each crossbridge generates enough force to slide the actin filament a few angstroms (1 angstrom (Å) equals 10^{-8} cm) longitudinally along the myosin filament. During contraction, filament sliding requires a cyclic "making" and "breaking" of crossbridges, with each cycle contributing a small amount to the total contraction.

The A band width is constant during all phases of muscle contraction, but the I band and H zone widths change. These widths are largest when the muscle is stretched (Figure 5-5a), and decrease as the muscle shortens (Figure 5-5b). In severely contracted muscle, the actin filaments meet in, or actually overlap, the center of the A band, and the Z lines may abut on the ends of the myosin filaments (Figure 5-5c). Under these conditions, the H zone and I band are not discernable in electron micrographs.

Shortening phenomena, resembling contraction in the living animal, can be observed in postmortem muscle before and during rigor development. These "contractions" can cause a marked toughening of meat, or a loss of natural meat juices. Normal rigor development leads to some shortening and tension development in muscle. The phenomenon of *cold shortening* occurs when prerigor muscle is exposed to temperatures between 0–15°C, and is probably caused by the cold induced release of Ca^{2+} ions from the sarcoplasmic reticulum. *Thaw rigor* occurs upon the thawing of muscle that has been frozen before rigor development (Chapter 7).

Muscle contraction requires an additional expenditure of energy, besides that normally consumed by the resting muscle. This energy is derived from ATP in a reaction, catalyzed by the enzyme myosin ATPase, in which the ATP is hydrolyzed to adenosine diphosphate (ADP) and inorganic phosphate. The enzyme system responsible for this reaction is located in the head region of the myosin molecule. The ATP splitting activity of myosin ATPase is greatly enhanced by the Ca^{2+} ions released into the sarcoplasm. Thus, the increased free Ca^{2+} ion concentration in the sarcoplasm promotes crossbridge formation between the actin and myosin filaments, and simultaneously increases ATP splitting, yielding the required chemical energy. The crossbridges between actin and myosin convert the chemical energy into mechanical energy, and initiate filament sliding, thus generating a contractile force.

RELAXATION OF SKELETAL MUSCLE

The relaxation of skeletal muscle is defined as the reestablishment of the resting state, and it can be measured by the decrease in muscle tension. In order for relaxation to occur, the conditions prevailing during the resting state must be reestablished; namely, a concentration of 10^{-7} moles/liter (or less) of intracellular free Ca^{2+} ions in the sarcoplasm, and a relatively high concentration of ATP. Therefore, relaxation can proceed only when the events that activated the contractile process are reversed. The first step in the relaxation process occurs in the repolarization half of the action potential, which returns the membrane potential to its resting value. Subsequent steps in the relaxation process are then able to occur.

The intracellular free Ca^{2+} ion concentration is next returned to its original low (rest) level by the action of the sarcoplasmic reticulum. This is done by the tubules of the sarcoplasmic reticulum, which remove the excess calcium from the sarcoplasmic fluid and bind it in an inactive form. This bound calcium in the tubules is then transferred to the terminal cisternae of the triads for storage and subsequent release when the next stimulus arrives. As the free Ca^{2+} ion concentration in the sarcoplasm decreases, the troponin molecules release the Ca^{2+} that had bound to them during initiation of the contraction. As the troponin loses this Ca^{2+}, it is again able to inhibit the formation of crossbridges. In the resulting absence of crossbridges, tension is not generated, and the filaments slide passively over one another. Thus troponin is a calcium ion sensitive protein. When the Ca^{2+} concentrations are low, troponin plus tropomyosin inhibits crossbridge formation between actin and myosin. When Ca^{2+} concentrations increase, in response to a stimulus, troponin binds Ca^{2+} and is no longer able to inhibit contraction.

Calcium accumulation by the sarcoplasmic reticulum must be accomplished against a calcium concentration gradient. Therefore, an active "pumping" process is involved, probably quite similar to the process that establishes and maintains the sodium and potassium gradients across nerve and muscle fiber membranes. An active pumping process of this type requires energy, in order to overcome the concentration gradient. This energy is furnished by the enzymatic hydrolysis of ATP to adenosine diphosphate (ADP) and inorganic phosphate. The enzyme responsible for ATP hydrolysis is not the myosin ATPase which liberates energy for contraction. Rather, ATP hydrolysis is accomplished by a separate enzyme system that is associated with the membranes of the sarcoplasmic reticulum. This enzyme system is sensitive to intracellular free Ca^{2+} concentration, and is activated by increases in free Ca^{2+}. Thus, the release of calcium from the terminal cisternae of the triads has a 2-fold effect: (1) it activates the contractile machinery to produce a contraction and, (2), it simultaneously activates the processes that pump calcium back into the terminal cisternae to end the contraction.

SOURCES OF ENERGY FOR MUSCLE CONTRACTION AND FUNCTION

From the discussion presented to this point, it should be evident that ATP is the ultimate source of energy for the contractile process, for the pumping of calcium during relaxation, and for maintaining the sodium and potassium gradients across the sarcolemma. Of these three uses of energy, contraction is by far the most expensive process. It has been estimated that during a single muscle twitch, the contraction uses 1000 times more energy than does reversal of the membrane potential, and at least 10 times more energy than does the calcium pump in the sarcoplasmic reticulum. Yet the amount of ATP present in the muscle is sufficient to supply energy for only a few twitches. Therefore, very rapid and efficient means must be available for the resynthesis of ATP within the living muscle.

When an animal is slaughtered, muscle does not instantaneously stop living and become meat. ATP continues to provide energy for all of the above muscle functions for a period of time. Pathways which provide for ATP synthesis by rephosphorylation in the living muscle also attempt to maintain the ATP level after death. The biochemical reactions that occur in muscles during their attempts in the initial postmortem period to maintain the conditions that prevailed during life, cause profound changes in their properties. These reactions constitute a major part of the processes described as the "conversion of muscle to meat" in Chapter 6.

The most immediate source of energy that can be mobilized for ATP synthesis is phosphocreatine. This is accomplished according to the reaction: ADP + phosphocreatine $\rightleftharpoons$ ATP + creatine. The enzyme that catalyzes this reaction is *creatine kinase.* As indicated by the double arrow, the reaction is reversible. However, the equilibrium lies quite strongly in the direction of ATP synthesis. Therefore, ATP broken down during a contraction is rapidly restored. In fact, this half of the equilibrium reaction occurs so rapidly that, unless special precautions are taken to block the reaction, ATP breakdown during a single muscle twitch cannot be demonstrated; only phosphocreatine breakdown is seen. The concentration of phosphocreatine in resting muscle is about twice that of the resting level of ATP. Therefore, it also is subject to depletion during extended periods of contraction and must be replenished during rest periods by other mechanisms.

The most efficient mechanism for ATP synthesis is a series of reactions collectively referred to as *aerobic metabolism,* in which food nutrients, such as carbohydrates, proteins, and lipids, are degraded to carbon dioxide and water, while part of the energy released is used to form ATP. The degradation products of these nutrients may all ultimately pass through this series of reactions. The degradation of the simple sugar, glucose, is used, in this section, to illustrate aerobic metabolism.

Recall, from your reading of Chapter 3, that glucose is stored in muscle as glycogen, and as such constitutes about 1 percent of the muscle by weight. The first part of the process, consisting of 12 sequential reactions, is called *glycolysis* (Figure 5-6). The entire process occurs in the sarcoplasm, and the enzymes that catalyze each step are soluble sarcoplasmic proteins. The glycogen is first split into glucose 1-phosphate units, each of which is then divided into two 3–carbon fragments; the end product of glucose breakdown is pyruvic acid:

$$H_3C-\overset{\overset{\displaystyle O}{\|}}{C}-COOH$$

The useful energy yield from glycolysis is 3 rephosphorylations (3 ADP → 3 ATP), and 4 hydrogen ions (H^+) per glucose 1-phosphate molecule split from glycogen. These ions are accepted by a carrier compound, nicotinamide adenine dinucleotide (NAD^+), and transported to the mitochondria for use in the third part of the mechanism, which is a further rephosphorylation.

The second part of the mechanism, a series of reactions called the *tricarboxylic acid cycle* (TCA cycle) occurs in the mitochondria. In this

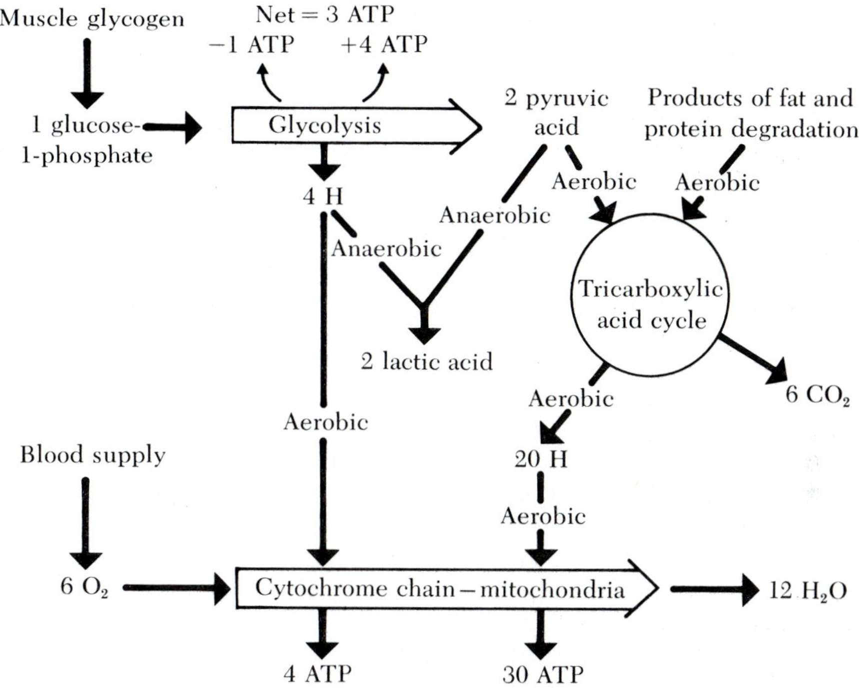

FIGURE 5-6
A diagrammatic illustration of the pathways that supply energy for muscle function. One molecule of glucose 1-phosphate split from glycogen, is degraded to CO_2 and H_2O in the glycolytic pathway, tricarboxylic acid cycle, and cytochrome chain. The energy yield (in terms of molecules of ATP produced) is indicated at each step. When oxygen is limited, energy may be supplied by glycolysis and conversion of pyruvic acid to lactic acid.

cycle, compounds derived from pyruvic acid (manufactured in the glycolysis process just explained) are sequentially broken down into carbon dioxide, and hydrogen ions. Carbon dioxide diffuses away, eventually entering the blood stream as waste, and the hydrogen ions are accepted by the carrier NAD^+, to form the compound NADH. The degradation products of fatty acids and proteins also enter into this cycle and are converted into useful energy (Figure 5-6).

Most of the useful rephosphorylation occurs in the third part of the mechanism, the *cytochrome chain*. The cytochromes are a group of iron containing enzymes that are located in the mitochondria, along with the TCA cycle enzymes. In the cytochrome chain, hydrogen ions from glycolysis and the TCA cycle are transferred from NAD^+, and are combined with molecular oxygen to form water. A large part of the energy released is used to rephosphorylate ADP, and the remainder is lost as heat. For each pair of hydrogen ions from the TCA cycle, 3 ATPs are produced.

In addition, 2 ATPs are produced for each pair of hydrogen ions that are released in glycolysis.

If one molecule of glucose is split from glycogen and carried through this entire sequence of reactions, the net ATP yield would be obtained as follows. From glycolysis, 3 ATP molecules are obtained, along with 4 hydrogen ions, which will yield 4 more ATP molecules in the cytochrome chain. At the end of glycolysis one glucose molecule will yield 2 pyruvic acid molecules. Each pyruvic acid molecule will yield 10 hydrogen atoms, for a total of 20 hydrogen atoms in the TCA cycle. These 20 hydrogen ions are converted, in turn, to 30 ATPs, in the cytochrome chain. Thus, when a single glucose molecule, derived from glycogen, is degraded to carbon dioxide and water, 37 ADP molecules are converted to 37 ATP molecules.

If a muscle is working slowly, and oxygen is supplied in adequate amounts, aerobic metabolism and phosphocreatine breakdown can adequately supply most of its energy requirements. However, when the muscle is contracting rapidly, its oxygen supply becomes inadequate for support of ATP resynthesis via aerobic metabolism. Under these conditions of oxygen shortage, a third mechanism, *anaerobic* metabolism, is able to supply energy for a short time. A major feature of anaerobic metabolism is the accumulation of lactic acid, which is explained in the following discussion. When the oxygen supply is inadequate, the hydrogen ions released in glycolysis and the TCA cycle cannot combine with oxygen at a sufficiently rapid rate. Thus, they tend to accumulate in the muscle. The excess hydrogens are then used to convert pyruvic acid to lactic acid, which permits glycolysis to proceed at a rapid rate. As stated above, each glucose yields 3 ATP molecules in glycolysis, so anaerobic metabolism can supply energy for muscle function. It is important to emphasize that the amount of energy available via this anaerobic route is limited. Lactic acid accumulation in the muscle lowers its pH, and at pH values of less than 6.0–6.5 the rate of glycolysis is drastically reduced, with a proportional reduction in ATP resynthesis. Under these conditions, fatigue develops quite rapidly, and the muscle is no longer able to contract due to insufficient energy, as well as because of the presence of the excess acidity (low pH).

During the muscle's recovery from fatigue, the lactic acid that has accumulated is transported out of the muscle via the blood stream, and is reconverted to glucose in the liver. ATP and phosphocreatine, the energy stores, are replenished by the processes of normal aerobic metabolism. The recovery process can occur quite rapidly for a slight fatigue, but can require extended periods if the fatigue is severe.

The cyclic process of energy provision for muscle contraction and function is summarized in Figure 5-7. To follow this process, begin with the gastrointestinal tract where nutrients (potential energy producing

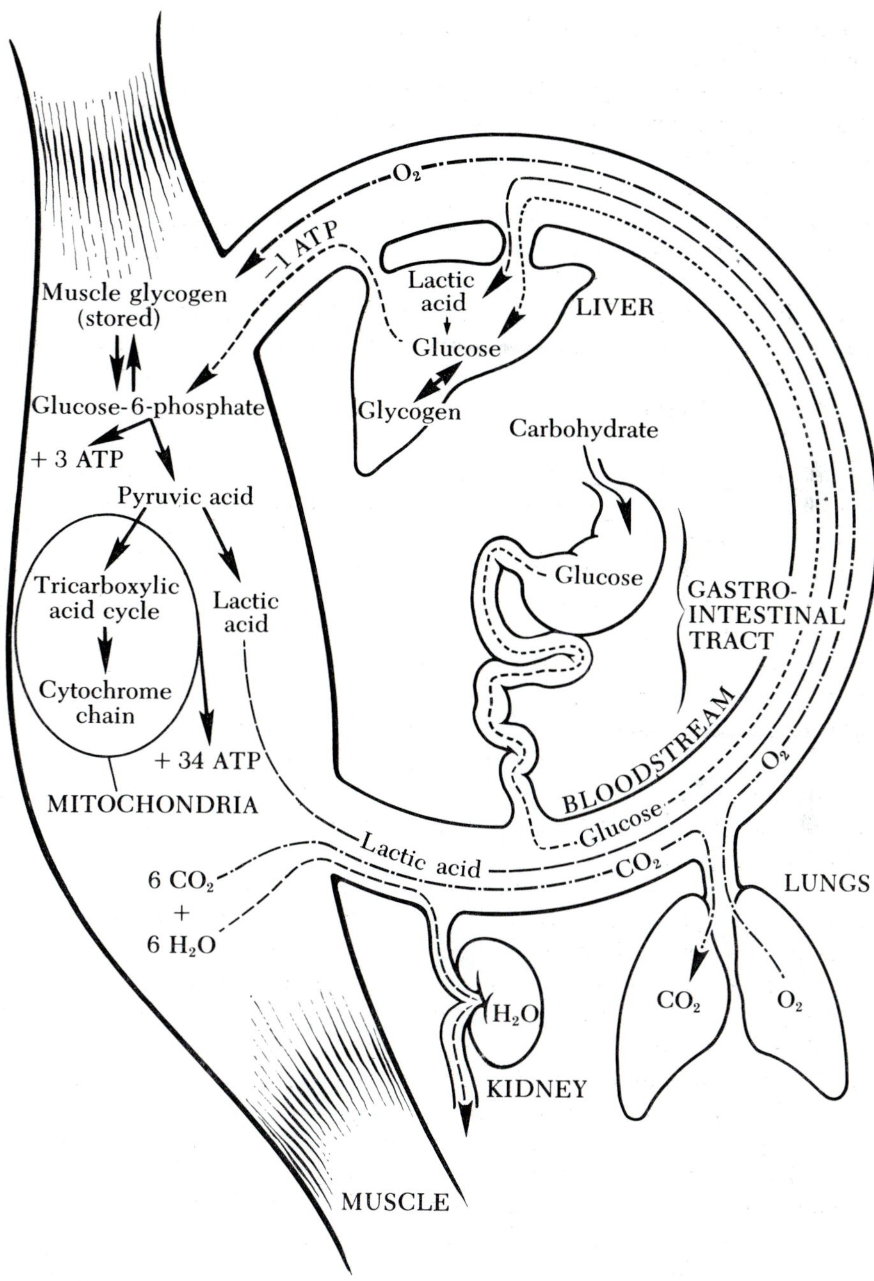

FIGURE 5-7
The cyclic nature of the pathways that provide energy for muscle contraction is summarized in this diagram.

compounds) are absorbed into the body; in the case illustrated, this is glucose. Glucose is carried by the circulatory system either to the liver (where it is converted to glycogen) for storage, or to the muscle where it may be metabolized for energy immediately, or stored as glycogen for future use. Liver glycogen can be hydrolyzed to glucose and carried to the muscle as needed. In the muscle, glycogen is metabolized to pyruvate in the glycolytic pathway described earlier (yielding 3 ATPs). Pyruvate is metabolized either in the TCA cycle and cytochrome chain to H_2O and CO_2 (yielding 34 ATPs), or it is converted to lactic acid. Lactic acid, H_2O, and CO_2 are removed from the muscle by the blood stream. Carbon dioxide is expelled from the body through the lungs, H_2O is excreted through the kidneys, and the lactic acid is resynthesized in the liver to glucose. Thus, we can see that a dynamic system is available for providing energy to muscle. It is only during periods of very rapid muscle contraction that this system is unable to keep pace with the energy demand. But when this does occur, fatigue develops rapidly, and the muscle must cease to contract for awhile to allow time for recovery.

REFERENCES

Guyton, A. C., *Textbook of Medical Physiology* (W. B. Saunders Co., Philadelphia), 1971.

Huxley, H. E., "The Mechanism of Muscular Contraction," Sci. Am., 213, 18 (1965). Offprint No. 1026, W. H. Freeman and Company, San Francisco.

Maruyama, K. and S. Ebashi, "Regulatory Proteins of Muscle," in *The Physiology and Biochemistry of Muscle as a Food, II*, E. J. Briskey, R. G. Cassens and B. B. Marsh, eds. (University of Wisconsin Press, Madison), 1970, pp. 373–381.

Murray, J. M. and A. Weber, "The Cooperative Action of Muscle Proteins," Sci. Am., 230, 58 (1974).

Wilkie, D. R., *Muscle* (St. Martin's Press, New York), 1968.

PART III

MEAT SCIENCE

CHAPTER 6

Conversion of Muscle to Meat

Muscle has evolved to the highly complex contractile system described in the previous chapter, and provides mobility for the animal kingdom. The details of muscle ultrastructure and its important life functions, contraction and relaxation, have been discussed in considerable depth in preparation for a more complete understanding of the process of conversion of muscle from its physiological locomotive function to its use as a food.

For many centuries people have used animal tissues for food without paying much attention either to their living functions, or to the changes that might take place in them before they are consumed. But, with the advent of the more centralized mass production techniques now used by the meat industry, the search for methods to control quality and uniformity in the end product has increased. This has led to a search for the causes of variation in meat quality with an eye toward quality improvement. A surprisingly large number of the variations in meat quality are now known to be caused by changes that take place after slaughter. The animal's muscle does not suddenly stop all its living functions and become meat. Instead, a number of physical and chemical changes take place over a period of several hours or even days. This process is referred to as the conversion of muscle to meat. In terms of the living state, this can be thought of as a gradual degradative process that, if it were allowed to contine indefinitely, would lead to a complete breakdown of the tissue to its constituent elements. Hence, there is a need for the preservative actions that are discussed in Chapters 10 and 11.

HOMEOSTASIS

An understanding of the living function of muscle is basic to the understanding of the changes which take place during its conversion to meat. In the living state, all organs and systems within the body cooperate to maintain an internal environment under which each can perform its function efficiently. Most organs in the body, including muscle, function efficiently only within a narrow range of physiological conditions (pH, temperature, oxygen concentration, and energy supply).

The maintenance of a physiologically balanced internal environment is termed *homeostasis*. This is a system of checks and balances that provides the body with a means of coping with the stresses that tend to alter the internal environment. Homeostatic regulation gives an organism the ability to survive under many different and sometimes adverse environmental conditions, including extreme variations in temperature, oxygen deficiency, and trauma.

The homeostatic mechanism is presided over by the nervous system and the endocrine glands. These two systems serve as communication and triggering mechanisms that coordinate adjustments in the function of various organs during periods of stress.

The concept of homeostasis becomes very important during the conversion of muscle to meat for two reasons: (1) many of the reactions and changes that occur during the conversion are a direct result of homeostasis (attempts to maintain life) and, (2) conditions in the immediate preslaughter period may alter postmortem changes and affect meat quality. This includes transportation of the animal to market, its handling during the marketing process, and its stunning or immobilization in preparation for slaughter. The consequences of these factors are discussed in Chapter 7.

EXSANGUINATION

The first step in conventional slaughter is to *exsanguinate* the animal, that is, to remove as much blood as possible from its body.

Exsanguination marks the beginning of a series of postmortem changes in the muscle. It is not hard to imagine that massive bleeding is an extreme stress. As soon as the blood pressure begins to drop, the circulatory system will adjust its functions in an attempt to maintain a blood supply to the vital organs. Heart pumping action will increase and peripheral vessels will constrict in an attempt to maintain pressure and hold blood in the vital organs. In fact, only about 50 percent of the total blood volume can be removed from the body, the remainder being held mainly in the vital organs. This in itself is an important consideration in meat production. Blood is an excellent medium for the growth of spoilage organisms, and excess blood in meat cuts is unappealing to the consumer. Therefore, a thorough bleeding is an essential beginning to the slaughter process.

CIRCULATORY FAILURE TO THE MUSCLES

What happens in a muscle as a result of the loss of its blood supply? The function of the circulatory system is to transport essential nutrients to the muscle, and carry waste products away from the muscle, either for excretion, or for further metabolism in other organs. Exsanguination eliminates this line of communication between the muscle and its external environment. Figure 5-7 diagrammatically shows the importance of the circulatory system as a main line of transport between the muscle and other organs.

Let us first consider the loss of the oxygen supply to the muscle. Muscles store only a limited amount of this important metabolite bound to the pigment myoglobin. In the living animal, oxygen is picked up in the lungs and carried to the cells of the body by the hemoglobin pigment of the blood. The myoglobin of the muscle cell has a greater attraction for oxygen than does hemoglobin, a characteristic that helps transfer oxygen from the blood to the muscle cells. The myoglobin provides a place for storing the oxygen until it is used by the cells for metabolism. The amount of oxygen stored in this manner is only sufficient to support oxidative reactions for a short period of time.

As the stored oxygen supply becomes depleted after exsanguination the aerobic pathway through the citrate cycle and the cytochrome system must stop functioning. Energy metabolism is then shifted to the anaerobic pathway, in much the same fashion as when there is insufficient oxygen for the living muscle during periods of heavy exercise and oxygen debt. Thus, we encounter another homeostatic mechanism in which an alternative source of energy is available to the muscle. Figure 5-6 shows that considerably less energy in the form of ATP is produced through the anaerobic path. However, the muscle is provided with a source of energy that will maintain the structural integrity and temperature of the cells for a while longer.

As was pointed out in Chapter 5, in the living animal the lactic acid produced by anaerobic metabolism is transported from the muscle to the liver where it us used in the synthesis of glucose and glycogen. These products can then be recycled to the muscle for energy when sufficient oxygen is again available. Since the circulatory system is no longer available to the exsanguinated animal, the lactic acid remains in the muscle and increases in concentration as metabolism proceeds. It will continue to accumulate until nearly all the original glycogen stored in the muscle is depleted, or until conditions are reached that slow or stop anaerobic glycolysis. Lactic acid accumulation causes a lowering of pH in the muscle. The ultimate pH of meat will depend, to a considerable extent, on the amount of glycogen contained in the muscle at the time of exsanguination.

POSTMORTEM pH DECLINE

The lowering of pH in muscle due to the accumulation of lactic acid is one of the most significant postmortem changes that occurs in a muscle during its conversion to meat. The rate at which the pH decline proceeds after the animal has been exsanguinated, and the extent of the total drop in pH are both highly variable. (See Chapter 7 for a discussion of the factors that cause variation in the rate and extent of pH decline.)

A normal pH decline pattern in pork muscle is represented (as shown in Figure 6-1) by a gradual decrease from approximately pH 7 in living muscle to a pH of about 5.6–5.7 within 6–8 hours postmortem, and then to an ultimate pH (reached approximately after 24 hours postmortem) of about 5.3–5.7. In some animals the pH drops only a few tenths of a unit during the first hour after slaughter and then remains stable at a relatively high level, giving an ultimate pH in the range of 6.5–6.8. In other animals, the pH will drop rapidly to around 5.4–5.5 during the first hour after exsanguination. Meat from these animals will ultimately develop a pH in the range of 5.3–5.6.

The accumulation of lactic acid early in the postmortem period can have an adverse effect on meat quality. The development of acidic (low pH) conditions in the muscle, before the natural body heat and the heat of the continuing metabolism have been dissipated through carcass chilling, causes denaturation of the muscle proteins. The amount of denaturation depends upon how high a temperature and low a level of pH is reached. Temperature appears to play a key role in denaturation, since muscle can attain a fairly low pH (5.2–5.4) after it has been thoroughly chilled without excessive denaturation occurring. The proteins in some species are more sensitive to this type of denaturation than are the proteins of others. For example, fish muscle proteins are more labile

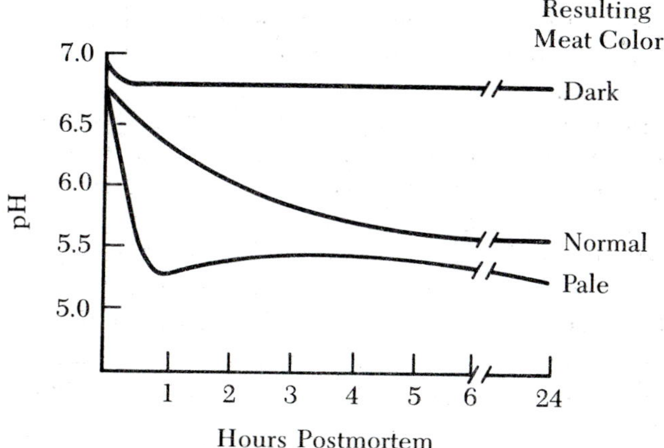

FIGURE 6-1
Postmortem pH decline curves. [Modified from Briskey, E. J., "Etiological Status and Associated Studies of Pale, Soft, Exudative Porcine Musculature," Adv. Food Research 13, 89 (1964) Academic Press, Inc.]

than most mammalian muscle and will become denatured at lower temperatures and a higher pH.

Denaturation of the proteins causes a loss of protein solubility, loss of water holding capacity, and a loss in the intensity of the muscle's pigment coloration. These changes are all undesirable, whether the muscle is going to be utilized as fresh meat or be subjected to further processing.

Muscles which have a very rapid and extensive pH decline will be pale in color, and have very low water holding capacity that will give a cut surface a very wet appearance. In severe cases, fluid will actually drip from the surface of the muscle. On the other hand, muscles that maintain a high pH during the conversion of muscle to meat will tend to be very dark in color, and very dry on the exposed cut surface, because the naturally occurring water is tightly bound to the proteins.

POSTMORTEM HEAT PRODUCTION AND DISSIPATION

As a result of exsanguination, an important temperature control mechanism in muscle, the circulatory system, is lost. No longer can heat from the deep parts of the body be carried rapidly to the lungs and other surface areas for dissipation. Therefore, a rise in muscle temperature occurs soon after exsanguination. How high the temperature rises depends upon the rate of metabolic heat production and on its duration. Figure

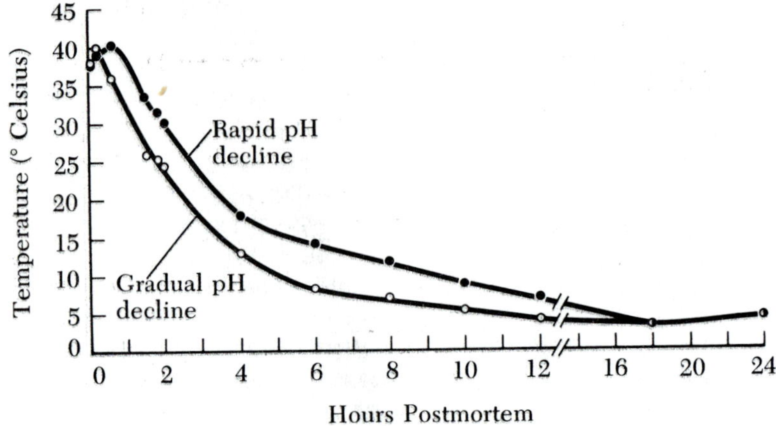

FIGURE 6-2
Postmortem temperature decline curves. [Modified from Briskey, E. J. and J. Wismer-Pedersen, "Biochemistry of Pork Muscle Structure; I. Rate of Anaerobic Glycolysis and Temperature Change versus Ultimate Muscle Structure," J. Food Sci. 26, 306 (1961).]

6-2 shows some of the variation in heat production and dissipation encountered in pork muscle that can be accounted for by metabolic differences. The size and location of a muscle within the body, and the amount of fat cover insulating it, will also influence the ultimate temperature rise and rate of heat dissipation. An important point to be made here is that the metabolic factors that cause the temperature to rise in postmortem muscle are the same ones that cause pH to decline. Therefore, denaturation becomes inevitable unless some means is employed to artificially remove heat from the muscle.

External factors associated with the slaughter process will also influence heat dissipation. The scalding and singeing processes to which some carcasses are subjected contribute to a delay in heat dissipation. The ambient temperature in the slaughter room, the length of the slaughter and dressing operation, and the temperature of the initial chill cooler will all have a considerable influence on the rate of carcass temperature decline.

RIGOR MORTIS

One of the most dramatic postmortem changes that occur during the conversion of muscle to meat is a stiffening of the muscles after death. The phenomenon of rigor mortis (Latin for "stiffness of death") has been recognized for hundreds of years. However, only during the past four or five decades has it been studied in connection with the conversion of muscle to meat.

The stiffening observed in rigor mortis is due to the formation of permanent crossbridges in the muscle between the actin and myosin filaments. This is the same chemical reaction that forms actomyosin during muscle contraction in life. The difference between the living state and the rigor state is that relaxation is impossible in the latter state, because no energy is available for breaking the actomyosin bond.

ATP complexed with Mg^{2+} (as noted in Chapter 5) is required for a muscle to be maintained in the relaxed state. As ATP is depleted from the muscle, permanent crossbridges begin to form. In the absence of ATP, this reaction is irreversible although it is recognized that muscles do not remain stiff indefinitely. The apparent "resolution" of rigor mortis is probably due to a physical degradation of the muscle structure. Current evidence indicates that this degradation occurs in the Z line area of the muscle.

Physical changes, such as a loss of elasticity and extensibility, shortening, and an increase in tension accompany the development of rigor mortis. The phenomenon of extensibility is the one that is most often used to follow the development of rigor mortis. Figure 6-3 shows a typical recorded pattern of the variation of muscle extension during the development of rigor mortis. During the period immediately following exsanguination muscle is quite extensible. If the muscle is loaded, or if a force is applied to it, the muscle will passively stretch; and, when the force is removed, the natural elasticity of the muscle will return it to its original length. At this time, few if any actomyosin bridges are present

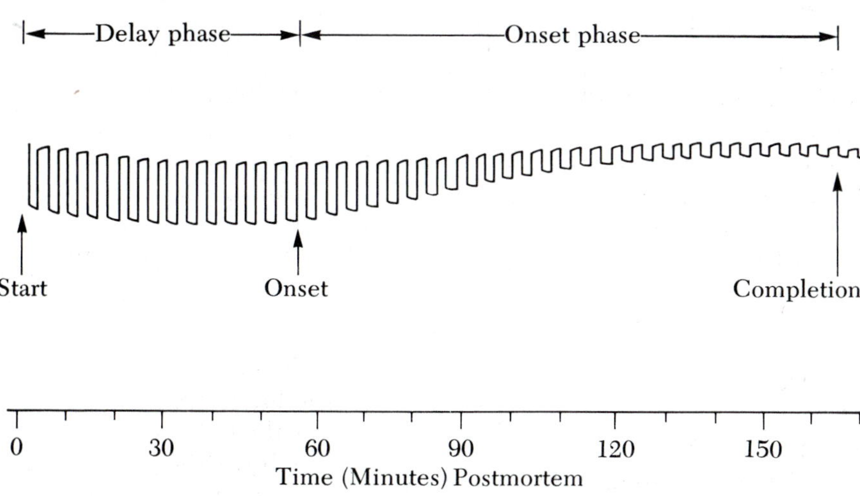

FIGURE 6-3
A typical pattern of extensibility change during the development of rigor mortis.

to prevent extension by a force. The period of time during which the muscle is relatively extensible and elastic is called the *delay phase* of rigor mortis.

After the muscle glycogen stores are depleted, the store of creatine phosphate (CP) is used for the rephosphorylation of ADP to ATP. As the stores of CP are depleted, the rephosphorylation of ATP becomes insufficient to maintain the muscle in a relaxed state. Actomyosin bridges begin to form, and the muscle gradually becomes less extensible under an externally applied force. The *onset phase* of rigor mortis begins when the muscle begins to lose extensibility, and this phase lasts until the *completion* of rigor mortis. When all the creatine phosphate is gone, and ATP can no longer be formed from ADP, the muscle becomes relatively inextensible. This signals the completion of rigor mortis.

The pattern of development of these physical changes varies from animal to animal, and even from muscle to muscle in the same animal. These variations are now known to be associated with properties of postmortem muscle that are important in the utilization of muscle as meat.

SHORTENING AND TENSION DEVELOPMENT

Since the actomyosin bond formed during the development of rigor mortis is the same as that formed during a muscle contraction, rigor mortis can therefore be thought of as an irreversible muscle contraction. Muscles shorten as permanent bonds form during the development of rigor mortis and, as a result, a tension develops within the muscle that contributes to the muscle stiffening. Rigor shortening differs from a normal contraction since more crossbridges are formed. During a normal contraction, crossbridges form at only about 20 percent of the possible binding sites, but in rigor nearly all the binding sites in the area of overlap between the actin amd myosin filaments will be utilized.

RELATION BETWEEN pH DECLINE AND THE DEVELOPMENT OF RIGOR MORTIS

The two major topics just discussed are closely correlated because both are related to energy metabolism, particularly to the metabolism of glycogen. Meat that undergoes the pH changes shown in the top and bottom curves of Figure 6-1 will have a rapid development of rigor mortis. In meat where the pH decline follows the top curve, the onset and completion of rigor mortis will be rapid because the initial energy supply is limited. In the case where the pH decline follows the bottom curve, the onset and completion of rigor will also be rapid because the energy supply is either rapidly metabolized, or low pH conditions will curtail important chemical reactions in the energy metabolism. The intermediate pH decline curve will be associated with a longer time of rigor mortis development than either of the two extremes.

LOSS OF HOMEOSTASIS

With increasing time postmortem, all homeostatic mechanisms are eventually lost. As the metabolites that are stored in a muscle are depleted, the temperature begins to decline because there are no mechanisms left for maintaining it at the original level. Nervous control from the central nervous system will be lost within 4–6 minutes after exsanguination. Uncontrolled nerve impulses may arise locally, causing muscles to twitch for a considerable time after exsanguination. The possibility that these local impulses affect postmortem metabolism is supported by the fact that muscles that have rapid delay and onset phases will not respond to external electrical stimulation for as long as muscles that have followed the normal (middle) pH decline curve shown in Figure 6-1.

LOSS OF PROTECTION FROM BACTERIAL INVASION

In the healthy living animal, the muscle is protected from invasions of microorganisms by a series of defenses, the first line of which is the tissues that cover the body and surround many of the internal organs. Connective tissues and cell membranes may also offer some protection. The lymphatic system and the circulating white blood cells are available to destroy organisms which enter the body. During the conversion of muscle to meat, membrane properties are altered and the muscle becomes susceptible to bacterial invasion. Since the circulatory and lymphatic systems are no longer operating, they cannot prevent the spread of microorganisms. Most of the changes that occur during the conversion of muscle to meat favor microorganism proliferation. However, the lowering of muscle pH will have an inhibitory effect on some microorganisms. Extra care must be exercised to prevent the contamination of meat with spoilage organisms during all postmortem handling and storage.

LOSS OF STRUCTURAL INTEGRITY

During the development of rigor mortis, the muscle structure will have much the same microscopic appearance as does an intact living muscle. However, it is almost certain that subtle degradative changes, such as the alteration of membrane properties, begin soon after exsanguination. The disintegration of Z line structure begins in some species soon after the completion of rigor mortis and, as stated earlier, may account for the loss of stiffness in muscles.

Electron photomicrographs taken at increasing lengths of time postmortem show a progressive disruption of the myofibrillar structure. This degradative change proceeds at different rates in different animals, and can be associated with quality differences in muscle when it is used as a food.

ENZYMATIC DEGRADATION

In muscle cells, proteolytic enzymes called *cathepsins* are held in an inactive state in organelles called *lysosomes*. As the pH of the muscle drops, these enzymes are released and begin to degrade the protein structure of the muscle. The release of cathepsins may be responsible for at least a portion of the observed postmortem structural changes.

The tenderization that occurs during the postmortem aging of beef muscle may be due (in part) to the breakdown of some of the collagen connective tissues of the muscle by cathepsins. In addition to proteolysis by enzymes, the muscle proteins may be subjected to a denaturation (as previously discussed). Collagenous connective tissues are also denatured as a result of a rapid pH drop in muscle. In this case, the connective tissues surrounding the loin eye muscle in the pork carcass have been degraded to the point where the muscle is no longer firmly attached to the skeletal structure of the carcass.

The above discussion is not intended to exclude the possibility that many of the changes attributed to protein denaturation may also be caused by enzymatic degradation, since the early pH drop would be favorable for the early release of the cathepsins. In this case structural changes in the proteins that are truly hydrolytic in nature have not been separated from those that would be classified as denaturation.

CHANGES IN THE PHYSICAL APPEARANCE OF THE MUSCLE

Color

In the living animal, muscle with a sufficient oxygen supply has a bright red appearance. If the muscle were in oxygen debt it would be darker red, or purple, in appearance.

In postmortem muscle, as the oxygen is used up, the muscle becomes dark purplish red in color. When fresh meat is first cut, the exposed surface will have this dark red appearance. Upon exposure to the atmosphere for a few minutes, the myoglobin will become oxygenated and change to a brighter red color. If the muscle has been subjected to severe denaturation, the color intensity will be considerably reduced and appear pale even when freshly cut. (The complete chemistry of myoglobin and its color changes is discussed in Chapter 8.)

Firmness

Living muscles maintain a certain amount of "tone", and are usually attached at both ends, either directly or indirectly, to some part of the skeleton, and are therefore relatively firm. As the muscles go into full rigor they become very firm and stiff. Later in the conversion process, as enzymatic degradation and protein denaturation proceed, the muscle becomes less firm. If the protein denaturation is extremely severe, the muscles will become very soft.

Water Binding Properties

Water accounts for 65–80 percent of the total muscle mass. In the living muscle cell, as in any other living cell, water plays a major role in cellular function. It acts as a solvent or carrier for substances that must be transported within the cell, is a lubricant, maintains the turgidity of the cell, and is an essential component of many chemical reactions. Much of the water in a muscle cell is bound rather tenaciously to the various proteins. If the proteins are not denatured they will continue to bind water during the conversion of muscle to meat, and (to a great extent) even through the cooking process. This retained water contributes to the juiciness and palatability of meat as a food.

The changes that occur in water binding, during the conversion from muscle to meat, depend upon the rate and extent of the pH drop and on the amount of protein denaturation. In cases when the pH of the postmortem muscle remains very high, the water binding properties of meat are similar to that of living muscle. When the pH drops rapidly during the conversion of muscle to meat, a low water binding capacity will result. The mechanisms by which water is held in meat are discussed in Chapter 8.

POSTMORTEM CHANGES AND MEAT UTILIZATION

Certain aspects of tenderness, juiciness, color, and flavor may all be influenced by the changes that occur during the conversion of muscle to meat. Important processing characteristics, such as emulsifying capacity, binding properties, cooking losses, and cooked meat color may also be affected by these changes. From the discussion thus far, it is obvious that postmortem changes are highly variable, and influence the way in which muscle can best be utilized as meat. These changes can be controlled, to a certain extent, to improve product quality. Therefore, the antemortem and postmortem factors affecting conversion that are discussed in Chapter 7 are an extremely important part of the overall effort to control and improve the quality of the meat supply.

REFERENCES

Bate-Smith, E. C. and J. R. Bendall, "Factors Determining the Time Course of Rigor Mortis," J. Physiol, 110, 47 (1949).

Bendall, J. R., "Postmortem Changes in Muscle," in *The Structure and Function of Muscle*, G. H. Bourne, ed. (Academic Press, New York), 1960, Vol. 3.

Briskey, E. J., "Etiological Status and Associated Studies of Pale, Soft, Exudative Porcine Musculature," Adv. Food Research, 13, 89 (1964).

Briskey, E. J., R. G. Cassens and J. C. Trautman, *The Physiology and Biochemistry of Muscle* (The University of Wisconsin Press, Madison, Milwaukee, and London), 1966.

Lawrie, R. A., *Meat Science* (Pergamon Press, London), 1966.

Price, J. F. and B. S. Schweigert, *The Science of Meat and Meat Products* (W. H. Freeman and Company, San Francisco), 1971.

CHAPTER 7

Factors Affecting Postmortem Change and the Ultimate Properties of Meat

When animals are moved to unfamiliar surroundings, they may become excited, fatigued, overheated, or chilled. All these conditions result from responses within the animal body that are caused by various factors in the new environment. When referring to the reactions of animals under these conditions, it is often noted that such animals are experiencing stress. The term *stress* is a general expression referring to the physiological adjustments, such as the changes in heart rate, respiration rate, body temperature, and blood pressure that occur during the exposure of the animal to adverse conditions. Such conditions, called *stressors*, occur when the environment becomes uncomfortable or hazardous to the animal.

PHYSIOLOGICAL RESPONSES DURING STRESS

The animal body has a store of many natural defenses against adverse conditions. These defenses attempt to maintain those internal conditions that enable the animal to continue its life processes (homeostasis). For example, low temperatures result in shivering and other heat producing

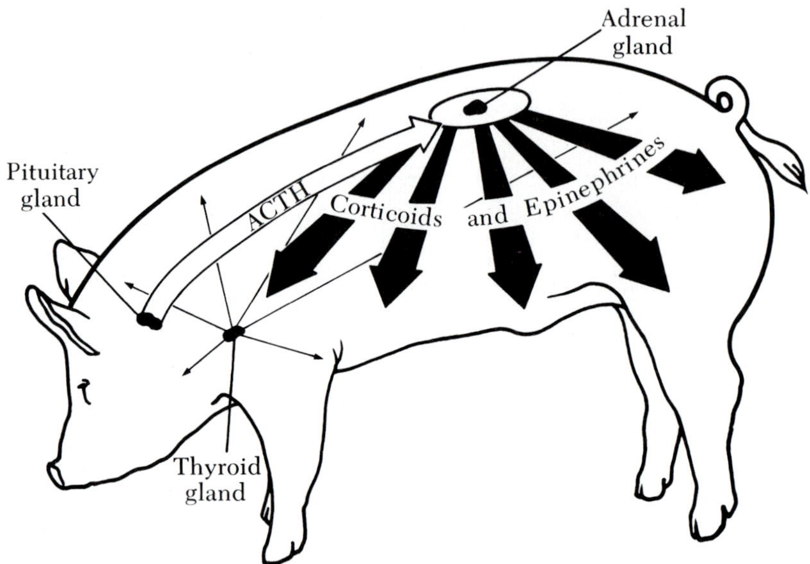

Influences of the "Stress" Hormones

ACTH – stimulation of corticoid release
Corticoids – synthesis of carbohydrates; maintenance of cell response and ion balance
Epinephrines – maintenance of circulation; breakdown of glycogen and lipid
Thyroid hormones – stimulation of oxidative metabolism

FIGURE 7-1
A simplified diagram of the sites of production, and the general actions of the major hormones that are related to stress reactions.

activities. Some of the energy stored in the form of carbohydrates or fat is released and converted to body heat by metabolic processes.

The adjustments in metabolism that occur during periods of stress are aided by the release of certain hormones (Figure 7-1). Hormones of importance in this regard are epinephrine and norepinephrine from the adrenal medulla, adrenal steroids from the adrenal cortex, and thyroid hormones from the thyroid gland. These hormones influence many chemical reactions, some of which come into play in almost every type of stress condition. In general, the adrenal hormones are those which provide stress resistance. Epinephrine helps break down the glycogen that is stored in the liver and muscles, as well as the fat that is stored in several locations in the body, in order to provide a ready source of energy. Epinephrine and norepinephrine also help to maintain proper blood circulation by their influence on the heart and blood vessels. Hormones from the adrenal cortex are also effective in reinforcing the ability of the tissues to respond during stress. Thyroid hormones increase metabolic rate, and thereby provide increased available energy to the animal.

The consequences of stress, and the metabolic adjustments associated with it, can be noted in the muscles. These adjustments usually place an increased demand on the muscles for contraction, so there is a need for an increased rate of blood flow in the muscles. Circulatory adjustments are therefore made on the basis of nutrient and oxygen demand, and on the need for temperature regulation. Many times the circulatory system is unable to provide the quantity of blood needed to maintain proper temperature and to support the contractions taking place in the muscles. At such times the muscle temperature will rise, and the muscle will become depleted to below normal levels of stored oxygen.

When the "stress" hormones are released, the muscles become prepared to meet the demand for contractions that an emergency might require. In this event the anaerobic pathway for glycogen breakdown, leading to the formation of lactic acid, is favored by the action of epinephrine. Consequently, a shift in type of metabolism occurs in the living animal in the same way as described for rapidly contracting muscle (Chapter 5) and postmortem muscle (Chapter 6).

When the demands on muscle tissue are so great that aerobic metabolism produces an inadequate amount of energy, and anaerobic breakdown of glycogen occurs, unusual conditions may develop within the muscles. Since the end point of glycogen breakdown under these conditions is lactic acid, there is a buildup of this metabolic product, particularly in muscles where white fibers predominate. Since lactic acid cannot be broken down in white skeletal muscle fibers, it must be removed and transported to the liver for transformation to glycogen, or to the heart, where it can be directly utilized for energy. Animals that accumulate excess lactic acid in their muscles require a cessation of (or adaptation to) the conditions causing the stress, and adequate time to rid their muscles of the excess lactic acid. If the quantities of lactic acid entering the blood stream are too great for the liver and heart to neutralize, a generalized acidosis will develop, and death may occur. In swine, this condition is the basis of many death losses from what is known as the *porcine stress syndrome.*

There is wide variation in the tendencies of animals to accumulate excess lactic acid, but the anaerobic breakdown of glycogen can occur in the muscles of all animals. If the circulatory system is capable of removing the lactic acid, the muscle simply develops a glycogen deficiency as a result of the stress. With adequate time and nutrition, the glycogen can be restored to the muscle in normal quantities (approximately 1 percent by weight). Variations in muscle glycogen level are directly associated with variations in the color and texture of meat that are described in this chapter under the head "Stress and Muscle Characteristics."

ENVIRONMENTAL ELEMENTS

Many elements of the environment can become stressful to animals under certain conditions. Extremes in temperature, humidity, light, sound, and space are effective in producing the changes already described in this chapter. However, they differ in their effects because the response that any one environmental condition will produce depends on the species, weight, age, sex, inherent stress resistance, and emotional state of the animal. The following discussion refers to the most common effects that the various stress factors produce in the muscles after the animal has been exposed to them for a short time.

Temperature

No stress factor causes a wider range of effects than does the environmental temperature. When temperatures become cooler than those to which the animals are acclimated, they begin to utilize additional processes to produce and conserve body heat. Shivering may cause a reduction in muscle glycogen levels, since this type of contraction is accompanied by an increased blood flow in the muscles and does not result in a buildup of lactic acid. On the other hand, unusually high temperatures place heavy demands on the animal's cooling mechanisms. Many times the muscle temperature will rise because the animal's body is unable to rid itself of heat rapidly enough. Animals with a poor ability to remove heat from their bodies can develop muscle temperatures of up to 42–43°C. Such high temperatures speed up the metabolic reactions, such as ATP splitting and glycolysis, that occur in times of stress, and may ultimately lead to death of the animal.

Humidity

The amount of moisture in the air (humidity) can influence the severity of the temperature effects just described. In fact, it is the interaction of these two environmental elements that largely determines the comfort of an animal. In general, high levels of humidity increase the discomfort of animals that are exposed to cold or hot temperatures for short periods. When animals require cooling, moisture in the air makes heat loss by breathing (evaporative cooling) more difficult. In cold environments, moisture in the air increases the rate of direct heat loss from the body.

A temperature–humidity index has been developed for the estimation of the combined effects of temperature and humidity (at warm tempera-

DRY BULB TEMP.	5	10	15	20	25	30	35	40	45	50	55	60	65	70	75	80	85	90	95	100
75									70	70	71	71	72	72	73	73	74	74	75	75
76							70	70	70	71	72	72	72	73	74	74	74	75	76	76
77						70	70	71	71	72	72	73	73	74	74	75	75	76	76	77
78					70	70	71	71	72	72	73	74	74	75	75	76	76	77	78	78
79				70	70	71	72	72	73	73	74	74	Alert		76	77	77	78	78	79
80			70	70	71	72	72	73	73	74	74	75		76	77	78	78	79	79	80
81		70	70	71	71	72	73	73	74	75	75	76	77	77	78	78	79	80	80	81
82		70	71	71	72	73	73	74	75	75	76	77	77	78	79	79	80	81	81	82
83	70	71	71	72	73	73	74	75	75	76	77	78	78	79	Danger		81	82	82	83
84	70	71	72	72	73	74	75	75	76	77	78	78	79	80	80	81	82	83	83	84
85	71	72	72	73	74	75	75	76	77	78	78	79	80	81	81	82	83	84	84	85
86	71	72	73	74	74	75	76	77	78	78	79	80	81	81	82	83	84	84	85	86
87	72	73	73	74	75	76	77	77	78	79	80	81	81	82	83	84	85			87
88	72	73	74	75	76	76	77	78	79	80	81	81	82	83	84	Emergency			87	88
89	73	74	74	75	76	77	78	79	80	80	81	82	83	84	85	86	86	87	88	89
90	73	74	75	76	77	78	79	79	80	81	82	83	84	85	86	87	87	88	89	90
91	74	75	76	76	77	78	79	80	81	82	83	84	85	86	86	87	88	89	90	91
92	74	75	76	77	78	79	80	81	82	83	84	84	85	86	87	88	89	90		
93	75	76	77	78	79	80	80	81	82	83	84	85	87	87	88	89	90			
94	75	76	77	78	79	80	81	82	83	84	85	86	87	88	89	90				
95	76	77	78	79	80	81	82	83	84	85	86	87	88	89	90					
96	76	77	78	79	80	81	82	84	84	86	87	88	89	90	91					
97	77	78	79	80	81	82	83	84	85	86	87	88	90	91						
98	77	78	79	80	82	83	84	85	86	87	88	89	90							
99	78	79	80	81	82	83	84	86	87	88	88	90								
100	78	79	80	82	83	84	85	86	87	88	90	91								
105	80	82	83	84	86	87	89	90	91											

RELATIVE HUMIDITY INTERVALS (%)

FIGURE 7-2
The temperature-humidity index (THI) for estimating the level of discomfort in animals. The numbers in the table are derived from dry bulb (T_{DB}) and wet bulb (T_{WB}) temperatures, according to the formula THI $= 0.4\ (T_{DB} + T_{WB}) + 15$. In the "discomfort index" shown in this figure, the chances of death loss during movement of livestock are specified as "alert," "danger," and "emergency". [From Livestock Conservation, Inc., Hinsdale, Illinois.]

tures) on animal comfort. Although the many species (and individuals) tolerate heat and humidity differently, the index is still useful to livestock handlers. It enables them to determine the approximate level of discomfort in the animals being handled or transported. Figure 7-2 illustrates the relationship of temperature (dry bulb) and relative humidity as they affect the *discomfort index*.

Light, Sound, and Space

Animal behavior is often influenced by such environmental elements as light, sound, and space. When confined in dark surroundings, animals tend to move toward sources of light. This probably reflects their preference for the unconfined conditions that the light represents. Some swine are particularly uncomfortable when restrained or confined in space that does not permit free movement. Unfamiliar noises are also stressful to many animals.

The environmental factors of light, space, and sound are particularly important during the marketing of livestock. Some animals become frightened in unfamiliar surroundings, while others become hostile. These differences in reaction are probably associated with many factors in the animal, such as hormone balance (epinephrine/norepinephrine ratio), fatigue, or previous handling experience. Because unpredictable emotional responses are elicited, highly variable responses are produced in the muscles.

STRESS AND MUSCLE CHARACTERISTICS

It is important to recognize that any of the environmental stress factors discussed can result in changes in the metabolites of muscle. These changes, in turn, are responsible for differences in the ultimate properties of meat. The nature of the changes depends on such factors as the duration or severity of the stress, and the level of the animal's stress resistance at the time of death.

Stress Resistance and Stress Susceptibility

Meat scientists have discovered that animals can be characterized as having, in variable degrees, *stress susceptibility* or *stress resistance.* If an animal is highly stress susceptible, it is likely to experience heat stroke, shock, and circulatory collapse when exposed to a stressor. Even a mild stress, not associated with high temperatures, may produce changes that result in death. In swine, the susceptible animals exhibit the porcine stress syndrome by such external signs as extreme muscularity, anxious behavior, muscle tremors, and reddening of the skin.

In general, stress susceptible animals have unusually high temperatures, rapid glycolysis (pH drop), and early postmortem onset of rigor mortis in their muscles. Although the postmortem changes are rapid, some degree of antemortem muscle temperature rise, lactic acid buildup, and depletion of ATP also occurs. This combination of conditions results in an exaggeration of the muscle to meat transformations (pH drop, protein denaturation) that normally occur. Table 7-1 shows some of the physiological indicators of stress in pigs, and the associated properties of muscle. In stress susceptible animals, the muscles usually

TABLE 7-1
Ultimate effects of high temperatures* on stress susceptible and stress resistant pigs

Factor	Stress susceptible	Stress resistant
Heart rate	Rapid increase	Gradual increase
Respiration rate	Rapid increase, followed by rapid decrease	Gradual increase
Body temperature	Marked increase	Limited increase
Blood CO_2 concentration	Increases	Decreases
Blood pH	Decreases	No change
Postmortem muscle pH	Rapid drop	Gradual drop
Rigor mortis	Rapid onset	Delay before onset

SOURCE: Summarized from Forrest *et al.*, (1968).
*42°C for 20–30 minutes.

become pale in color, soft in texture, and moist or exudative after a normal 18–24 hour chilling period. The *pale, soft, exudative* (PSE) condition is evident in some muscles of approximately 20 percent of slaughter swine and is associated with lowered processing yields, increased cooking losses, and reduced juiciness. It is usually restricted to the muscles of the ham and pork loin, but it is occasionally seen in the dark muscles of pork shoulder, as well as in some muscles of beef, lamb, and poultry. A dark, firm, dry muscle condition can also be produced in meat from animals with a degree of stress susceptibility, if they have survived a stress of sufficient duration to deplete their glycogen reserves.

Stress resistant animals are able to maintain normal temperature and homeostatic conditions in their muscles in spite of relatively severe stress. However, they may accomplish this at the expense of muscle glycogen stores. Glycogen deficiency usually occurs when animals survive stress, such as that associated with fatigue, exercise, fasting, excitement, fighting, restraint, electrical shock, or adrenalin injection, but are slaughtered before they have sufficient time to replenish their muscle glycogen stores. Muscle glycogen deficiency in these animals causes a slow rate (and limited extent) of glycolysis after death. The resultant high pH in such tissue minimizes the color change that otherwise occurs in the postmortem period, i.e., the muscle reflects less total light due to minimal structural change and the pigments themselves reflect light of a dark red or purplish color. Tissue of this type is also dry or sticky in texture because of its excellent water binding capacity. (Fresh meat color and water holding properties are discussed more fully in Chapter 8.) *Dark cutting meat* occurs frequently in beef (approximately 3 percent), pork, and lamb. Its major disadvantages are an unattractive dark, firm, dry appearance and a favorable pH for bacterial growth. These conditions result in monetary loss, particularly in the merchandising of the fresh beef.

There are many variations between the extremes of PSE and dark cutting muscle. In general, the conditions can be described by the rate and extent of pH decline immediately postmortem. Figure 6-1 illustrates some of the possible pH patterns, along with the resulting conditions of the meat. Thus, one can visualize the biological variation in animal stress resistance in terms of the variation in the ultimate quality of the muscle as food.

ANIMAL PRODUCTION FACTORS THAT AFFECT MUSCLE PROPERTIES

Heritability

Table 7-2 shows heritability estimates for some of the physical properties of muscle observed after the postmortem changes have occurred. These estimates indicate that the physical properties of muscle are at least moderately heritable. The heritability estimates for color, firmness, and structure may be interpreted as being indicative of the heritability of stress resistance or susceptibility, since these muscle properties are associated with stress responses. Livestock producers therefore could make important improvements in the acceptability of muscle as food by selecting breeding animals whose close relatives had muscles with normal color, firmness, structure, adequate intramuscular fat, and low resistance to shearing force (high degree of tenderness). The importance of these physical properties of meat is described in Chapter 8.

Some physical properties of muscle are associated with an animal's breed or strain. Although breed or strain differences exist in all species,

TABLE 7-2
Heritability estimates for physical properties of meat

Characteristic	Species	Heritability estimate, approximate average percentage
Color score or color–firmness–structure score	Swine	30
	Cattle	30
Marbling score (Intramuscular fat)	Swine	25
Resistance to shear or tenderness score	Swine	30
	Cattle	60

the available research suggests that these influences are particularly noticeable in swine. The differences are largely the result of variations among breeds in postmortem muscle metabolism rates.

Diet

The influence of diet on the physical properties of muscle is of minor importance, as long as there are no serious nutritional deficiencies. However, it should be noted that the growth patterns of body tissues can be changed by controlling dietary energy, thus affecting the composition of muscle. These dietary effects are mostly the result of age differences among market animals that are slaughtered at uniform weights. (The influence of age on muscle composition has been reviewed in Chapter 4.)

Growing Environment

The growing environment of meat animals exerts important influences on the properties of muscle. The overall level of stress caused by the environment is of major importance, although variable end results are observed as resulting from the same stress level. Among animals which are stress susceptible, the long-term mild stress caused by some growing environments (such as close confinement, hard or slippery floors, chronic disease, and fluctuating temperatures) probably exhausts their ability to resist the severe stress experienced during marketing. Surveys, showing that confinement reared pigs are more susceptible to stress than those reared in the natural environment, suggest that modern production facilities must be carefully designed, in order to avoid a reduction of pork muscle quality. Other surveys have shown that pigs marketed during periods of highly variable temperatures produce a high incidence of PSE muscle. When pigs are reared in carefully controlled research environments their muscles tend to become PSE after slaughter, if the rearing environment had fluctuating temperatures.

Stress resistant animals are capable of adjusting to certain levels of discomfort in the rearing environment, and consequently retain a high degree of stress resistance. It is possible that the handling, exercise, and other stressors imposed during marketing do not cause serious carcass effects in such animals because they have a reserve of stress resistance. Or, they just might not become highly excited upon experiencing these treatments. It is interesting to note that cattle raised in particularly comfortable environments, such as show cattle, have a higher incidence of dark cutting muscle than those reared under feedlot conditions. In this instance, marketing stress results in muscle glycogen deficiency, high pH, and dark color. In resistant animals, some stress in the rearing environment tends to reduce the chances that abnormal postmortem muscle conditions will be present. Yet, this same level of stress is undesirable if the animals are inherently stress susceptible. Of course, the variable degree of stress susceptibility results in highly variable responses to the environment, and a wide range of muscle conditions after slaughter.

PRESLAUGHTER HANDLING

The procedures necessary to convert the tissues of a living animal into edible food are necessarily stressful. The animal is exposed to a combination of environmental stimuli. The several steps involved in the marketing process may include sorting, loading onto a truck, transportation, weighing, driving, feeding, fasting, water spraying, and stunning. The severity of the effects of these treatments depends upon the climate, equipment that is used, personnel, and many other factors.

Transportation and Holding

Generally speaking, the transportation phase of livestock marketing can be one of the most severe segments of the process. It is during transit that most death losses and tissue bruisings occur. Improperly ventilated trucks, or warm climatic conditions, can result in extreme discomfort for the animals. Experiments using simulated conditions of transit have shown that some animals will die within 20–30 minutes of continuous exposure to these adverse conditions. In addition to possible death losses, muscle tissue shrinkage and reduction of weight of the dressed carcass can result from severe or prolonged transportation. However, under normal marketing conditions the weight of the muscles is not affected, even though animals usually lose 2–5 percent of their live weight in the marketing process, due to loss of gastrointestinal tract contents.

Holding livestock in a stockyard prior to slaughter provides opportunity for resting and feeding. In addition to improving the ability of the animal to withstand later handling, this interval can influence the level of energy stores in the muscles. Starchy feeds, and sugar especially, will restore depleted muscle glycogen levels, thus permitting the development of a normal postmortem pH. However, it is advisable to withhold feed for 24 hours prior to slaughter in order to facilitate the process of evisceration and minimize the chances for microbial contamination of the carcass from the gastrointestinal tract.

Immobilization and Bleeding

In the United States, the immobilization of meat animals is accomplished by certain methods that are accepted under the meat inspection laws. These include carbon dioxide immobilization, electric shock, and captive bolt (or projectile) stunning. Animals slaughtered for Kosher meat (see Chapter 13) are not immobilized before exsanguination (bleeding).

Although the immobilization process is not free of stress, it probably reduces stress responses compared to bleeding without immobilization. However, the overall effectiveness of these procedures is dependent on the careful design and operation of the equipment that is used. The animals should be rendered unconscious without stopping the action of the heart, so that it can aid the bleeding process.

The properties and composition of the muscles are influenced by the type and effectiveness of the immobilization procedure. The severity of this process is usually expressed in the muscles by the degree of glycogen depletion. Figure 7-3 shows the glycogen levels in the breast muscles of chickens bled under anesthesia, after electrical immobilization, and without immobilization. These differences in glycogen would result in differences in ultimate pH and physical properties of the muscles.

Immobilization should be followed, as quickly as possible, by a rapid bleeding to prevent the animal from regaining consciousness, and to

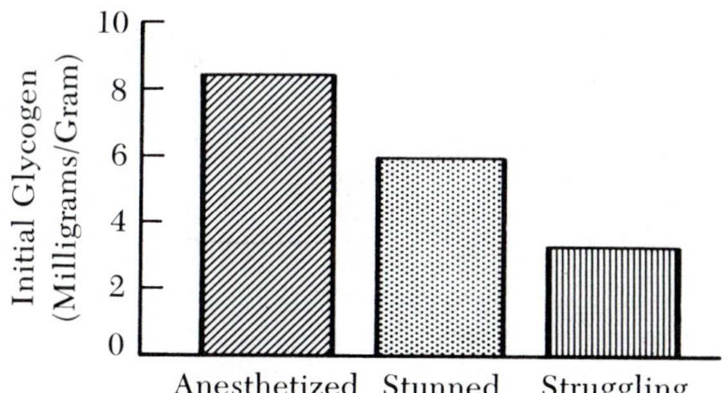

FIGURE 7-3
Glycogen levels in chicken breast muscles immediately after death: (left) with bleeding under anesthesia, (middle) after electrical stunning, and (right) without stunning. [From DeFremery, D., "Some Aspects of Postmortem Changes in Poultry Muscle," in *The Physiology and Biochemistry of Muscle as a Food*, E. J. Briskey, et al., eds. (University of Wisconsin Press, Madison), 1966, p. 205.]

release the pressure of the blood. Some immobilization systems, particularly those employing electric shock, raise the blood pressure to the point that hemorrhaging can occur in the muscles. Unless bleeding is accomplished within a few seconds of immobilization, the meat may exhibit blood spots that cannot be removed. These problems can be minimized by the use of proper voltages, and the correct placement of the electrodes.

POSTMORTEM TEMPERATURE

The temperature at which freshly slaughtered animal carcasses are stored can bring about distinct changes in the rate of chemical reactions that occur in muscle tissue. Reactions catalyzed by the enzymes in muscles are particularly temperature sensitive. Temperature differences of 10°C can cause the rates of these reactions to change by a factor of three or more. Consequently, it is considered desirable to reduce muscle temperature after death as quickly as possible, in order to minimize the protein denaturation that occurs in this period and to inhibit the growth of microorganisms. On the other hand, extremely rapid reduction of muscle temperature in the postmortem period can cause undesirable consequences.

Thaw Rigor and Cold Shortening

Two conditions, known as *thaw rigor* and *cold shortening*, have been recognized in recent years as resulting from a low temperature in the muscle before the onset of rigor mortis.

Thaw rigor is a severe type of rigor mortis that develops when muscle that has been frozen prerigor is thawed. The contraction produced may cause a physical shortening of 80 percent of the original length of the unrestrained muscle. However, a more common degree of thaw rigor shortening is approximately 60 percent of the original length. The contraction is accompanied by the release of large quantities of the meat juices and a severe toughening. Figure 7-4 illustrates the severe physical shortening that occurs in thaw rigor. From it one can visualize the extent of muscle distortion that occurs when meat is frozen prerigor. Although the extent of shortening of muscles attached to the skeleton would be less than in unrestrained muscle, a loss of tenderness and other qualities can take place.

Muscle need not be frozen in order for some of the undesirable attributes of thaw rigor to develop. Temperatures above 0°C, but below 15–16°C, if attained prerigor, cause a type of contraction known as cold

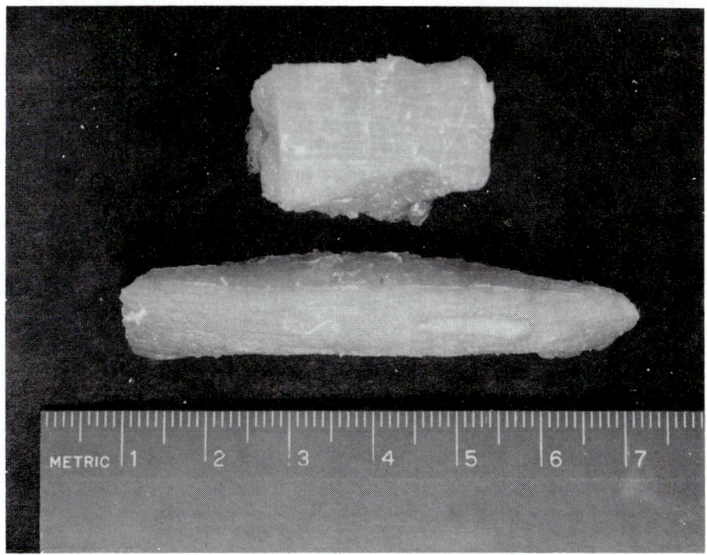

FIGURE 7-4
Thaw rigor shortening. A freshly excised muscle sample (bottom) is shown in comparison with an identical sample that was frozen prerigor and thawed (top). The sample that has undergone thaw rigor (top) is only 42 percent of its original length.

shortening. Although such shortening is less severe than that of thaw rigor, the underlying cause relates to the release of calcium ions and the resulting muscle rigidity. The practical implications are also important, since many commercial chilling systems remove body heat from the superficial muscles so rapidly that cold shortening is induced. Such shortening is especially severe in unrestrained muscles, but it also occurs in localized areas of muscles that are attached to the skeleton.

Severe shortening and an early onset of rigor mortis may be produced by maintaining muscle at relatively high temperatures. It is likely that this effect is the result of a rapid depletion of ATP stores. Consequently, there appears to be an optimum temperature at which muscle should be held during the onset of rigor mortis to minimize shortening, toughening, and other undesirable carcass effects of the rigor process. Although rigor mortis cannot be prevented, carcass chilling systems that maintain temperatures of 15–16°C in the muscles during rigor onset probably minimize the severity of rigor associated changes. The time at which rigor begins to develop is approximately 6–12 hours postmortem for beef and lamb, 15 minutes–3 hours for pork, and 5 minutes–1 hour for poultry. Figure 7-5 shows the relation between extent of shortening and the

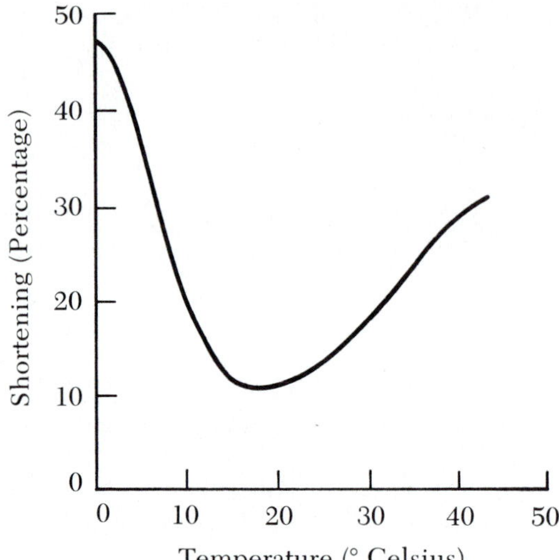

FIGURE 7-5
The effect of temperature on the degree of shortening in unrestrained bovine muscle. [Summarized from the data of Locker and Hagyard, J. Sci. Food Agric., 14, 787 (1963).]

temperature of the muscle during rigor onset. In addition to this relation, in the shortening range of 20–50 percent, tenderness decreases as the degree of shortening increases.

POSTMORTEM HANDLING PRACTICES

Many of the postmortem meat preparation processes can alter the rate or extent of change that takes place in muscle. Obviously, scalding, dehairing, skinning, eviscerating, and other procedures will influence carcass temperature and induce such changes as were previously described. The following discussion includes specific examples of processes that influence postmortem events through other than temperature change.

Carcass Suspension

The suspension of animal carcasses from overhead rails places tension on some muscles and leaves others unrestrained. Studies have shown that the amount of tension partially controls the amount of rigor shortening. When sarcomere lengths are measured, those muscles with the greatest tension during rigor have the longest sarcomeres, and are therefore in a reduced state of postrigor contracture. Such muscles are more tender than those allowed to shorten freely. In young animals, much of the muscle to muscle variation in tenderness is the consequence of the varying degrees of muscle tension present during rigor onset.

Figure 7-6 shows two methods of beef carcass suspension which differ in the resultant tension developed in certain muscles. When carcasses are suspended from the achilles tendon, as is the normal commercial

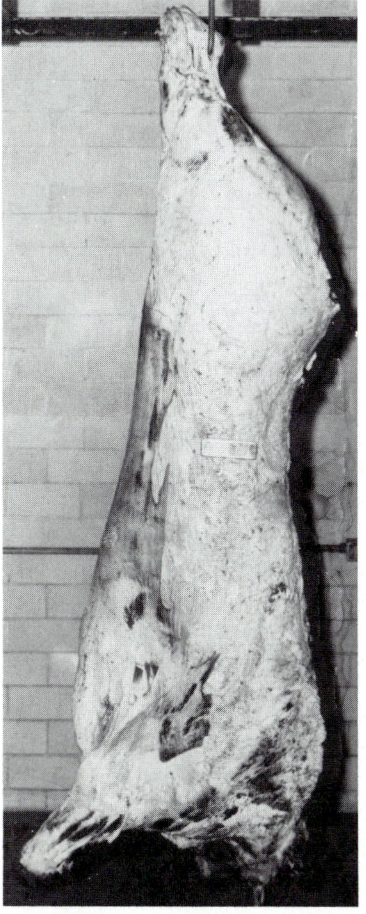

(A)

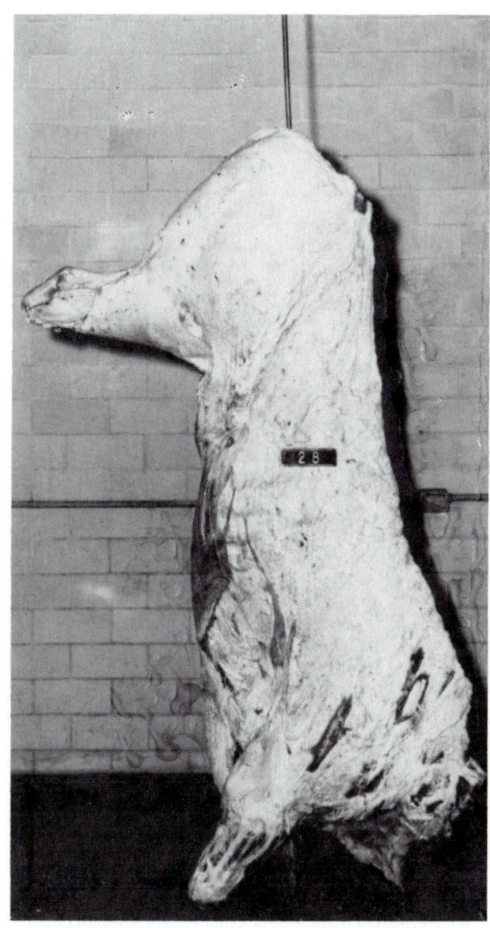

(B)

FIGURE 7-6
Carcass suspension (A) from the Achilles tendon, and (B) from the pelvic bone produces different degrees of muscle tension. [From Hostetler, R. L., B. A. Link, W. A. Landmann and H. A. Fitzhugh, Jr., "Effect of Carcass Suspension on Sarcomere Length and Shear Force of Some Major Bovine Muscles," J. Food Sci., 37, 132 (1972). Copyright © by Institute of Food Technologists.]

TABLE 7-3
Sarcomere length and shear force of muscles after achilles tendon and pelvic bone suspension of beef carcasses

Muscle	Sarcomere length*		Shear force**	
	Achilles	Pelvic	Achilles	Pelvic
Longissimus (loin)	1.9	2.3	6.0	4.9
Semimembranosus (round)	1.9	2.6	5.0	4.3
Semitendinosus (round)	2.4	2.9	5.3	4.9
Biceps femoris (round)	1.9	3.2	4.2	5.0
Rectus femoris (round or loin tip)	2.2	2.5	4.7	3.8
Adductor (round)	2.0	3.4	4.7	4.0
Gluteus medius (sirloin)	2.0	2.8	4.9	3.7
Psoas major (loin)	3.6	2.4	3.7	4.2
Triceps brachii (chuck)	2.3	2.3	4.7	4.8
Mean	2.2	2.7	4.8	4.4

SOURCE: Hostetler, R. L., B. A. Link, W. A. Landmann and H. A. Fitzhugh, Jr., 1972, Effect of Carcass Suspension on Sarcomere Length and Shear Force of Some Major Bovine Muscles. J. Food Sci. 37:132.
*micrometers
**Warner-Bratzler shear, 1.27 cm diameter muscle cylinders.

practice, a maximum amount of tension develops in the tenderloin (psoas) muscle and it consequently is extremely tender. When the carcass is suspended from the pelvic (aitch) bone, the tension is increased in several muscles of the loin and round, making them more tender than they are in commercially suspended carcasses.

Table 7-3 shows that suspension from the pelvic bone increases the sarcomere length of several muscles, but decreases it in the psoas major. Tenderness, measured as shear force, was improved due to the muscle stretching.

Prerigor Processing

Sausage makers know that the interval of time between slaughter and meat grinding affects the physical properties of their finished products. In general, meat that is ground prerigor, and mixed with curing ingredients, including salt, has superior water binding properties and maximum juiciness. The advantage in grinding or processing the prerigor meat stems from the early exposure of undenatured muscle proteins to the action of salt. These proteins become solubilized before the denaturing effects of the postmortem events can take place. Therefore, they do not undergo the severe transformations to which the proteins in intact muscles are subject.

The feasibility of processing some intact cuts of pork prior to completion of rigor has been demonstrated. The introduction of curing ingredients before the major losses of protein solubility have taken place results in cured cuts that are superior in color and moisture retention. The *hot processing* of pork has not been widely adopted by industry because of the extra costs associated with labor and handling problems.

REFERENCES

Bray, R. W., "Variation of Quality and Quantity Factors Within and Between Breeds," in, *The Pork Industry: Problems and Progress*, D. G. Topel, ed. (Iowa State University Press, Ames), p. 136, 1968.

Briskey, E. J., "Etiological Status and Associated Studies of Pale, Soft, Exudative Porcine Musculature," Adv. Food Research 13, 89 (1964).

Forrest, J. C., J. A. Will, G. R. Schmidt, M. D. Judge and E. J. Briskey, "Homeostasis in Animals (Sus domesticus) During Exposure to a Warm Environment," J. Appl. Physiol. 24, 33 (1968).

Hedrick, H. B., "Influence of Ante-Mortem Stress on Meat Palatability," J. Anim. Sci. 24, 255 (1965).

Judge, M. D., "Environmental Stress and Meat Quality," J. Anim. Sci. 28, 755 (1969).

Topel, D. G., E. J. Bicknell, K. S. Preston, L. L. Christian and G. Y. Matsushima, "Porcine Stress Syndrome," Mod. Vet. Practice 49, 40 (1968).

CHAPTER 8

Properties of Fresh Meat

In this text, the term "fresh meat" is used in a special context to include that product which has undergone the chemical and physical changes which follow slaughter but has not been further processed by freezing, curing, smoking, etc. The properties of fresh meat dictate its usefulness to the merchandiser, its appeal to the purchaser or consumer and its adaptability for further processing. Of particular importance are the characteristics of water holding capacity, color, structure, firmness, and texture.

WATER HOLDING CAPACITY

Water holding capacity is defined as the ability of meat to retain its water during application of external forces such as cutting, heating, grinding, or pressing. However, some loss of moisture usually occurs even during the mildest application of these treatments because a portion of the water present is in the *free* form. Many of the physical properties of meat (including the color, texture, and firmness of raw meat, and the juiciness and tenderness of cooked meat) are partially dependent on water holding capacity. The basis for these relationships is discussed at various points in this chapter and in Chapter 6.

The water holding capacity of muscle tissue has a direct effect on the shrinkage of meat during storage. When the tissues have poor water holding properties, the loss of moisture and, consequently, the loss of weight during storage (shrink) is great. This moisture loss occurs from the exposed muscle surfaces of carcasses during storage. After wholesale or retail cuts are made, an even greater opportunity for moisture loss exists because of the large muscle surface exposed by cutting. Retail cuts are wrapped with materials having a low *water vapor transmission rate* to minimize this loss.

Retail meat cuts may lose some of their moisture even after packaging in moisture-proof wrapping. The free water (and associated soluble proteins) may exude from the cut surfaces and accumulate around the meat causing a wet, unattractive retail package. This production of visible meat juice is known as *weep*. It occurs during display of retail cuts, during shipment of wholesale cuts, and during storage of cuts prior to shipment. These losses in weight, palatability, and nutritive value are serious industry problems. When large quantities of meat juices escape, the products lose weight and value to the wholesaler and retailer. These products also lose palatability, as well as some of their soluble proteins, vitamins, and minerals.

Weep is particularly evident in fresh cuts of pork and poultry and is discriminated against by consumers. Pale, soft, exudative (PSE) pork has a high percentage of free water that accumulates on the surface of the meat soon after the cuts are packaged. This condition is the most serious problem associated with PSE pork when it is merchandised on a fresh meat basis. Similarly, the rapid accumulation of juice in packages of cut poultry has forced retailers to perform the cutting operation, even though the cutting and packaging of poultry parts could be done more efficiently by processors.

Water holding capacity is especially critical in the meat ingredients of manufactured products that are subjected to heating, grinding, and other processes. Weight losses during the manufacturing processes are largely the result of water evaporation. Achieving proper protein/water ratio is important for palatability and adequate yield of finished product weight.

Chemical Basis of Water Holding Capacity

Although there are no clear lines of division among the types of water binding in muscle, one may consider that the water exists in the *bound*, *immobilized*, and *free* forms. Due to the distribution of their electrons, water molecules are not electrically neutral, but have a positive and a negatively charged "end" (they are *polar*). Thus they can become associated with electrically charged reactive groups on the muscle proteins. Of the total water in muscle, 4–5 percent is so located, and is known as bound water. It remains tightly bound even during the application of a severe mechanical or other physical force. Other water molecules are subsequently attracted to the bound molecules in layers that become successively weaker as the distance from the reactive group on the protein becomes greater. Such water may be termed immobilized water, but the quantity so immobilized depends on the amount of force exerted physically on the muscle. Water that is held only by surface forces is known as free water. Figure 8-1 is a diagram of the principal mechanisms by which the muscle proteins retain water.

Several factors influence the number of reactive groups on the proteins and their availability for binding water. In the immediate surroundings of the muscle proteins, these conditions are largely the result of the postmortem changes discussed in Chapter 6. The specific conditions are dependent on the production of lactic acid, loss of ATP, onset of rigor mortis, and the changes of cell structure associated with proteolytic enzyme activity.

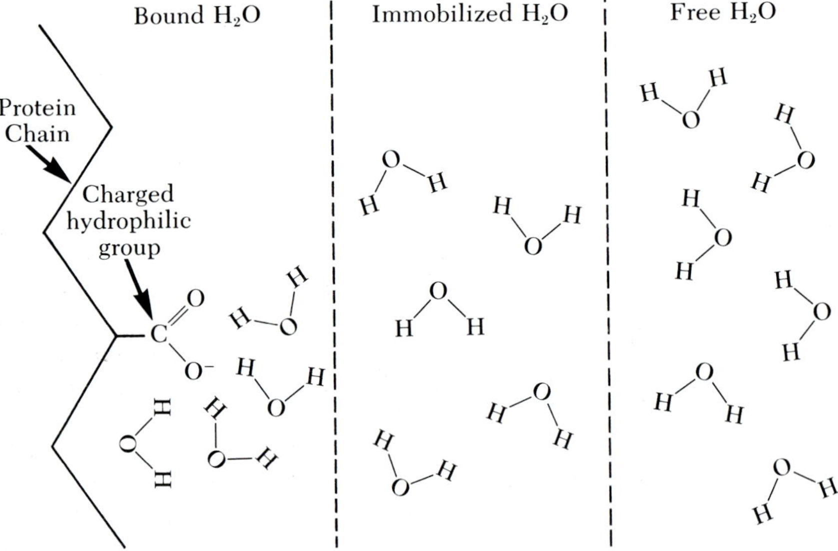

FIGURE 8-1
Charged hydrophilic groups on the muscle proteins attract water, forming a tightly bound layer (left) the molecules of which are oriented by their own polarity and that of the charged group. An immobilized layer (middle) is formed that has a less orderly molecular orientation toward the charged group. The free water molecules (right) are held only by capillary forces, and their orientation is independent of the charged group.

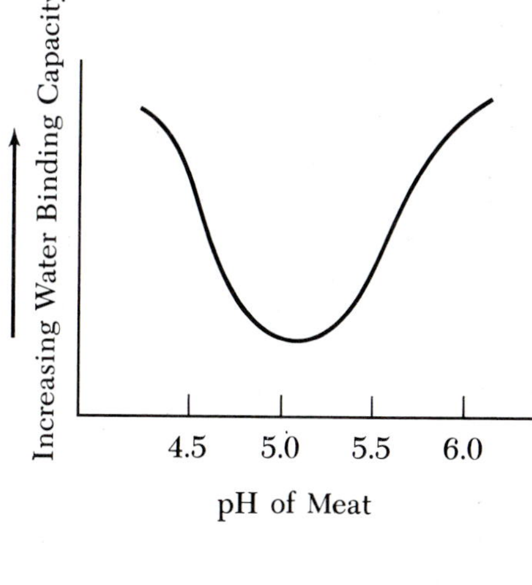

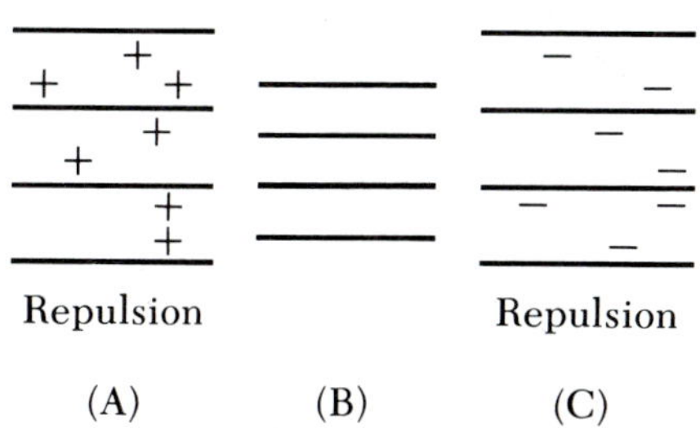

FIGURE 8-2
Effect of pH on the amount of immobilized water present in meat, due to its impact on the distribution of charged groups on the myofilaments and the amount of space between them. (A) Excess positive charges on the filaments. (B) Balance of positive and negative charges. (C) Excess negative charges on the filaments. [From Wismer-Pedersen, J., "Chemistry of Animal Tissues—Water," in *The Science of Meat and Meat Products*, 2nd ed. J. F. Price and B. S. Schweigert, eds. W. H. Freeman and Co. San Francisco. Copyright © 1971.]

NET CHARGE EFFECT. The formation of lactic acid and the resultant drop in pH in the postmortem period are responsible for an overall reduction of reactive groups available for water binding on the protein. This change results in a varying amount of denaturation and loss of solubility in the proteins. The reduction of reactive groups occurs because the pH approaches the *isoelectric point* of the muscle proteins: that pH at which the number of positively and negatively charged groups is equal. Consequently, these groups tend to be attracted to each other, and only those "left over" are available to attract water. This influence of pH is called the *net charge effect.* Figure 8-2 shows the effect that a pH other than the isoelectric point pH (5.0–5.4) has. At the pH values that are possible in meat (5.2–6.8), it is obvious that higher values are associated with a greater net charge on the protein and a greater percentage of immobilized water.

STERIC EFFECTS. Research has shown that, in normal meat, only about one-third of the loss of water holding capacity postmortem is due to the drop in pH. Other changes associated with the onset of rigor mortis also reduce water binding properties. The breakdown of ATP and the protein interactions associated with rigor mortis are largely responsible for the formation of a tight network within the contractile proteins. Certain ions, principally divalent cations such as calcium (Ca^{2+}) or magnesium (Mg^{2+}), have the ability to combine with two of the negatively charged reactive groups on the proteins. This tends to pull the protein chains close together and prevent those reactive groups that are still available from binding water. The lack of space for water molecules within the protein structure is known as a *steric effect* on water binding.

Steric effects are produced in muscle proteins in direct proportion to the breakdown of ATP in the postmortem period. ATP degradation is associated with a release of ions into the network of peptide chains. It is possible that these ions are liberated by a loss of integrity in the membranes of the sarcoplasmic reticulum.

AGING. Some of the loss of protein hydration caused by steric effects and pH decline may be recovered if meat is permitted to *age*. The proteolytic enzymes, such as the cathepsins, are probably responsible for some subtle changes in cell membrane permeability, or in their protein structure. This enzymatic degradation of muscle membrane structure, although not shown by experimentation, may allow some diffusion of ions into the areas surrounding the muscle proteins. A redistribution of ions occurs, resulting in the replacement of some divalent ions on the protein chains with monovalent ions. Consequently, for every divalent cation that is replaced, one reactive group on a protein is freed to bind water, and the forces pulling the protein chains together are lessened. This *exchange of ions* in muscle proteins during aging results in an improved water binding capacity. A limited improvement in water holding capacity also occurs in the aging period due to a slight rise in pH, but this accounts for only a small proportion of the total change.

COLOR

A color, as detected by the eye, is the result of a combination of several factors. Any specific color has three attributes, known as *hue*, *chroma*, and *value*. Hue describes that which one normally thinks of as a color—yellow, green, blue, or red (the wave length of the light radiation). Chroma (purity, or saturation) are terms which describe the intensity of the fundamental color with respect to the amount of white light that is mixed in with it. The value of a color is an indication of the overall reflectance (brightness) of the color.

The most important contributors to meat color are the pigments that absorb certain wave lengths of light and reflect others. However, other factors influence and modify the manner in which color is visually perceived. Meat color is the total impression seen by the eye, and is influenced by the viewing conditions. There are also marked differences among individuals in their color perceptions. The structure and texture of the muscles involved also influence the reflection and absorption of light.

Pigments

The pigments in meat consist largely of two proteins; *hemoglobin*, the pigment of the blood, and *myoglobin*, the pigment of the muscles. In well bled muscle tissue, myoglobin constitutes 80–90 percent of the total pigment and is much more abundant than hemoglobin. Such pigments as the catalase and cytochrome enzymes may also be present, but their contribution to color is minor.

The two major pigments are similar in structure, except that the myoglobin molecule is one-fourth as large as the hemoglobin molecule. Myoglobin consists of a protein portion called a *globin* (since it is a globular protein), and a nonprotein portion called a *heme ring* (Figure 8-3). The heme portion of the pigment is of special interest because the

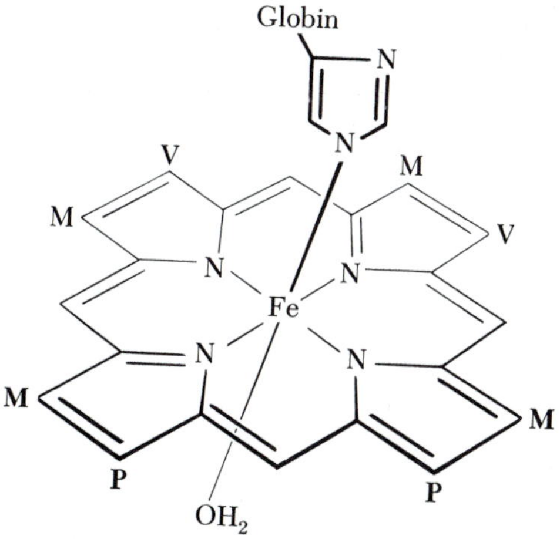

FIGURE 8-3
Schematic representation of the heme complex of myoglobin. The globin and water are not part of the planar heme complex. M, V, and P stand for methyl, vinyl, and propyl radicals attached to the porphyrin ring that surrounds the iron atom. [From Bodwell, C. E. and P. E. McClain, "Chemistry of Animal Tissues—Proteins," in *The Science of Meat and Meat Products*, 2nd ed. J. F. Price and B. S. Schweigert, eds. W. H. Freeman and Co., San Francisco. Copyright © 1971.]

color of meat is partially dependent on the chemical state (oxidation state) of the iron within the heme ring.

Myoglobin quantity varies with species, age, sex, muscle, and physical activity. This accounts for much of the variability in meat color. Species differences will become apparent when the light color of pork is compared with the bright red color of beef. The pale muscles of veal carcasses are indicative of the fact that the muscles of immature animals have a lower myoglobin content than those of more mature individuals. The intact male has muscles that contain more myoglobin than do those of the female or the castrate at comparable ages. Because of their differences in myoglobin content, the light breast muscles of the chicken contrast strongly with the dark muscles of the leg and thigh. Game animals have muscles that are darker than those of domestic animals partially because of the effect of a higher level of physical activity on their myoglobin content. In general, beef and lamb (or mutton) have more myoglobin than pork, veal, fish, or poultry. The following list shows the most typical color of meat from the various species:

Beef—Bright, cherry red
Fish—Gray-white to dark red
Horse—Dark red
Lamb and Mutton—light red to brick red
Pork—Grayish pink
Poultry—Gray-white to dull red
Veal—Brownish pink

Muscle to muscle differences in myoglobin content (and much of the variation among species) are due to the type of muscle fibers that are present. Those muscles with relatively high proportions (30–40 percent) of red fibers appear darker red in color. However, when they are viewed histologically, these myoglobin-rich fibers are still seen to be mixed with easily distinguishable white fibers. Thus, dark muscle color is often simply a consequence of the relatively high frequency of red fibers (Figure 8-4).

CHEMICAL STATE. The reaction of pigments with any of several materials can result in color changes in meat. However, the ability of the pigment to combine with or tie up a molecule depends on the iron in the heme ring being in the proper chemical state. Figure 8-5 depicts two chemical states for myoglobin, based on the valence of iron in the heme ring. When the iron is *oxidized* (in the ferric state), it cannot combine with other molecules, including molecular oxygen. When it has been *reduced* (is in the ferrous state) it will readily combine with water (as in uncut meat), or with oxygen (as in meat exposed to the air). The key to

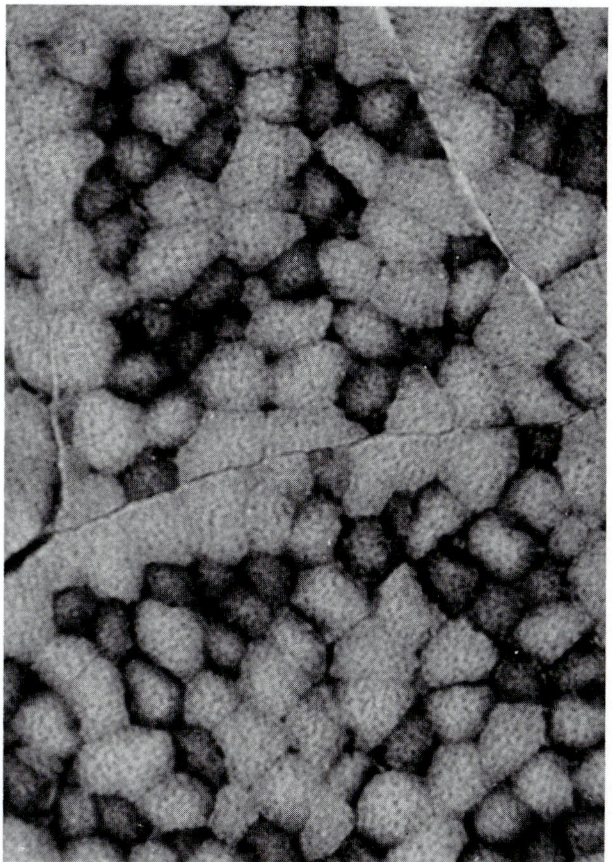

FIGURE 8-4
The color of meat depends upon the proportion of red and white muscle fibers. In this photomicrograph the dark staining fibers are the myoglobin rich red fibers. Bovine muscle (× 140).

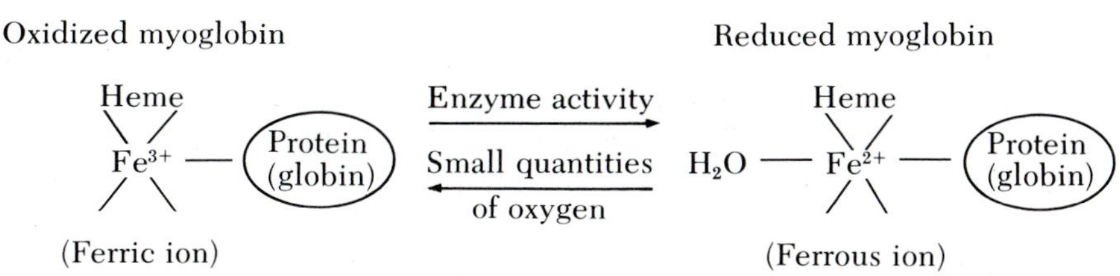

Oxidized myoglobin		Reduced myoglobin
Heme — Fe^{3+} — Protein (globin)	Enzyme activity → ← Small quantities of oxygen	H_2O — Fe^{2+} (Heme) — Protein (globin)
(Ferric ion)		(Ferrous ion)

FIGURE 8-5
The oxidation state (valence) of the central iron atom in the heme group of myoglobin under different conditions.

the maintenance of the pigment's ability to react with other molecules is to encourage the presence of reducing conditions in the muscle tissue. This is important because molecular oxygen reacts with the reduced iron of myoglobin, a reaction which provides the desirable red color of fresh meat.

Reducing conditions in meat can occur naturally, as a result of the normal enzyme activity (the so-called *electron transport chain*) that takes place continuously. These enzymes use all of the oxygen available in the interior of the muscles. Consequently, the pigment in uncut meat is in the reduced form and has only water with which to react. Such pigment is purple in color, and is called reduced myoglobin (Figure 8-6).

Upon cutting, grinding, or exposure to air, the pigments in meat undergo color changes due to their reaction with oxygen. If only small quantities of oxygen are present, such as in a partial vacuum or a sealed semipermeable package, the iron portion of the pigment becomes oxidized (Figure 8-5) and changes to a brown color. In this oxidized state the pigment is called *metmyoglobin* (Figure 8-6). The formation of this color is a serious problem in the merchandising of meat because most customers associate it with a product that has been stored too long, even

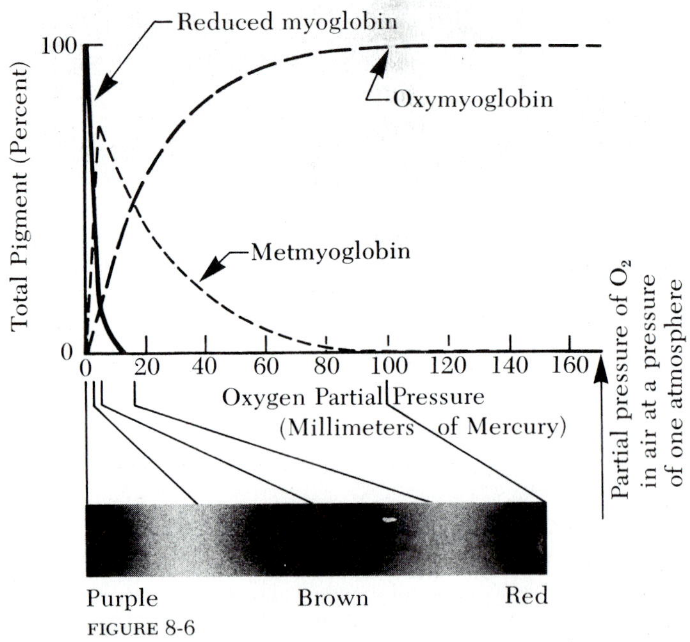

FIGURE 8-6
Relation of oxygen partial pressure in the atmosphere to pigment chemical states and color.

though it can be formed in a matter of minutes. It is especially troublesome because the meat remains brown in color indefinitely, or for as long as it is exposed to the air. Only by removing the oxygen and developing reducing conditions can it be converted back to a desirable color.

The prevention of metmyoglobin formation is an important requirement for fresh meat merchandising. Meat discoloration by metmyoglobin formation can occur during several phases of meat processing. When a product is permitted to remain in contact with flat surfaces, such as pans or tables, the oxygen supply may drop to a level that favors the development of the brown color. In other instances, the meat may be wrapped in a paper that only permits the passage of small quantities of oxygen to the surface of the meat.

When meat is allowed to come in full contact with the air, the reduced pigments will react with molecular oxygen and form a relatively stable pigment called *oxymyoglobin* (Figure 8-6). This pigment is responsible for the bright red color that consumers expect in fresh meat. Oxymyoglobin will form within 30–45 minutes after exposure to the air. This bright red color development is known as *bloom*. In this reaction, myoglobin (purple) is *oxygenated* (atmospheric or molecular oxygen is added). Metmyoglobin (brown) is the *oxidized* form of the pigment (the chemical state of the iron ion is changed). Under atmospheric conditions, the oxymyoglobin (*oxygenated* pigment) is stable and not easily *oxidized* to metmyoglobin.

The formation of oxymyoglobin occurs spontaneously when meat is exposed to the air, but its stability depends on a continuing supply of oxygen since the enzymes involved in oxidative metabolism in the tissues rapidly use up the available oxygen. As gases in the air diffuse inward, an oxygen gradient (and consequently a color gradient) is established from the surface of the muscle inward. The bright red color on the surface depends on the availability of oxygen in the superficial layers of the tissues.

Muscles differ in their rates of enzyme activity which, in turn, regulates the amount of oxygen available in the outermost layers of tissue. As the pH and temperature of the tissues increase, the enzymes become more active and the oxygen content is reduced. Consequently, maintaining the temperature of fresh retail meat near the freezing point minimizes the rate of enzyme activity and oxygen utilization and helps maintain a bright red color for the maximum possible time.

Fresh meat wrapping materials are designed to provide an abundant amount of oxygen at the muscle surface. Clear films are available that have high oxygen and low water permeabilities. Cellophane, polyvinyl chloride, and polyethylene provide the *oxygen transmission rate* that is necessary for the retention of a red meat color.

Discoloration

Discoloration or unusual color development can occur in meat in several ways, some of which are unrelated to the normal chemical reactions of the pigments. Fresh retail cuts of meat usually retain an attractive color for approximately 72 hours if good merchandising practices are followed. However, in addition to the proper wrapping materials mentioned previously, a long shelf life is dependent upon proper sanitation, a stable, low temperature (0°C or slightly lower), and proper lighting conditions.

The PSE and dark cutting conditions in meat (Chapter 7) are partially the result of an unusual amount of water binding in the muscles, and its influence on light reflection, rather than resulting from the chemical state of the pigment. The paleness of PSE pork is believed to be the result of a high proportion of free water in the tissues, compounded by the direct effect of the low pH on the pigments. The free water in PSE tissue probably influences color because it is located between the muscle cells rather than within them. Tissues containing a great amount of *extracellular* water have many reflecting surfaces that totally reflect light, but have only a limited light absorption capability. The color intensity is therefore greatly reduced. The pigments of PSE muscle might also appear light in color either because of a possible denaturation during the early postmortem period, or because of a direct effect of the low pH on the light reflecting properties of the pigments.

In dark cutting meat, the high water binding capacity maintains an unusually large proportion of water as *intracellular* water. Because of this, white light reflections between the cells are minimized. In addition, color absorbtion is enhanced. Dark cutting tissue also has a high rate of oxygen-using enzyme activity, due to its high pH. This reduces the proportion of the pigment in the red oxygenated (oxymyoglobin) state.

Discoloration of meat may occur as a result of myoglobin destruction due to bacterial growth. In this case the organisms use the pigment as a nutrient, and the heme ring is split from the protein part of the pigment. This usually causes a green color to develop. Greening on the surface of meat is almost always the result of bacterial contamination. This should not be confused with the green color that is occasionally seen when a thin film of fat covering the meat surface diffracts white light, and makes visible all the colors of the spectrum. Browning may also occur as a result of a reduction in oxygen tension on the meat surface by bacteria, and the subsequent formation of metmyoglobin.

A discoloration, in the form of a darkening of the meat, may also occur after cuts have been exposed to the air for a long time. In this instance, even though there is an adequate amount of oxygen available to maintain a bright color, the darkening occurs because the cut surface is drying out. As the drying progresses and the pigments become concentrated, the color becomes a deeper red.

STRUCTURE, FIRMNESS, AND TEXTURE

Some of the physical properties of fresh meat, such as structure, firmness, and texture, are difficult to measure objectively. These factors are usually evaluated by consumers with their visual, tactile, and gustatory senses. However, these properties are no less important than many other more easily measured properties of meat. Figure 8-7 illustrates the visual contrast between two meat cuts that differ in structure, firmness, and texture. The extent to which these cuts conform to the surface on

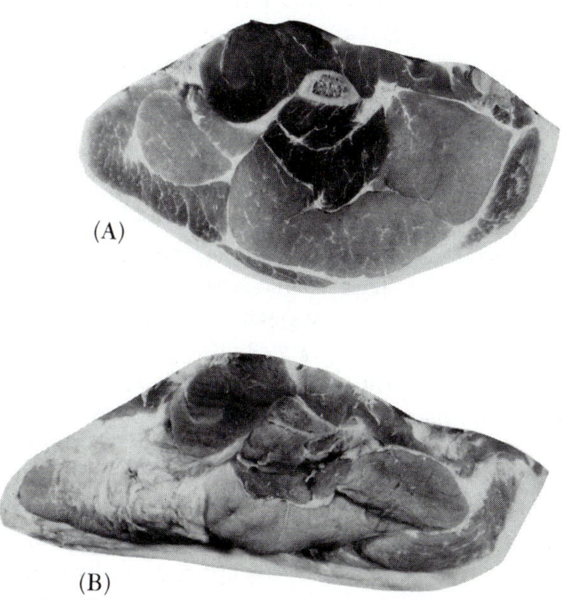

FIGURE 8-7
An example of the varying firmness of pork muscle; butt end views of two fresh hams. (A) A firm, rigid structure. (B) A soft structure with pronounced muscle separation. [From Briskey, E. J. and R. G. Kauffman, "Quality Characteristics of Muscle as a Food," in *The Science of Meat and Meat Products*, 2nd ed. J. F. Price and B. S. Schweigert, eds. W. H. Freeman and Co., San Francisco. Copyright © 1971.]

which they rest, the degree of separation between their muscles, and the prominence of the bundle divisions make a composite impression that can be either attractive or unattractive. Many factors, such as rigor state and associated water holding properties, intramuscular fat, connective tissue, and bundle size contribute to these physical properties. Several of these factors have been referred to in discussions of water holding capacity, color, and postmortem changes.

Rigor State

If, during the course of carcass chilling, muscle firmness is compared to that of the carcass of a freshly slaughtered animal, there are noticeable differences. The change is said to result from the carcass "setting-up" during chilling. This increased firmness develops from the loss of extensibility that accompanies the completion of rigor mortis, and the solidification of fat within and surrounding the muscles. In some muscles the sarcomeres are very short (approximately $2\mu m$), indicating that the filaments have a high degree of overlap, and the microstructure is very tight and dense.

During the storage and processing of fresh meat, the muscles retain a firm, somewhat moist condition unless they have undergone unusual rates of postmortem change, as described in Chapter 7. Yet, certain subtle changes occur that result in an improved palatability, especially in tenderness. These changes are partially responsible for the improvements in palatability associated with *aging*, and they are sometimes referred to as the factors that are responsible for the *"resolution" of rigor mortis.*

"Resolution," or the softening of rigor mortis, is believed to result from alterations in the structure of the myofilaments and their crossbridges. Some destruction of Z line proteins (Chapter 3) occurs which reduces the rigidity of the myofibrils (Figure 8-8). There is also a change in the amount of crossbridging between filaments that allows sarcomere lengthening and other modifications. It is possible, although not certain, that these subtle but important structural changes are caused by the action of cathepsin enzymes. These structural alterations may either relate directly to the tenderization accompanying aging, or permit the shift of ions necessary for improved water holding capacity, or both.

The degree of water holding capacity associated with each rigor stage, or with the rate of postmortem change is observable because of large scale effects on firmness, structure, and texture. Those muscles with an extremely high proportion of bound water are firm, have a tight structure, and a dry or sticky texture. Conversely, tissues with poor water binding ability are soft, have a loose structure, and a wet or grainy texture.

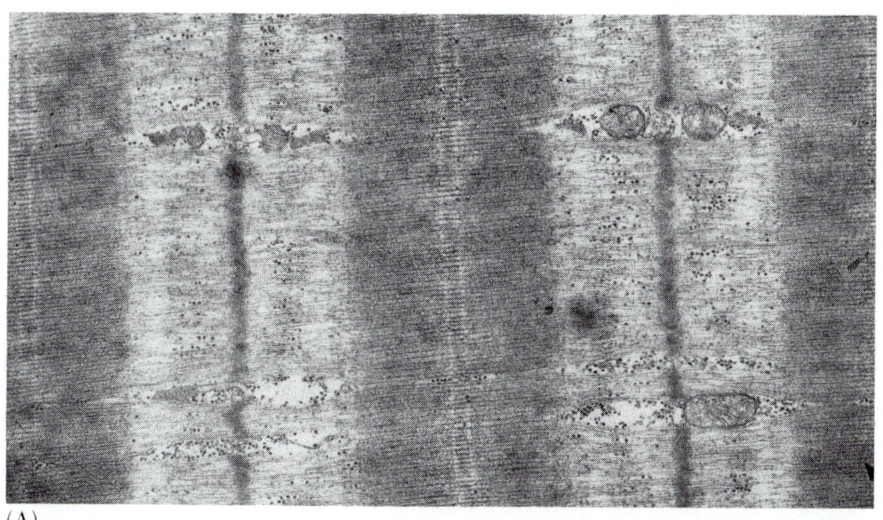

(A)

(B)

FIGURE 8-8
Degradation of the Z line occurs during the aging of meat. (A) Bovine muscle at death (× 14,800). (B) Bovine muscle after 24 hours of storage at 25°C (× 31,300). [From Henderson, D. W., Goll, D. E. and Stromer, M. H., "A Comparison of Shortening and Z line Degradation in Postmortem Bovine, Porcine, and Rabbit Muscle," Am. J. Anat. 128, 117 (1970).]

Intramuscular Fat

Intramuscular fat (marbling) contributes to the firmness of refrigerated meat. The solidification of fat that occurs during chilling increases its firmness. Thus, fat has an important merchandising influence. It helps retail cuts, such as steaks and chops, retain a uniform thickness and characteristic shape during handling and storage.

Connective Tissue

The amount of connective tissue in muscle affects the texture of meat. Those muscles which perform vigorously during the life of the animal, such as the biceps femoris or semitendinosus of the rear leg, tend to appear coarse in texture. On the other hand, a muscle such as the infrequently used psoas major of the loin, appears fine in texture. Of course, heavily used muscles develop extensive amounts of connective tissue to support their activity. The coarse textured muscles are less tender than the fine textured ones, unless special methods of cookery are used to break down the connective tissue.

The quantity of connective tissues in the muscles dictates what type of cutting is used, and how the meat will be utilized. Most meat cutting methods incorporate the objective of separating tender from less tender cuts so that maximum palatability is realized, and the usefulness of each cut is not limited by a large internal variation in connective tissue or tenderness.

A coarse texture is sometimes apparent in muscles from older animals. Although the amount of connective tissue does not increase with advanced age, its prominence and strength become greater. If protein repair and replacement were to occur at different rates in muscle fibers and connective tissues, it would be possible for the latter tissue to become more prominent in older animals. These changes in texture may also result partially from the increasing muscle fiber diameters that develop as an animal ages.

The size of the muscle fibers and bundles has been associated with the texture of meat. Large bundles and fibers, and abundant connective tissues surrounding the bundles, generally indicate coarse textured meat and a need for special care in the preparation of the meat to insure that it will be tender.

Absorption of Off-Flavors

The texture and consistency of meat render it highly susceptible to the absorption of volatile materials. Aromatic compounds from other foods, such as apples or onions, are readily absorbed by meat tissues. Consequently, off-flavors may occur when meat is stored in the presence of such products.

REFERENCES

Dean, R. W. and C. O. Ball, "Analysis of the Myoglobin Fractions on the Surfaces of Beef Cuts," Food Technol., 14, 271 (1960).

Hamm, R., "Biochemistry of Meat Hydration," Adv. Food Research, 10, 355 (1960).

Goll, D. E., N. Arakawa, M. H. Stromer, W. A. Busch and R. M. Robson, "Chemistry of Muscle Proteins as a Food," in *The Physiology and Biochemistry of Muscle as a Food, 2*, E. J. Briskey et al., eds. (University of Wisconsin Press, Madison), p. 755, 1970.

Khan, A. E. and L. van den Berg, "Some Protein Changes During Postmortem Tenderization in Poultry Meat," J. Food Sci., 29, 597 (1964).

Whitaker, J. R., "Chemical Changes with Aging of Meat with Emphasis on the Proteins," Adv. Food Research, 9, 1 (1959).

Watts, B. M., J. Kendrick, M. W. Zipser, B. Hutchins and B. Saleh, "Enzymatic Reducing Pathways in Meat," J. Food Sci., 31, 855 (1966).

CHAPTER 9

Principles of Meat Processing

Processed meat products are defined as those in which the properties of fresh meat have been modified by the use of one or more procedures, such as grinding or chopping, addition of seasonings, alteration of color, or heat treatment. Typical processed meat products include items such as cured ham, bacon, corned beef, and an almost endless variety of sausages. Most of these products are subjected to a combination of several basic processing steps before reaching their final form. Since there are hundreds of different processed meat products, each with its own characteristics, it is impossible to discuss completely all of the procedures followed in their manufacture. However, most products undergo certain basic processing steps in common, and it is our purpose to discuss these basic procedures in this chapter. Although each processed product has its own specific characteristics and methods of preparation, they all can be classified as either *comminuted* or *noncomminuted* products.

Typical noncomminuted products include hams of all types, bacons, Canadian bacon, and corned beef. In the meat industry, many of these products are commonly referred to as *smoked meats*. Their distinguishing characteristic is that they are prepared from whole, intact cuts of meat (with the bone removed in some cases). These products usually are cured, seasoned, heat processed, and smoked, and often they are molded or formed.

Comminution involves subdividing the raw meat materials, so that the product consists of small meat pieces, chunks, chips, or slices. Most comminuted products may be classed as sausages. *Sausages* are comminuted, seasoned meat products that may also be cured, smoked, molded, and heat processed. The degree of comminution varies widely. Some sausages are very coarsely comminuted; examples are salami, pork sausage, and summer sausage. In other sausages, the meat can be so finely subdivided that the sausage mix is a viscous mass with many characteristics of an emulsion. These are referred to as *emulsified sausages*; examples are frankfurters and bologna. All sausages can be classified into one of six categories, depending upon the processing methods that are used in their manufacture. The six classes, and examples of each, are: (1) *fresh*—fresh pork sausage; (2) *uncooked, smoked*—smoked pork sausage, mettwurst, Italian pork sausage; (3) *cooked, smoked*—frankfurter, bologna, knackwurst, mortadella, berliner; (4) *cooked*—liver sausage, braunschweiger, beer salami, cooked salami; (5) *dry or fermented*—summer sausage, cervelat, dry salamis, cappicola, pepperoni; and (6) *cooked meat specialties*—luncheon meats and loaves, sandwich spreads, jellied products.

There are also many comminuted products that are not classed as sausages. Hamburger and ground beef are probably the most common. Many commercial products begin as ground meat or meat chunks, pieces, or slices that are then formed into patties by machines using sufficient pressure to force the meat tightly together. These products are often marketed under the general titles of "burgers" or "steaks". They may even be breaded and precooked. Boneless poultry meat pieces, either chicken or turkey can be formed into roll type products by using a suitable binder to hold the meat pieces together, and a casing, which gives the product a desired shape.

Many processed products do not fit any of the above categories. Such products as breaded or frozen fish sticks or patties, or shrimp must be considered as processed meat. All canned meat products are classed as processed meat, including canned hams requiring refrigeration because their heat treatment is sufficient only to cook and pasteurize the product, to those fully sterilized meat products that need not be refrigerated. Precooked meat of any type used in TV dinners, or in institutional serving, must also be included as processed products. Thus, it is apparent that processed meat products take many shapes, sizes, and varieties. Their number and variety is limited only by the processor's imagination. Some processed products are listed in Table 9-1 and the reader will recognize many of the names; however, this listing is certainly not complete.

TABLE 9-1
Processed meat products

Noncomminuted		
Bacon Beef bacon Canadian bacon Cappicola Butt Ham Corned beef Cured pork loin	Hams Boneless Canned Cooked Country cured Prosciutti Regular Semiboneless Tonino Water added	Picnic shoulder Pigs' feet Pork hocks Smoked butt Smoked spare ribs
Comminuted		
Berliner Blood sausage Bologna Bratwurst Braunschweiger Cervelat Chorizo Consenza Country style pork sausage Farmer sausage Frankfurter Fresh pork sausage Göteberg Gothaer Head cheese Holsteiner Knackwurst (knoblauch) Kolbassi Lebanon bologna Liver sausage Luncheon loaves Cheese Chicken roll Dutch Ham Head cheese Honey loaf	Luncheon loaves (continued) Jellied corned beef Jellied tongue Liver and bacon Liverwurst Macaroni and cheese Meat and cheese Meat loaf Olive loaf Pickle and pimento Poultry roll Souse Spiced veal Tongue Turkey roll Veal Luncheon roll Messina Mettwurst Minced ham Mortadella Pepperoni Pescara Piccolo Polish sausage (kielbasa) Pork roll	Salami B.C. salami Beer Capri Catania Cooked Corti Cotto De Lusso genoa Genoa German Golden west milano Italian Kosher Luguria Lombardia Sicilian Sorrento Savona genoa Sandwich spreads Scrapple Smoked pork sausage Strassburger Summer sausage Swedish sausage Thuringer (fresh, smoked) Veal sausage Vienna Wiener wurst

HISTORY OF MEAT PROCESSING

Meat processing originated in prehistoric times, and no doubt developed soon after people became hunters. Probably the first type of processed product was sun dried meat, and only later was meat dried over a slow burning wood fire to give a dried, smoked meat similar to jerky. The salting and smoking of meat was an ancient practice even in the time of Homer, 850 B.C. These early processed meat products were prepared for one purpose, their preservation for use at some future time. People had learned at a very early time that dried or heavily salted meat would not spoil as easily as the fresh product. Meat processing probably developed out of this knowledge, coupled with the necessity for storing meat for future use. With advances in preservation technology, especially in refrigeration and packaging, meat processors were no longer tied primarily to preservation by high salt concentration or drying. They were free to experiment with lower salt and higher moisture levels in the finished product, and with new seasonings and combinations of meat ingredients, thereby creating many new processed products. Reasons for preparation of modern processed meat products include development of unique flavors and forms of product, provision of a variety of products, and development of new products in addition to preservation of meat.

Many of our present day meat products were known to the ancient Egyptians and Romans. Roman butchers prepared cracklings, bacon, tenderloins, oxtails, pigs feet, salt pork, meat balls, and sausage of many varieties. A book dating from the reign of Augustus (63 B.C.–14 A.D.) contains directions for the preservation of meat with honey (using no salt), and for the preservation of cooked meats in a brine solution containing water, mustard, vinegar, salt, and honey. Recipes are also given for liver sausage, pork sausage, and for a "round" sausage of chopped pork, bacon, garlic, onions, and pepper that was stuffed in a casing and smoked until the meat was pink.

BASIC PROCESSING PROCEDURES

Curing

HISTORY AND PRESENT APPLICATION. Meat curing is the application of salt, color fixing ingredients, and seasonings to meat in order to impart unique properties to the end product. Cured meat products were originally prepared by the addition of salt at concentrations that were high enough to preserve the meat. Salt inhibits spoiling largely by reducing the amount of water available for microbial growth. Since a high salt concentration promotes oxidation of myoglobin molecules, meat preserved by salting has an unattractive gray color. The use of nitrate to "fix" the red color of cured meat probably evolved more by accident than by design. Potassium nitrate (saltpeter) was most likely present as an impurity in the salt used as a preservative, and early meat processors eventually learned that when it was present, a more pleasing bright reddish color resulted. With the development of refrigeration and freezing and their application to the preservation of meat, the main purpose of meat curing changed from preservation to the development of unique color, flavor, texture, and palatability properties. The most notable difference between present day cured meats and those of the past is the lower salt level and blander flavor of the modern product. Today, color development is equally as important as flavor and texture changes.

MEAT CURING INGREDIENTS. Two main ingredients must be used in order to cure meat; *salt*, and *nitrite*. However, other substances are added to accelerate curing, stabilize color, modify flavor and texture, and reduce shrinkage during processing. Salt (sodium chloride) is included in all meat curing formulas. Since it is not generally used at levels high enough to effect preservation, its main function is as a flavoring agent. However, even in low concentration, salt has some preservative action. Nitrite, either as a potassium or a sodium salt, is used to develop cured meat color. They impart a bright reddish pink color, which is desirable in a cured product. Sodium or potassium nitrates were the first compounds used for this purpose. However, it was discovered that the use of nitrite would accelerate color development and thus, nitrate and nitrite were often used in combination. Nitrates are now being removed from use in meat curing since nitrite will accomplish the desired reaction more rapidly, even in the absence of nitrate, and also because recent Food and Drug Administration regulations prohibit its use in many products. Several *reductants* (compounds capable of donating electrons) are incorporated in meat curing mixtures in order to accelerate color development. Nitrite must be reduced to nitric oxide before proper color development occurs and reductants speed this reaction. The most commonly used reductant is a sodium salt of ascorbic acid (vitamin C) or of an isomer, isoascorbic (erythorbic) acid. Other factors that contribute to reduction reactions are discussed in a subsequent section. Alkaline *phosphates* are often incorporated into curing mixtures. Although they do not enter directly into the curing reaction, they do increase the water binding capacity of meat and reduce shrinkage of meat products during subsequent processing. Phosphates also retard development of oxidative rancidity, and may improve texture. *Sea-*

sonings, including spices, herbs, vegetables, and sweeteners, are often incorporated in meat along with curing ingredients. They do not enter into the curing reaction, but do impart unique flavors.

METHODS FOR THE INCORPORATION OF CURE INGREDIENTS. Several techniques are utilized for incorporating the curing mixture into meat products. Irrespective of the method used, the important requirement is to distribute the cure ingredients throughout the entire product. Inadequate or uneven distribution will result in poor color development, with the attendant possibility of spoilage in areas that are not penetrated by the curing mixture. "Bone sour" in hams, and gray areas in the interior of other products are examples of problems resulting from improper distribution of the curing mixture.

Cure ingredients are incorporated into sausage products during the mixing or comminution processes. The ingredients are added in dry form, or as a concentrated solution, and are uniformly distributed through the product during the blending and comminution steps.

In the preparation of smoked meat products, curing agents may be incorporated by several methods into meat cuts such as hams, bacons, picnics, or briskets. These cuts are sometimes immersed in a *brine*, a solution composed of curing ingredients dissolved in water. This is a slow method of curing, since extended periods of time are required for the brine to diffuse through the entire product. In addition, microbial growth and spoilage can occur during curing, even though the product is refrigerated. At present, only specialty products are cured in this manner.

The curing procedure can be shortened greatly by the injection of brine directly into the cuts using any one of several methods, all of which achieve a more rapid and uniform distribution of brine throughout the tissues. In cuts that have an intact vascular system, a brine solution can be pumped directly into the artery (*artery pumping*). The successful use of artery pumping requires care, so that the blood vessels are not ruptured by an excessive pumping pressure. A very popular method of pumping that can be used on any type of meat cut is called *stitch pumping*. In stitch pumping, brine is injected (through a hollow needle) into various parts of the cut, especially the thickest part, and near joints. The newest method of pumping, *multiple injection*, is a widely used variation of stitch pumping. With this technique, brine is injected simultaneously and automatically through a series of hollow needles, the shafts of which have numerous evenly spaced holes along their lengths.

An older method is the *dry cure*, in which the curing agents are rubbed in dry form over the surface of a cut. The cuts are then placed on a shelf or in a box and allowed to cure. For large cuts, the cure ingredients must be applied several times during the curing period. Since the dry curing procedure is slow, and requires a large amount of hand labor, it is now used only on specialty items such as country cured hams and bacons.

CHEMISTRY OF CURED MEAT COLOR. Since a primary purpose of curing meat is to develop an attractive, stable color in the processed product, the chemistry of color fixation is discussed here in some detail.

The basic reaction occurring during color development is represented by the following equation.

$$\text{myoglobin} + \text{nitric oxide} \rightarrow \text{nitric oxide myoglobin} \xrightarrow{\text{heat}} \text{nitrosyl hemochrome}$$

Nitric oxide myoglobin has an attractive, bright red color, and is the pigment present in the cured product prior to heat processing. The color is stabilized by the denaturation of the protein portion of myoglobin, for example, by heating. The resulting pigment is nitrosyl hemochrome which is responsible for the bright pink color characteristic of cured meat.

Cured meat pigment can be rapidly produced by exposing myoglobin directly to nitric oxide, a colorless gas that is slightly soluble in water. However, procedures for the direct incorporation of nitric oxide into meat products have not been adapted on a commercial scale. In commercial curing procedures, nitric oxide is presently produced by the reduction of nitrate or nitrite. If nitrate is used in curing, it must first be converted to nitrite, which is in turn reduced to nitric oxide. Nitrate-reducing bacteria are responsible for the conversion of nitrate to nitrite. If nitrite is added directly to the cure, the necessity for nitrate reduction is eliminated and color development is much more rapid. Nitrite is added to nearly all commercial curing mixtures while nitrate may or may not be included. Meat inspection regulations limit the amount of nitrite permitted in the finished product to 200 parts per million (ppm). Also, the amount of nitrate must not exceed 500 ppm in the product because it represents a potential source of excess nitrite. The maximum amounts of nitrite, in the form of either sodium or potassium salts, that may be used in meat curing are: 2 lbs. in 100 gallons of brine (239.7 grams/100 liters), 1 oz. for each 100 lbs. of meat in dry cure (62.8 grams/100 kilograms), or 1/4 oz. per 100 lbs. of chopped meat for sausage (15.7 grams/100 kilograms). The residual nitrite levels in the finished product are much lower than the levels that are initially added.

There are several mechanisms by which nitrite can be converted to nitric oxide. In water solution, at the pH of meat (5.5–6.0), a portion of the nitrite is present as nitrous acid (HNO_2). At these pH values, nitrous acid decomposes to nitric oxide, as shown by the following equation.

$$3\ HNO_2 \rightleftharpoons HNO_3 + 2\ NO + H_2O.$$

This is a rather slow pathway for nitric oxide production and thus its importance as a source of nitric oxide is probably minimal in present rapid curing methods.

Nitrite can also be reduced to nitric oxide by the natural reducing activity of post-mortem muscle tissue. Many substrates and enzymes, notably those of the tricarboxylic acid cycle, are present and active in meat and can furnish reducing equivalents (hydrogen atoms and electrons) as NADH. (NADH is the reduced form of nicotinamide adenine dinucleotide [NAD^+].) Under anaerobic conditions, these reducing equivalents are used by the electron transport chain of the mitochondia to reduce nitrite. This endogenous reduction may supply part of the nitric oxide necessary for color development. However, the process is slow, and can produce significant quantities of nitric oxide only in curing procedures of longer duration.

As mentioned previously, nitric oxide formation can be greatly accelerated by the addition of reducing agents to the curing mixture. The chemistry of this process is not well understood. However, it seems to involve a series of equilibrium reactions in which the reductant donates electrons to nitrite, leading to formation of nitric oxide. The sodium salts of ascorbic acid or erythorbic acid are most widely used as reductants. Δ-Gluconolactone can be added to accelerate reduction and stabilize the color slightly. This compound forms gluconic acid when it is added to meat, thus lowering the pH to some extent.

Figure 9-1 illustrates some of the reactions that heme pigments might undergo during color development in cured meat. Since nitrite is a very efficient oxidizing agent for myoglobin, the initial reaction is probably the conversion of *myoglobin* and *oxymyoglobin* to *metmyoglobin*. Nitric oxide then combines with the heme portion of metmyoglobin to produce *nitric oxide metmyoglobin*. Nitric oxide metmyoglobin must be reduced to *nitric oxide myoglobin*, the desired cured meat pigment. Reduction of nitric oxide metmyoglobin involves addition of an electron to Fe^{3+} of the heme converting it to Fe^{2+}. This reduction might be accomplished either naturally in the meat, or by reductants included in the cure. Natural reduction is a slow process, and plays a significant role only in lengthy curing processes. The reducing activity of ascorbic acid salts (ascorbates) accelerates nitric oxide metmyoglobin reduction by donating electrons to the ferric state (Fe^{3+}) of the heme. Sulfhydryl groups (—SH) released during the heat processing of cured meat are also very strong reducing compounds, and can contribute significantly to the reduction of metmyoglobin or nitric oxide metmyoglobin. When reductants are used, the time for cured meat color development can be shortened to several hours instead of several days, as would be required if natural reducing activity were to be solely responsible for nitric oxide production and nitric oxide metmyoglobin reduction. The use of reductants in

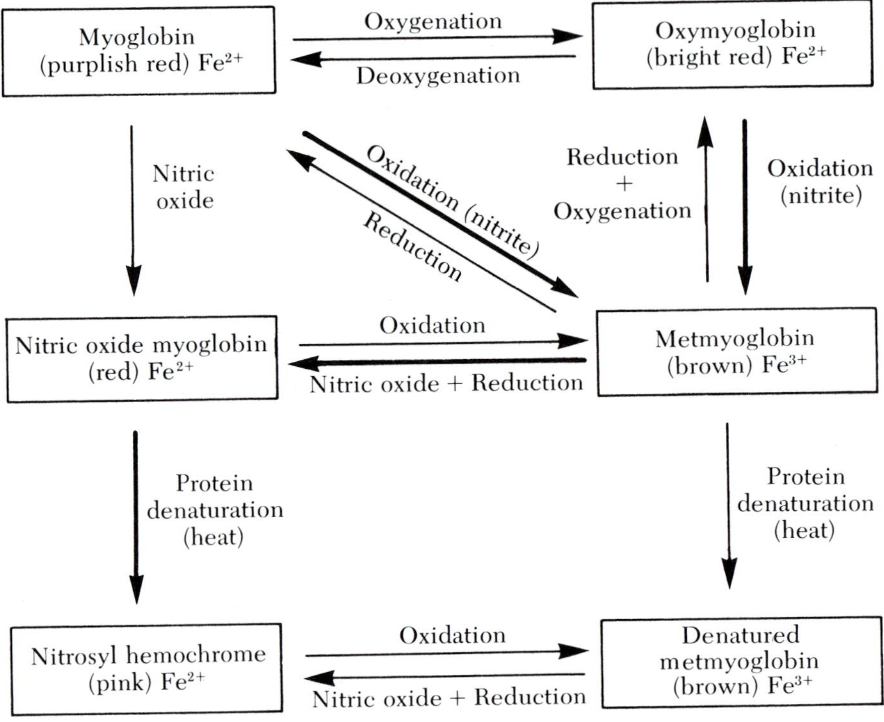

FIGURE 9-1
Chemical changes that can occur to myoglobin during the development of the final pigment in cured meat.

meat curing forms the basis for modern, continuous-process frankfurter production, and accelerated ham and bacon processes.

The final reaction shown in Figure 9-1, the formation of *nitrosyl hemochrome*, involves denaturation of the protein portion of myoglobin, but leaves the heme structure intact with the nitric oxide attached. Denaturation is caused by the heat of the cooking process. The color of nitrosyl hemochrome is pink, in contrast to the more reddish color of nitric oxide myoglobin. The color of the denatured pigment is more stable than the color of the native pigment.

STABILITY OF CURED MEAT PIGMENT. Nitrosyl hemochrome is a heat stable pigment, so it does not undergo further color changes upon any additional cooking of the cured product. However, cured meat pigments can undergo other reactions that result in color changes. Often these changes include a loss of color. Nitric oxide myoglobin and nitrosyl hemochrome are very susceptible to *light fading*. This effect can become important if meat cuts are displayed under strong fluorescent illumination while they are also exposed to the air. Under these conditions the surface color of cured meats will fade in about one hour. Under the same conditions fresh meat will hold its color for three days or longer. Light fading is a two step process that includes: (1) the dissociation of nitric oxide from the heme catalyzed by light, followed by (2) oxidation of the nitric oxide by oxygen. The heme groups can now also be oxidized by oxygen. A brownish gray color develops on the exposed meat surfaces during light fading because the faded pigment (sometimes called *hemichrome*) now has its heme group in the ferric (Fe^{3+}) state.

The most effective way to prevent light fading is to exclude oxygen from contact with the meat surfaces. This can be done either by the vacuum packaging of meat, or by packaging it in oxygen impermeable films. In either case, clear films and illuminated display methods can be used without any significant fading occurring. If no oxygen is present, nitric oxide will not be oxidized, and it can then recombine with the heme. In some instances, the application of ascorbates either in curing, or by being sprayed on the meat surface or wrapping material will also retard light fading. It is assumed that ascorbates provide a continuous production of nitric oxide from residual nitrite in the meat. The use of opaque packaging materials is also effective in prevention of fading since the process requires both light and oxygen.

The presence of rancid fat in cured meat products can also result in color instability. Unsaturated fatty acids, like cured meat pigments, oxidize in the presence of oxygen and, in addition, the oxidation of one will accelerate the oxidation of the other. This is referred to as *co-oxidation.*

Cured meat pigment is also susceptible to bacterial discoloration, during which a green color develops on the surface of the product. This greening usually results from poor sanitation or improper storage conditions where the product is contaminated by equipment surfaces and then is held under favorable conditions for microbial growth. Under aerobic conditions, the bacteria responsible for the greening effect produce hydrogen peroxide, which directly oxidizes the cured meat pigment.

Excessive amounts of nitrite in the cure can also cause a greening of cured meat pigment, called *nitrite burn.* Since nitrite is highly reactive in an acid environment, nitrite burn is especially a problem in fermented sausages and pickled pigs feet, because of their high acidity. The greening due to nitrite burn results from an oxidation of the cured meat pigment, but the exact details of the reaction are unknown.

PUBLIC HEALTH ASPECTS OF NITRITE USAGE. Nitrite is toxic if consumed in excessive amounts. A *single dose* of nitrite in excess of 15–20 mg/kg of body weight may be lethal. However, the maximum level of nitrite now permitted in cured meat products is 20–40 times *below* this lethal dose. Thus, there is essentially no problem with nitrite toxicity, as long as the recommended levels are followed.

Since about 1960, there has been increasing concern about a certain class of cancer producing (*carcinogenic*) chemical compounds, generally referred to as *nitrosamines*. Under certain conditions, these compounds can be formed in food products by reactions between nitrite and secondary amines ($—NR_2H$, where R is any substituent group). In general, nitrosamines are known to be carcinogenic when ingested by animals. However, for specific nitrosamines, the carcinogenic properties are quite variable among various species. In addition, some nitrosamines are general carcinogens, but others specifically affect certain organs only.

The Delaney Amendment to the Food, Drug, and Cosmetic Act of 1958 stipulates that known carcinogens shall not be permitted in food. Therefore, the enforcement of this law would mean that at least one of the reactants that produces nitrosamines should be eliminated from foods. Amines are impossible to eliminate, since they occur as breakdown products of natural precursors present in foods. Therefore, it has been proposed that nitrite be ruled out as a food additive.

Since the major use of nitrite in foods is in meat curing, it has been suggested that the elimination of nitrite from cured meat would be a solution to the nitrosamine problem. Nitrosamines have been isolated from several cured meat products, generally at less than 50 parts per billion (ppb), but the large majority of samples tested have been negative.

It is not a simple matter to eliminate nitrite from cured meat. The characteristic color that is produced when nitrite is added to meat has already been discussed. Also of great importance is the fact that nitrite inhibits the germination of *Clostridium botulinum* spores, and prevents toxin production in cured meat products that are heat-pasteurized during processing and require post-processing refrigeration. Weiners, bologna, bacon, and canned hams are examples of such products. It is generally recognized that 150 ppm of nitrite is the minimum level that is required to inhibit *Clostridium botulinum*. Without nitrite, the hazard of botulism food poisoning would forbid preparation of these products as they are now manufactured. Possibly higher salt concentration or a greater heat treatment would give the needed protection, but consumer acceptance of such products has not been established.

Some effort has been expended in searching for a nitrite substitute in meat curing, and a number of patents have been granted. Nicotinic acid, nicotinamide, and their derivatives are examples of compounds that form complexes of the desired color with myoglobin and have a potential as nitrite substitutes. A common problem with these substitutes is that the complex formed with myoglobin is less stable, and more susceptible to oxidation, than is the nitric oxide complex. Many potential substitutes also have serious side effects in the human body. For example, substances derived from nicotinic acid are potent vasodilators; that is, they cause dilation of the small blood vessels. Antimicrobial action of potential substitutes has not been established. In summary, a substance equal, or superior, to nitrite for color development and antimicrobial action has not been found.

Comminution, Blending, and Emulsification

COMMINUTION. The process by which particle size is reduced for incorporation of meat into sausage type products is called *comminution*. The degree of comminution (or particle size) differs greatly between various processed products, and is often a unique characteristic of a particular product. Some items are very coarsely comminuted, but other products are so finely divided that they form a meat emulsion. Two main advantages are gained from all comminution processes. These are an improved uniformity of product due to a more uniform particle size and distribution of ingredients, and an increase in tenderness as the meat is subdivided into smaller particles.

Equipment commonly used for comminution includes the meat grinder, silent cutter, emulsion mills, and flaking machines. Grinders are usually employed for the first step in the comminution of sausage type products. For nonemulsified sausages, grinding is often the only form of comminution employed. In the past, the silent cutter was used to form meat emulsions, but it is now usually used to reduce the particle size of meat and fat, and for mixing ingredients, prior to their emulsification in an emulsion mill. Compared to the silent cutter, emulsion mills operate at much higher speeds, form emulsions in much less time, and produce emulsions that have a smaller fat particle size. Since emulsion mills operate at high speed, the meat materials are subjected to considerable friction. This results in a higher emulsion temperature in a mill than in a silent cutter. Excessively high temperatures can reduce the stability of the resulting emulsion.

BLENDING. As a separate processing step, *blending* refers to an additional mixing to which comminuted products are subjected prior to further processing. The purpose of a separate blending step is to insure a more uniform distribution of ingredients, especially of the cure and seasoning, than could be achieved just with grinding. Coarsely ground sausages are blended prior to being stuffed into casings. Large batch blending of meat, seasonings, and other ingredients is a common procedure prior to emulsification.

EMULSIFICATION. An *emulsion* is defined as a mixture of two immiscible liquids, one of which is dispersed in the form of small droplets or globules in the other liquid. The liquid that forms the small droplets is called the *dispersed phase*, and the liquid in which the droplets are dispersed is called the *continuous phase*. The size of the dispersed phase droplets ranges from 0.1 to 5.0 micrometers (μm) in diameter.

Two typical emulsions are mayonnaise and homogenized milk; in both, fat droplets are dispersed in an aqueous medium. Meat emulsions are a two-phase system, with the dispersed phase consisting of either solid or liquid fat particles, and the continuous phase being water containing dissolved and suspended salts and proteins. Thus, they can be classified as oil in water emulsions, as illustrated in Figure 9-2. Many fat particles in commercial meat emulsions are larger than 50 μm in diameter, and thus they do not conform to one requirement of a "classical" emulsion.

Emulsions are generally unstable unless another component, known as an *emulsifying* or *stabilizing agent* is present. When fat is in contact with water there is a high interfacial tension between the two phases. An emulsifying agent functions to reduce this interfacial tension, thus permitting the formation of an emulsion with less energy input as well as increasing its overall stability. A distinguishing characteristic of emulsifying agents is that their molecules have an affinity for both water

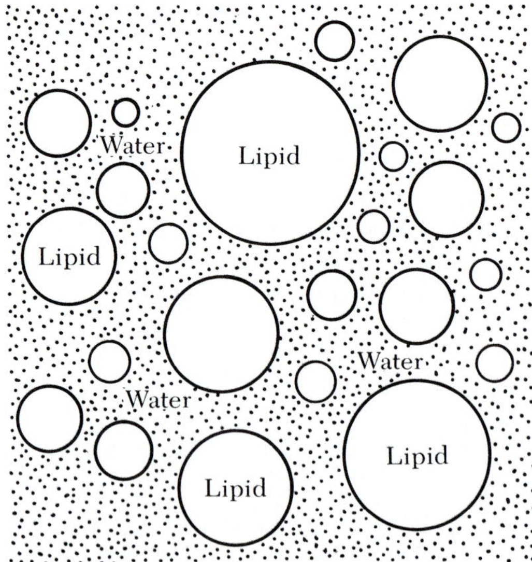

FIGURE 9-2
Schematic drawing of an oil in water emulsion showing the dispersion of lipid droplets (dispersed phase) in water (continuous phase).

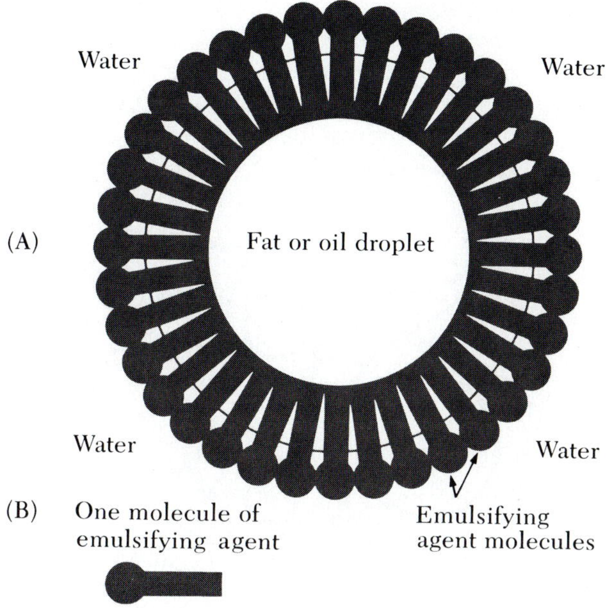

FIGURE 9-3
(A) The emulsifying agent is located at the interface between the lipid and water phases. These molecules are oriented so that their hydrophilic portions are in contact with the water phase while their hydrophobic portions are in contact with the lipid phase. The emulsifying agent forms a monomolecular layer surrounding the lipid droplet. (B) A molecule of emulsifying agent with its hydrophilic and hydrophobic portions indicated.

and fat. The *hydrophilic* (water loving) portions of such molecules have an affinity for water, while their *hydrophobic* (water hating) portions have more of an affinity for the fat. These affinities are best satisfied when the hydrophobic and hydrophilic portions of the emulsifying agent can align themselves between both the lipid and aqueous phases. If enough of the emulsifying agent is present, it will form a continuous layer between the two phases (Figure 9-3), thereby helping to stabilize the emulsion by separating the two phases.

The components of a meat emulsion are shown diagramatically in Figure 9-4. The emulsion consists of a matrix of muscle and connective tissue fibers, or segments of fibers, suspended in an aqueous medium containing soluble proteins and other soluble muscle constituents. Spherical fat particles coated with soluble protein are dispersed in the

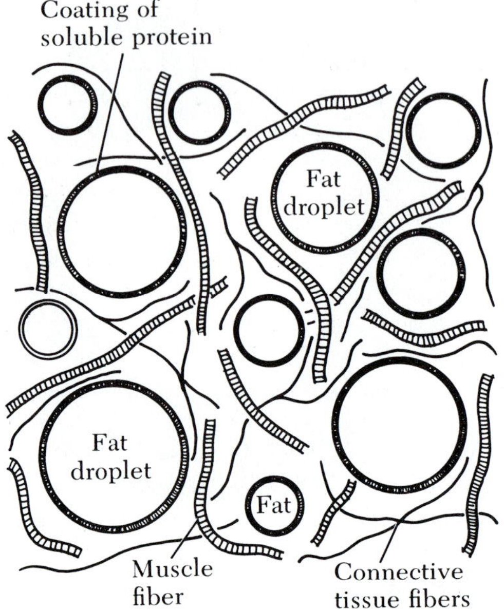

FIGURE 9-4
A diagrammatic illustration of a meat emulsion. Fat droplets are dispersed in an aqueous medium that contains soluble proteins, other soluble muscle constituents, and segments of muscle fibers and connective tissue fibers. Note that each fat droplet is coated with a thin layer of soluble protein (emulsifying agent) which has been released into the aqueous medium from the muscle fibers.

matrix. In sausage emulsions, soluble proteins dissolved in the aqueous phase act as emulsifying agents by coating all surfaces of the dispersed fat particles. The soluble proteins may be either sarcoplasmic or myofibrillar. However, myofibrillar proteins are much more efficient emulsifying agents and thus contribute more to emulsion stability. It was noted in Chapter 3 that the myofibrillar proteins, actin and myosin, are insoluble in water and dilute salt solutions, but are soluble in more concentrated salt solutions. One major function of salt in sausage emulsions is to solubilize these proteins into the aqueous phase so that they become available for coating fat particles.

FACTORS AFFECTING EMULSION FORMATION AND STABILITY. The extent to which fat is incorporated into a stable emulsion is influenced by several factors, including the temperature during emulsification, fat particle size, pH, amount and type of soluble protein, and emulsion viscosity. During chopping and emulsification, the emulsion temperature increases due to friction in the chopper and emulsion mill. Some warming is beneficial, in that it helps to release soluble protein, accelerates cured color development, and improves flow characteristics. However, if the temperature becomes too high during emulsification, emulsion breakdown can occur during subsequent heat processing. The maximum permissible temperature depends upon the type of equipment used for emulsification. Final emulsion temperatures of 20–25°C can be reached without a deleterious effect on stability, when high speed emulsion mills are used. Lower temperatures must be maintained when emulsions are formed with the slower speed silent cutter. There are several possible explanations for the deleterious effects of excessive temperature. These include a denaturation of soluble proteins, a decrease in emulsion viscosity and the melting of fat particles. Since the soluble proteins serve as emulsifying agents, their denaturation during emulsification can cause breakdown during heat processing. As emulsion viscosity decreases, the decreasing emulsion stability can be explained as follows. Since the dispersed fat particles are less dense than the continuous aqueous phase, they tend to float to the surface. Their rapid migration is opposed by the viscous nature of the continuous phase. Therefore, as the continuous phase becomes less viscous, the tendency toward separation increases. Melted fat particles are more easily dispersed into smaller particles that can affect emulsion stability, as is discussed later in this section. The temperature increases can be controlled or reduced by adding ice, rather than water to the meat ingredients during chopping and emulsification. Ice is superior to water because of the additional latent heat of fusion it must absorb, in order to melt. Approximately 80 calories are needed to convert one gram of ice at 0°C to water at 0°C. From this point on, only about one calorie of heat per gram of water is needed to raise its temperature one degree Celsius. Thus, the amount of heat used in melting a gram of ice is sufficient to raise the temperature of one gram of water 80°C. Other effective means for reducing temperature during emulsification include the addition of carbon dioxide "snow," or the use of partially frozen meat ingredients.

During emulsification, fat that is present in the meat ingredients must be subdivided into smaller and smaller particles until an emulsion is formed. However, as fat particle size decreases, there is a proportional increase in the total surface area of fat particles. Every 1-fold decrease in average fat particle diameter causes an increase in total surface area. For example, if a sphere of fat with a diameter of 50 μm is chopped into spheres with a diameter of 10 μm, 125 fat particles will result. However, their total surface area will have increased from 7850 sq. μm to 39,250 sq. μm. This 5-fold increase in surface area will necessitate that much more soluble protein is needed in order to completely coat the surfaces of the smaller particles. Thus, the overchopping of emulsions can result in so large a fat particle surface area that the soluble protein present cannot adequately stabilize the emulsion. Uncoated, or partially coated, fat particles do not form a stable emulsion. Generally, as fat particle size decreases, more soluble protein is required to form a stable emulsion.

In the preparation of meat emulsions, lean meat ingredients are chopped with salt in order to facilitate protein extraction prior to the addition of ingredients having a high fat content. Emulsion stability increases as the amount of soluble protein available to act as an emulsifying agent increases. The amount of protein extracted is affected by several factors. More protein is extracted as the pH of muscle increases. This may partially explain the more stable emulsions formed at higher pH values. The rigor state of meat ingredients also influences emulsification. Prerigor meat is superior to postrigor meat, because as much as 50 percent more salt soluble protein can be extracted. As was stated earlier, the salt soluble myofibrillar proteins are superior to the sarcoplasmic proteins as emulsifying agents. Therefore, more fat can be emulsified with the protein extracted from prerigor meat than with the same amount extracted postrigor. The amount of fat emulsified per unit of protein is referred to as the *emulsification capacity.* The meat processor's decision as to whether prerigor meat should be used will depend upon whether the advantages gained justify the additional cost involved in producing prerigor meat for emulsion manufacture. In addition, most of the advantages of prerigor meat can be gained by the preblending of postrigor meat prior to emulsification. In this process the meat is chopped with ice, salt, and cure ingredients, and held at 0–4°C for up to 12 hours before emulsification, thereby allowing more efficient protein extraction.

Meat emulsion breakdown is actually a reaggregation (coalescence) of the finely dispersed fat particles into larger, easily visible fat particles. Of course, emulsion breakdown is undesirable and must be avoided, if high quality emulsified products are to be prepared. If breakdown occurs, it usually takes place during heat processing. During heating, the dispersed fat particles melt, and in the liquid state they exhibit a greater tendency toward coalescence. Particles that are completely coated with soluble protein (emulsifying agent) are much less likely to coalesce than those having either no coat, or a partial coat. The enlarged fat particles can appear as *fat pockets* on the surface of the product, or in its interior (Figure 9-5). Large fat particles can also coalesce at the end of a sausage into what is known as a *fat cap.* However, less dramatic fat separation, such as greasy surfaces on the finished product are more common, and can cause significant problems in sausage manufacture.

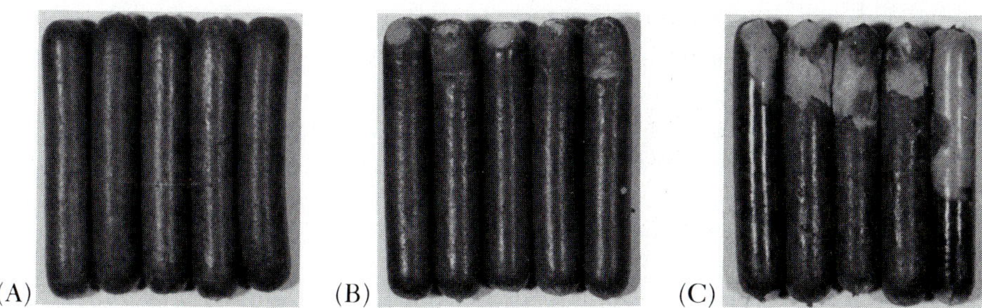

FIGURE 9-5
Emulsion breakdown in frankfurters. (A) A product produced from a stable emulsion. (B) and (C) are examples of fat caps and fat pockets that developed during heat processing due to emulsion breakdown. Note the shiny surfaces of (B) and (C). This indicates that grease is present on the surface of the franks.

Formulation

Many different ingredients are incorporated into processed meat products including meat, curing mixtures, seasonings, binders and fillers, and water. In the preparation of a particular product, the manufacturer is not restricted to a set recipe. The specific ingredients, and their respective amounts, can be varied to arrive at the desired product, much as a diet can be balanced by varying foods and their relative proportions. In the process of formulation, the meat processor selects the ingredients that are to be included, and the amounts that will be used. The first goal of a formulation is to produce products of uniform appearance, composition, taste, and physical properties from batch to batch, day to day, week to week, and month to month. Successful formulation depends upon the availability of accurate information about the properties and composition of the potential raw materials that may be included in the product. For example, meat ingredients vary in composition, color, and chemical and physical properties. Spices vary in purity and strength. The amount of binders and fillers that can be included is restricted by meat inspection regulations, as are the water and fat content. The second goal of a formulation is to produce a product which meets the preset quality standards at the least cost in raw materials. Because of fluctuations in the cost of the various meat ingredients, it is often economically desirable to partially or completely substitute one ingredient for another. The process of formulation must determine to what extent substitutions can be made, and when it would be best, in terms of economics, to do so. Linear programming procedures for least cost formulation are widely used in the meat industry to help accomplish these goals.

MEAT INGREDIENTS. A basic requirement for producing uniform processed meat products is the proper selection of meat ingredients. Animal tissues vary widely in moisture, protein and fat content, in pigmentation, and in ability to bind water and fat. Thus, the processor must know the properties and composition of the various available meat tissues in order to arrive at the correct meat formulation. Meat tissues that are used in sausages are listed in Table 9-2. Note that, in addition to skeletal muscle meat, by-product (variety) meat of nonskeletal muscle

TABLE 9-2
Meat raw materials for sausage type products

Cattle	Hogs
Skeletal muscle meat	Skeletal muscle meat
Boneless, primal cuts, such as chucks, plates, flanks, navels	Boneless primal cuts
Boneless bull meat	Trimmings from primal cuts
Boneless cow meat	Trimmings from blade and neck bones
Cheek meat	Pork jowls and butts
Head meat	Head meat
Beef trimmings	Cheek meat
By-product (variety) meat	By-product (variety) meat
Hearts	Hearts
Tripe	Livers
Livers	Tongues
Tongues	Stomachs
Ox lips	Gelatin skins
Weasand meat (esophagus muscle)	Snouts
Giblet meat	Lips
Blood	Backfat
Brains	Caul fat
Lungs	Ears
Udders (nonlactating)	Diaphragm meat
Spleens	Spleens
Tongue trimmings	Weasand meat (esophagus muscle)
Suet, cod, and brisket fat	
Calves	**Sheep**
Skeletal muscle meat	Skeletal muscle meat
Boneless primal cuts	Boneless primal cuts
Cheek meat	Cheek meat
By-product meat	By-product meat
Tongues	Hearts
	Tongues
Poultry	
Skeletal muscle meat	
Deboned chicken	
Deboned turkey	

SOURCE: MacKenzie, Donald S., *Prepared Meat Product Manufacturing* (AMI Center for Continuing Education, American Meat Institute, Chicago, Illinois), p. 26, 1966.

origin is a potential sausage ingredient, and is often used in cooked sausages. Not all sausages contain variety meat but those that do, have an equivalent, and frequently a superior, nutritional value to those containing only skeletal meat. Also, the use of by-products reduces formulation and finished product costs, compared to products containing only skeletal muscle meat. However, since some consumers object to consuming by-products for esthetic reasons, meat inspection regulations require that cooked sausages containing these meat ingredients be clearly labeled; for example, "Frankfurters with By-products."

In the meat processing industry, the terms *bind* or *binding* refer to the water retention capabilities of lean meat, or to the surface cohesion of meat chunks to each other. In sausage emulsions, the term *bind* includes both water retention and the ability to emulsify fat. Some meat ingredients have a very high binding capacity, and others are inferior. Thus, meat raw materials can be (and are) classified by their binding capacity (Table 9-3). Beef skeletal muscle is an example of meat with a high binding capacity; head and cheek meat, on the other hand, are intermediate in binding ability. Meat with a high fat content, high amount of connective tissue, or a large proportion of smooth muscle, is low in binding ability; for example, regular (50 percent fat) pork trimmings, beef briskets, and tongue trimmings. Collagen, the connective tissue protein, can contribute to water binding during sausage mixing or emulsification. However, during heat processing, collagen shrinks and partially gelatinizes. Shrunken and gelatinized collagen has a good water binding capacity, but is poor in fat emulsification ability. Tripe, lips, stomachs, skin, and snouts may be included in sausage, but the amount must be limited since these materials have very inferior binding proper-

TABLE 9-3
Binding classification of meat raw materials

High Binding Meats	Intermediate Binding Meats	Low Binding Meats	Filler Meats
Bull meat	Beef cheek and head meat	Regular pork trimmings (50 percent fat)	Ox lips
Cow meat	Pork cheek and head meat	Pork jowls	Tripe
Beef chucks	Beef flanks, plates, navels	Beef briskets	Pork stomachs
Boneless pork shoulders	Beef shanks	Hearts	Skin
Lean pork trimmings (80 percent lean)		Beef hanging tender	Snouts
Poultry meat (without skin)		Weasand meat (esophagus muscle)	Lips
		Giblets (diaphragm)	Livers
		Tongue trimmings	
		Deboned poultry backs	

ties. (Commercially, they are referred to as *filler meats.*) A newer meat ingredient for emulsions is machine deboned poultry meat. The composition and binding ability of this product depends upon the particular parts of the carcass that are used. For example, meat from necks and backs contains a high proportion of connective tissue and fat from the skin. On an equal weight basis, deboned poultry meat usually ranks slightly below cow meat in fat emulsifying ability.

Proteins originating from the meat ingredients are responsible for water binding and fat emulsification. Variation in the ability of animal tissues to bind or emulsify fat in meat emulsions is due to: (1) the amount of soluble protein potentially available, and (2) the emulsifying capacity of the protein (the efficiency of soluble protein in emulsifying fat). For example, cow meat has a slightly lower salt soluble protein content than does pork cheek meat. However, cow meat is superior in actual use, since its lower protein content is more than offset by its greater emulsifying capacity. Its emulsifying capacity is higher because it has less connective tissue and greater myofibrillar protein content than does pork cheek meat. Emulsification values have been determined for many of the common meat ingredients by combining factors (1) and (2) above into a single value. Tables of these values may then be used for linear programming of sausage emulsion formulations, if the total amount of protein in the meat batch is known.

Moisture/protein ratios of various tissues provide some guide for the prediction of a product's final composition. The ratios presented for the more common sausage ingredients (Table 9-4) are approximate and should only be used as rough guides. However, upon comparing the binding properties and moisture/protein ratios, it becomes apparent that meat with lower ratios generally performs better in sausage formulations.

Fat is an important constituent of processed meat products, and makes a large contribution to their palatability. Fat content affects the tenderness and juiciness of sausages. It also serves as the dispersed phase in meat emulsions. The occurrence of nonemulsified fat in emulsion type products is often a problem to the processor. The fat content of meat ingredients varies more widely than does the moisture/protein ratio, and depends primarily upon the type of cut or trimmings as well as on the carcass grade. In preparing Table 9-4, the fat contents were approximated by difference, allowing 1 percent for ash content. Since meat inspection regulations limit fat content to a maximum of 30 percent in cooked sausages, knowledge of fat content in meat ingredients is important.

MOISTURE. Moisture accounts for 45–60 percent of the finished weight of processed meat products, more than any other single com-

TABLE 9-4
Moisture, protein, and fat content and moisture/protein ratios of some common sausage meats

Meat Ingredient	Moisture %	Protein %	Fat %	Moisture/Protein Ratio
Bull meat	70.7	20.8	87.5	3.40
Chicken white meat (hand deboned)	73.8	23.3	1.2	3.16
Chucks, boneless	69.6	19.5	10.0	3.57
Chicken dark meat (hand deboned)	73.1	18.5	6.4	3.95
Beef trimmings, lean, 75/85	57.6	16.9	25.0	3.41
Beef cheek meat	66.4	18.5	14.5	3.59
Boneless picnics	60.0	15.6	23.4	3.84
Pork head meat	57.9	16.1	25.0	3.60
Beef flanks	35.0	9.9	54.0	3.54
Regular pork trimmings	36.0	9.6	54.0	3.77
Pork jowls (skinned)	23.4	6.3	70.0	3.72
Broiler backs and necks (machine deboned)	66.6	14.5	17.6	4.59
Beef hearts	64.1	14.9	20.0	4.30
Turkey frame meat (machine deboned)	73.3	13.5	11.5	5.43
Pork backfat (untrimmed)	16.1	4.2	79.0	3.83
Beef tripe	75.5	12.8	11.0	5.90

SOURCES: Froning, G. W., "Poultry Meat Sources and Their Emulsifying Charcateristics as Related to Processing Variables," Poultry Sci. 49, 1625 (1970); Kramlich, W. E., A. M. Pearson and F. W. Tauber, *Processed Meats* (AVI Publishing Company, Inc., Westport, Connecticut), 1973.

ponent. Most of the moisture originates from the lean meat ingredients. However, the processor adds additional water to many products as part of the formulation. There are several reasons for adding water. Many products would be dry and unpalatable if only the moisture contained in the meat ingredients were present in the final product. Additional water improves their tenderness and juiciness. As previously mentioned, moisture, added as ice, also helps to keep product temperatures down during emulsification. Water serves as the carrier for distributing the curing ingredients into noncomminuted smoked meat items, such as hams and bacons. The moisture added during pumping, immersion of cuts in brine, or that is added to sausage formulations also serves to replace moisture that will be lost during subsequent processing operations, particularly in heat processing. Thus, by adding water, the yield of finished product can be improved. The amount of moisture that is added during processing depends upon the composition and properties of the raw materials, the losses anticipated during all stages of processing, and the moisture level desired in the final product.

Processed meat products must comply with federal meat inspection regulations with respect to moisture content. For example, moisture content of cooked sausage products must not exceed 4 times the meat protein content (by analysis) plus 10 percent. In other words, if the product contains 12 percent meat protein, then the maximum allowable amount of water is 58 percent of the final product $[(4 \times 12\%) + 10\% = 58\%]$. In fresh sausage that is not heat processed, a maximum of 3 percent moisture may be added to facilitate processing. Regulations also state that the finished weight of smoked meat products such as hams, picnics, or pork butts must not exceed the weight of the fresh, uncured product. Therefore, the water added during pumping must be sufficient only to replace that which will be lost during the remainder of the processing operation. If more than this amount of water is added, and the finished product weight is greater than the fresh, uncured product weight, it must be labeled "water added." Finished weight of "water added" smoked meats must not exceed 110 percent of the fresh weight (commonly referred to as 10 percent added water). "Water added" products are generally more tender and juicy, and have met with consumer acceptance. Laboratory enforcement procedures for the maximum allowable moisture content of fresh sausage, hams, picnics, butts, and water added products are based on the meat protein content of the respective products, as is the case for the cooked sausages described above. Although the multiplier is 4.0 times the protein content for all cooked sausages, it varies for other products, generally being between 3.79 and 4.0, depending upon the natural moisture/protein ratio for a given type of meat.

EXTENDERS, BINDERS, AND FILLERS. A variety of nonmeat products are incorporated into sausage and loaf items. These materials are commonly referred to as extenders, binders, and fillers. They are included in sausage formulations for one or more of the following reasons: (1) to improve emulsion stability, (2) to improve water binding capacity, (3) to enhance flavor, (4) to reduce shrinkage during cooking, (5) to improve slicing characteristics and, (6) to reduce formulation costs.

We have previously used the terms bind or binding to refer to both water retention and fat emulsification in emulsified products, to the water retention ability of lean meat, and to the ability of meat chunks to adhere to one another in nonemulsified products. Therefore, *binders* are those nonmeat materials that contribute to both water binding and fat emulsification. *Fillers* are able to bind large amounts of water, but contribute little to emulsification. The nonspecific term *extender*, is used to describe any nonmeat ingredient, except for water, salt, and seasoning, added in sufficient quantity to increase the bulk or change the composition of sausages.

Binders commonly used in sausage formulations are characterized by high protein content, and are either dried milk or soybean products. Soy products include soy flour, soy grits, soy protein concentrate, and isolated soy protein. The grits and flour contain 40–60 percent protein and are identical, except for the fact that the flour is finely ground and the grits are coarsely ground. Both products impart a distinctive flavor to meat items which has limited their application. On the other hand, soy protein concentrate contains about 70 percent protein and has a bland flavor. Grits, flour, and concentrate may be used to bind meat patties and loaves. Isolated soy protein contains approximately 90 percent protein and, besides having a very bland flavor, is dispersible in water, and possesses good water and fat binding capabilities. Acceptable nonmeat frankfurter and bologna type sausages have been prepared using isolated soy protein as the sole source of protein.

Manufactured vegetable protein (MVP) is the newest nonmeat ingredient that has been developed as a binder and extender for processed meat products. MVP is generally produced by simultaneously cooking and extruding a mixture of soy flour and other ingredients, such as flavoring and coloring, into particles whose size can be varied from a millimeter to a centimeter, or more, in diameter. Unflavored MVP has a typically bland, toasted flavor, and contains about 50 percent protein and 7 percent moisture.

Major uses of MVP are as an extender for ground beef, and as a binder and extender in meat patties. A typical formulation is to add a mixture of 3 parts water and 1 part MVP to ground beef, at a rate of 20–25 percent extender mixture and 75–80 percent ground beef. The use of vegetable protein in this manner significantly reduces formulation costs. The extruded protein products have a meatlike texture, hydrate rapidly, and have an affinity for juice retention. Meat patties containing MVP shrink less during cooking than do all meat patties, due to their extra juice retaining ability. Textured protein products simulating beef, ham, chicken, and bacon bits are found in such items as salad dressings, seasoning mixes, casseroles, and many types of prepared foods and snacks.

Nonfat dried milk solids (NFDM), calcium reduced nonfat dried milk, and dried whey are products derived from milk that are used as binders. NFDM contains approximately 35 percent protein, of which 80 percent is casein and the remainder is largely β-lactoglobulin and lactalbumin. NFDM has a limited ability to emulsify fat because the casein is combined with large amounts of calcium, making it poorly soluble in water. You will recall that proteins must be solubilized in order to function as emulsifying agents. If sodium replaces most of the calcium, water solubility and emulsification capacity of the casein is improved. This sodium replacement product is called calcium reduced NFDM. Dried whey, from which a part of the lactose has been removed, may have a higher emulsifying capacity than NFDM, since casein is not present and β-lactoglobulin and lactalbumin are readily soluble proteins.

The common filler ingredients used in sausages and loaves include: (1) cereal flours obtained from wheat, rye, barley, corn, or rice, (2) starch extracted from these flours, or from potatoes and, (3) corn syrup or corn syrup solids. These flours are high in starch content but relatively low in protein. Therefore, they are able to bind large amounts of water but are poor in emulsification ability. Corn syrup and corn syrup solids are derived from corn starch by a partial degradation of the starch into glucose, maltose, and dextrin units. They are less sweet than sucrose and are added for flavor and textural properties.

The amount of extenders permissible in sausages and loaves is specified by meat inspection regulations. Soy products (with the exception of isolated soy protein), cereal flours and starches, and NFDM or calcium reduced NFDM may be added to cooked sausages, either singly or in combination, to a maximum of 3.5 percent by weight of the finished product. Isolated soy protein may account for no more than 2 percent of the sausage. Sausages containing more extenders than these limits must be labeled as "imitation." These limits also apply to loaf items identified as meat loaves. Nonspecific loaves that do not include the word meat in the product name are not limited with respect to extender content: examples are pickle and pimento loaf, and macaroni and cheese loaf.

SEASONINGS. *Seasoning* is a general term applied to any ingredient that is added to improve or modify the flavor of processed meat products. Thus, the obvious reason for incorporation of seasonings into processed meat products is to create distinctive flavors. The use of seasonings allows the processor to create new products and provide variety in existing products. A great amount of artistry is required for imaginative and successful use of seasonings, and meat processors guard the secrecy of their seasoning formulations very closely. In addition to flavor, seasonings contribute somewhat to the preservation of meat. The preservative action of salt, especially in high concentrations, has been mentioned previously. Certain spices possess antioxidant properties, and thereby reduce the rate of oxidative rancidity development. On the other hand, spices may carry excessively high bacterial loads that will shorten the shelflife of products. Table 9-5 lists many of the common spices used in meat products, their origin, and some typical examples of their use in processed meat.

TABLE 9-5
Spices commonly used in processed meats, and their origin

Spice	Part of Plant	Region Grown	Use
Allspice	Dried, nearly ripe fruit of *Pimenta officinalis*	Jamaica, Cuba, Haiti, The Republic of Trinidad and Tobago	Bologna, pickled pigs feet, head cheese
Anise (seed)	Dried ripe fruit of *Pimpinella anisum*	Russia, Germany, Scandinavia, Czechoslovakia, France, Netherlands, Spain	Dry sausage, mortadella, pepperoni
Bay leaves (laurel leaves)	Dried leaf of *Laurus nobilis*	Mediterranean region, Greece, Italy, Great Britain	Pickle for pigs feet, lamb, pork tongue
Cardamom	Dried ripe seeds of *Elettaria cardamomum*	Malabar Coast of India, Sri Lanka (Ceylon), Guatamala	Frankfurters, liver sausage, head cheese
Cassia	Dried bark of *Cinnamomum cassia*, *C. loureirii*, *C. burmanni*	People's Republic of China, India, Indo-China	Bologna, blood sausage
Celery Seed	Dried ripe fruit of *Apium graveolens*	Southern Europe, India	Pork sausage
Cinnamon	Dried bark of *Cinnamomum zeylanicum* and *C. loureirii*	Sri Lanka (Ceylon), Sumatra, Java, Vietnam, Borneo, Malabar Coast of India	Bologna, head sausage
Clove	Dried flower buds of *Eugenia caryophyllata*	Brazil, Sri Lanka (Ceylon), Tanzania (Zanzibar), Malagasy Republic (Madagascar)	Bologna, head cheese, liver sausage
Coriander (seed)	Dried ripe fruit of *Coriandrum sativum*	England, Germany, Czechoslovakia, Hungary, Russia, Morocco, Malta, India	Frankfurters, bologna, Polish sausage, luncheon specialities
Cumin	Dried ripe seeds of *Cuminum cyminum*	Southern Europe, India, Mediterranean areas of North Africa, Saudi Arabia, India, and People's Republic of China	Curry powder
Garlic	Fresh bulb of *Allium sativum*	Sicily, Italy, Southern France, Mexico, South America, India, U.S.A.	Polish sausage, many types of smoked sausage
Ginger	Dried rhizome of *Zingiber officinale*	Jamaica, West Africa, West Indies	Pork sausage, frankfurters, corned beef
Mace	Dried waxy covering (aril) that partly encloses the seed of the nutmeg *Myristica fragrans*	Southern Asia, The Malay Archipelago (Indonesia and East Malaysia)	Veal sausage, liver sausage, frankfurters, bologna

(*continued*)

TABLE 9-5
Spices commonly used in processed meats, and their origin (*continued*)

Spice	Part of Plant	Region Grown	Use
Marjoram	Dried leaf (with or without flowering tops) of the *Marjorana hortensis* or *Origanum vulgare*	Northern Africa, Greece, and other Mediterranean countries	Liver sausage, Polish sausage, head cheese
Mustard (black, white, yellow, brown, red)	Dried ripe seed of *Brassica nigra, B. juncea, B. alba*	People's Republic of China, Japan, India, Italy, Russia, The Netherlands, England, U.S.A.	Good in all sausage
Nutmeg	The dried and ground ripe seed of *Myristica fragrans*	Southern Asia, Malay Archipelago (Indonesia and East Malaysia)	Veal sausage, bologna, frankfurters, liver sausage, head cheese
Onion	Fresh bulb of *Allium cepa*	U.S.A., worldwide	Liver sausage, head cheese, baked loaf
Paprika	Dried, ripe fruit of *Capsicum annuum*	Hungary, Spain, U.S.A., Ethiopia	Frankfurters, Mexican sausage, dry sausage
Pepper (black)	Dried, unripe fruit of *Piper nigrum*	The Republic of Singapore, Lampung, Sumatra, Penang, Sarawak, Thailand, India, Philippines, Indonesia, Tanzania (Zanzibar)	Bologna, Polish sausage, head cheese
Pepper (Cayenne or red) or red)	Dried ripe fruit of *Capsicum frutescens*	Tanzania (Zanzibar), Japan, Mexico, U.S.A.	Frankfurters, bologna, veal sausage, smoked sausage, smoked country sausage
Pepper (white)	Dried, ripe, decorticated fruit of *Piper nigrum*	The Republic of Singapore, Thailand	Good in all sausage
Pimento (pimiento)	Ripe, undried fruit of *Capsicum annuum*	Spain, U.S.A.	Loaves
Sage	Dried leaves of *Salvia officinalis*	Dalmatian Coast, U.S.A.	Pork sausage, baked loaf
Savory	Dried leaves of either *Satureja hortensis* or *S. montana*	Mediterranean countries, U.S.A.	Good in all sausages
Thyme	Dried leaves and flowering tops of *Thymus vulgaris*	Mediterranean coast	Good in all sausages

Salt and pepper form the basis for sausage seasoning formulas. All other seasoning ingredients are supplementary to salt and pepper, but are very necessary to obtain the distinctive flavor associated with various products. These seasonings include spices, herbs, vegetables, sweeteners, and other ingredients, such as monosodium glutamate, that contribute to flavor enhancement.

Spices are aromatic substances of vegetable origin. Although spices originate in all parts of the world, most are grown in the East and West Indies, Malaysia, India, The People's Republic of China, and Japan, and some in Africa and Europe. The parts of the plant used to obtain the spice varies. For example, the fruit is used for pepper, allspice, and nutmeg, the flower bud for clove, the aril (fleshy seed covering) for mace, the rhizome for ginger, and the bark for cinnamon and cassia. *Aromatic seeds* such as cardamom, dill, mustard, and coriander are also classed as spices. *Herbs* are the dried leaves of plants, and those used in sausages include sage, savory, thyme, and marjoram. Seasonings originating from vegetable bulbs are onion and garlic. Since all the above seasonings are natural products, they are quite variable in flavor, strength, and quality due to variation in weather during growth, the local soil fertility, and storage conditions. Thus, a large amount of empirical experience is necessary for the proper selection of natural seasonings for use in processed meat.

Natural seasonings can be used whole; for example, whole pepper corns are included in certain dry sausages. But usually they are used in a processed form, either ground up or as essential oils and oleoresins. Ground spices are available to the processor in several degrees of fineness. The most finely ground (*microground*) spices are sometimes preferred because they disperse more completely in the product, are invisible, and do not detract from product appearance. *Extractives*, the essential oils and oleoresins, have some advantages over ground spices. They are free of microbial contamination and are invisible in the finished meat product. Extractives are obtained from the natural spices by pressing, distillation, or solvent extraction. They can be blended in the same manner as ground spices. The extractives are usually combined with a carrier, such as salt or dextrose, for incorporation into processed meat.

The sweeteners used in meat products are sucrose, dextrose, corn syrup or corn syrup solids, and lactose. Lactose has little sweetening ability and is present in sausages only when nonfat dried milk is included in the formulation. Nonfat dried milk contains approximately 50 percent lactose. Corn syrup and corn syrup solids are only 40 percent as sweet as sucrose, and they are therefore often classified as fillers. Sucrose and dextrose are used primarily for their sweetening ability. In addition, they are readily available for fermentation by the lactic acid producing organisms used in preparing some dry sausages. The fermentation products are responsible for the characteristic flavors of these products.

OTHER INGREDIENTS FOR PROCESSED MEAT FORMULATION. Several other ingredients are used in processed meat formulations. The use of nitrite and nitrate for cured meat color development is discussed extensively earlier in this chapter, as is the use of ascorbate or erythrobate as reducing agents for the acceleration of color development.

Alkaline phosphates are widely used in meat curing brines at present. These compounds had been used unknowingly in meat preparation for many years through use of the "stock pot." It is known that the phosphate content of the stock increases each time it is used. It is widely recognized that flavor is added to meat cooked in such stock, and to gravy that is prepared from it. Compounds such as disodium phosphate, hexametaphosphate, sodium tripolyphosphate, and sodium pyrophosphate are approved by United States meat inspection regulations for use in the pumping brine for hams, picnics, and similar products. Phosphates cannot be added to sausage and other prepared meat products in the United States, but they are used extensively in these products in several European countries.

Several beneficial effects of phosphates in meat curing can be cited. Phosphates increase the water holding capacity (water retention ability) of meat. When phosphates are used, there is less moisture loss during cooking. For example, there is less *purge* (water and gelatin released from meat during cooking) in processing canned hams and less loss in fully cooked hams (hams reaching an internal temperature of 68°C or greater). There is some increase in tenderness and juiciness of the cured product when phosphates are used. There is also an improvement in cured meat color acceptability, uniformity, and stability. Phosphates offer protection against browning during storage, and act synergistically with ascorbates to protect against rancidity in cured meat (i.e., the protection offered by both compounds is greater than the sum of the separate protection offered by each, alone).

Forming Processed Meat Products

Most processed meat products are formed at some point in processing, in order to give each product a uniform or characteristic shape. Since sausages are comminuted products, they must be placed in some type of forming device or covering to give them shape, to hold them together during further processing, and for protection. The shape or form of particular types of sausage is now dictated by tradition, after having been in use for many years (see Figure 9-6). Many noncomminuted smoked meat products are also formed during processing. In general, these meat products are formed in either *molds* or *casings*. Metal molds are often used for loaf type items that are sliced prior to merchandising. The product is placed in the mold after comminution and blending, and is then cooked to "set" its shape. The formed loaf is removed from the mold before slicing. Baked loaves are usually formed in metal pans. Casings are more widely used as forms and containers for sausages. The process of placing meat products, either comminuted or noncomminuted, into casings is referred to as *stuffing*. Two types of casings are in general use: (1) natural, and (2) manufactured.

NATURAL CASINGS. Prior to the development of manufactured casings, only natural casings were available to meat processors. They are derived almost exclusively from the gastro-intestinal tracts of swine, cattle, and sheep. Hog casings are prepared from the stomach, small intestine (smalls), large intestine (middles), and terminal end (colon) of the large intestine (bungs). The parts of cattle used for beef casings are the esophagus (weasands), small intestine (rounds), large intestine (middles), bung, and bladder. Only the intestines of sheep are used to produce sheep casings.

Natural casings are very permeable to moisture and smoke. One of their most important characteristics is that they shrink and thereby remain in close contact with the surface of a sausage as it loses moisture. Thus, they are often used in dry sausage manufacture. Most natural casings are digestible and can be eaten.

MANUFACTURED CASINGS. Four classes of manufactured casings are available: (1) cellulose, (2) inedible collagen, (3) edible collagen, and (4) plastic. Cellulose casings are prepared from cotton linters (the short fibers that are closest to the seed). These are first dissolved and then regenerated into casings. Other sources of cellulose have been successfully used. Cellulose casings are manufactured in sizes ranging from 1.5 centimeters in diameter, for small sausages, up to 15 centimeters for large sausages such as bologna. They are manufactured with stretch and shrink characteristics similar to those of natural casings. The inner surface of the casing is sometimes coated with an edible, water soluble dye that transfers to the sausage surface and artificially colors the product. The advantages of cellulose casings include their ease of use, the variety of sizes that are available, their uniformity of size, greater

FIGURE 9-6
Sausages stuffed in either natural or artificial casings. The shape and size of each sausage is dictated in part by the type of natural casing traditionally used for the sausage. The sausages are numbered in the drawing, and are identified as follows:

Sausage	*Casing or Container*
1. Ring bologna	Natural – beef small intestine
2. Smoked sausage	Natural – hog small intestine
3. Bologna	Artificial – cellulose tube

(*continued*)

strength, and low microbial levels. Their strength is especially important in view of the widespread use of automated processing procedures. Fibrous cellulose casings, consisting of cellulose extruded on a paper base material, are very strong and are used for large sausages (bologna) and roll-type items (turkey rolls).

Both edible and inedible collagen casings are regenerated from collagen extracted from skins and hides. The inedible collagen casings combine some of the advantages of both cellulose and natural casings; especially their strength, uniformity, and shrink characteristics. They must be removed prior to consumption of the products, as must cellulose casings. Edible collagen casings are used largely for fresh pork sausage and frankfurters. They are very uniform in physical characteristics and have greater strength than natural casings.

Plastic tubes or bags are used as sausage containers in certain applications. They are impermeable to smoke and moisture. Therefore, they are used with products which are not smoked, such as fresh pork sausage or liver sausage, or with products which are heat processed in hot water.

FORMING NONCOMMINUTED SMOKED MEAT. The forming of smoked meat products differs in several ways from the forming of sausage type products. Natural and collagen casings are generally not used as containers for noncomminuted meat, but (in certain instances) fibrous cellulose casings, plastic tubes, or bags are used. For example, boneless hams can be placed in a fibrous cellulose or plastic container after curing and boning. The enclosed product can then be flattened in a spring-loaded frame and heat processed in the frame device. Bone-in hams and picnics are placed in cloth stockinettes prior to smoking and heat processing to protect the product and absorb moisture and melted fat which

FIGURE 9-6 (*continued*)

4. Bung bologna	Natural – beef bung
5. Beer salami	Natural – beef bladder
6. Mettwurst	Natural – beef small intestine
7. Liver sausage	Natural – sewn hog bung
8. B. C. salami	Natural – beef large intestine
9. Smoked sausage	Natural – hog small intestine
10. Knackwurst	Natural – beef small intestine
11. Luncheon salami	Artificial – cellulose tube
12. Pickle and pimento loaf	Artificial – metal loaf pan
13. Luncheon salami	Artificial – cellulose tube
14. Head cheese	Natural – hog stomach
15. Salami	Artificial – fibrous cellulose
16. Pepper loaf	Artificial – metal loaf pan
17. Frankfurter	Artificial – cellulose (removed)

[Photograph courtesy of National Live Stock and Meat Board.]

cook out during heat processing. The stockinette also serves to shape the ham or picnic. Bacons are formed in a high pressure bacon press after they are smoked and before being sliced, in order to shape or "square" them to uniform dimensions. This facilitates high speed mechanical slicing, and results in more uniformity between slices, as well as a more uniform number of slices per pound.

Smoking and Heat Processing

The smoking and heating (cooking) of processed meats can be considered as two separate processing steps. They are discussed together since, in most products, the two processes occur simultaneously or in immediate succession, so that variables in one process affect the other. In modern processing methods, the same facilities are used to accomplish both processes. Most products are both smoked and cooked (frankfurters, bologna, and many hams). However, a few products are only smoked with a minimum of heating (mettwurst, some Polish sausage, and bacon) while others are cooked but not smoked (liver sausage).

HEAT PROCESSING. Typical heat processed meat products are cooked until internal temperatures of 65–75°C are reached. This is sufficient to kill most of the microorganisms present, including the trichinae that are occasionally found in pork. The product is thereby pasteurized, and its shelf life significantly extended. Pasteurization is one important function of heat processing. Products are seldom heated to the extent that they are sterile and stable at room temperature. Thermal processing as a preservative technique is discussed in Chapter 11.

In addition to pasteurization, other important changes result from heat processing. Of special significance is the firm, set structure that develops as a result of protein denaturation, coagulation, and partial dehydration. For example, an emulsified product stuffed into a cellulose casing has no definite shape. The hardening and firming that occurs during cooking sets the structure so that when the casing is removed, the product's shape and form is retained. Similar changes occurring in smoked meat products, such as hams, give the product a more rigid structure, so that their shapes are retained during further handling, packaging, and distribution. Textural changes, increased tenderness, and browning also occur during heating. All of these factors are discussed in more detail in Chapter 12. A third important purpose of heat processing is to fix the cured meat pigment by the denaturation of nitric oxide myoglobin, as previously discussed. The degree of heating used in a particular process is dependent upon the federally and state approved labels, specifications, and standards of identity under which a product is manufactured. Some products, such as bacon, are heated only to an internal temperature of 52°C, at which temperature color fixation occurs (by the formation of nitrosyl hemochrome). Such products are not considered as being cooked meat products and, except for dry or semidry sausages and smoked meats, require further cooking before they are eaten. Meat inspection regulations require that products labeled as cooked (ready to eat hams, luncheon meat, frankfurters, and many others) must be heated to an internal temperature of 65–68°C.

SMOKING. The smoking of meat is the process of exposing a product to wood smoke at some point during its manufacture. Smoking methods originated simply as a result of meat being dried over wood fires. The development of specific flavors, and the improvement of appearance are the main reasons for smoking meat today, even though smoking provides a preservative effect. More than 200 individual compounds have been identified in wood smoke. The classes of chemical compounds that are present include aldehydes, ketones, alcohols, phenols, organic acids, cresols, and acyclic hydrocarbons. Although most of these compounds exhibit either bacteriostatic or bactericidal properties it is believed that formaldehyde accounts for most of the preservative action of smoke. In addition, phenols also have an antioxidant activity that retards the onset of oxidative rancidity. All of the compounds listed above probably contribute to the characteristic flavor of smoked meat.

In most present day processed meats, smoking contributes little if any preservative action. Smoke components are absorbed by surface and interstitial water in the product, but in no case do they penetrate more than a few millimeters. In products where the surface remains intact, a preservative effect will persist. But if the surface is disrupted, as in slicing or casing removal, the bacteriostatic effect is essentially lost. Since few products today reach the consumer with the surface intact, and since less smoke is applied than was formerly the case, the present day purpose of smoking meat is mainly to develop a distinctive flavor and a surface appearance that is attractive to the consumer. However, a few other advantages do accrue from the smoking of meats. For example, it aids in the development of a smooth surface or skin beneath the cellulose casing of frankfurters that facilitates peeling of the casing prior to packaging.

Several methods are used to apply smoke. The traditional and most widely used method is in a smokehouse. The product is hung from racks or trees which, in turn, are placed in the sealed house. Smoke is generated outside of the house by the controlled combustion of moist sawdust, or by the friction of a rapidly moving steel plate against the end of a log or board. Smoke is carried into the house by a system of fans. Because of their low resin content, hardwoods, usually oak or hickory, are most commonly used to generate the smoke. Softwoods are also used in order to achieve special flavor effects.

The modern smokehouse is equipped to heat process meat products as well as to apply smoke. In order to accomplish both processes, the temperature, density of the smoke, and the relative humidity are carefully controlled. In many cases, smoke is applied during the initial phase of cooking, or even before cooking the product. The smoke density determines the length of time that products must be smoked in order to achieve the desired level of smoke deposition. This is a very important factor in modern continuous process frankfurter production, where the franks must be exposed to large volumes of smoke in a short period of time. Continuous smokehouses are now in use that are capable of smoking and heat processing frankfurters in 30–60 minutes.

Humidity control in the smokehouse is important for several reasons, the most important of which is to insure a high yield of cooked product. At the start of a cooking schedule, relative humidity can be high, but as the temperature increases, it will decrease, and a significant loss in product weight, as moisture, can occur. Acceptable weight losses range from 5–10 percent, depending somewhat upon the particular product. In addition to decreasing shrinkage, a high relative humidity during cooking makes the casings more permeable to smoke, speeds cooking, and eases casing removal after chilling. However, excessively high relapative humidity can contribute to emulsion breakdown, and to the appearance of surface grease or, in more extreme cases, fat caps and jelly pockets.

Electrostatic smoking processes have been developed in an attempt to speed smoke deposition, but these processes have not achieved widespread commercial application. Liquid smoke preparations have been developed as an attempt to eliminate the smoking process. Liquid smoke is prepared by the condensation and fractional distillation of wood smoke. It is usually applied as an aqueous spray on the product surface. Liquid smoke preparations are free of carcinogenic compounds, such as 3,4-benzopyrene, that have been discovered in low levels in natural wood smoke. The presence and amount of carcinogens in natural smoke depends somewhat upon the temperature at which the smoke is generated. The danger of carcinogenesis from natural smoke in meat products seems negligible. However, if it becomes necessary to eliminate the use of natural smoke, the use of liquid smokes will probably increase. These preparations are widely used in Europe at present, but have only seen limited application in North America.

Dehydration

Dehydration of meat products is one of the several basic processing steps. However, few meat products are dehydrated as a separate process. In those cases where drying is a separate step, the objective is primarily preservation. Drying to preserve the product can be accomplished by freeze dehydration or by the application of heat. These preservative processes are discussed in Chapter 11. In many meat products with a reduced moisture content, drying occurs simultaneously with the next processing step—aging.

Aging

This process involves keeping the manufactured product for varying periods under controlled temperature and humidity conditions. There are several purposes for aging processed meat, including: (1) flavor development, (2) textural changes, (3) completion of the various curing reactions, and (4) the drying and hardening of the product. The development of a distinctive flavor often results from microbial fermentation in the product. The organisms responsible for fermentation are usually lactic acid producing bacteria that can enter the product from the plant environment and processing equipment. If this were always the case, only chance would dictate whether the product would encounter the correct numbers and types of organism. To eliminate the element of chance, and to achieve uniform quality in various fermented products, many processors now add specific lactic acid producing microorganisms as a starter culture. Examples of such starter culture organisms are *Lactobacillus plantarum* and *Pediococcus cerevisiae.*

The aging period is also necessary for proper cure development. This is especially true when the curing mixture contains only nitrate, since time must be allowed for the growth of nitrate reducing bacteria, and for the conversion of nitrate to nitrite. Some lactic acid producing bacteria do not reduce nitrate, so when they are used as starter cultures to accelerate fermentation, nitrite must also be included in the cure. Two types of textural changes may occur. Tenderization may occur due to the action of autolytic enzymes (cathepsins) that are present in muscle (see Chapter 8). More commonly, the product will become firmer or harder due to a loss of moisture during drying. Varying amounts of moisture loss are desired, depending upon the product. However, in all cases the rate of drying is closely controlled. Too rapid a drying rate, especially initially, will result in loss of moisture primarily from the surface, and the development of a hard exterior that will retard or prevent the proper drying of the interior. Theoretically, the drying rate at the surface should be only slightly greater than that required to remove moisture which migrates to the surface from the interior of the product. Conversely, if the drying rate is too slow, the surface will be moist enough to support excessive mold, bacterial, or yeast growth.

Aging can either follow or precede the smoking process, depending upon the particular product. Quite often, aged products are not fully heat processed, but are only subjected to a cold smoke. For example, semidry fermented sausages are heat processed at a minimum temperature of 56.5°C, whereas dry sausages, such as summer sausage and salami, never exceed a temperature of 32°C during their manufacture. In addition to semidry and dry fermented sausages, country cured hams, which are widely produced in the southeastern United States, are an example of an aged processed meat product. The length of the aging period for these products varies from 1 week to several months, depending largely upon the amount of moisture loss that is desired. Country cured hams are aged for 3–6 months, during which time they will shrink at least 18 percent due to moisture loss. Genoa salami is aged for an average of 90 days, and will sustain an average moisture loss of 25 percent during this time.

REFERENCES

Fox, J. B., Jr., "The Chemistry of Meat Pigments," Agric. Food Chem. 14, 207 (1966).

Mackintosh, D. L., *Pork Operations in the Meat Industry* (American Meat Institute, Chicago, Illinois), 1967.

MacKinzie, D. S., *Prepared Meat Product Manufacturing* (American Meat Institute, Chicago, Illinois), 1966.

Merory, J., *Food Flavorings, Composition, Manufacture and Use* (AVI Publishing Company, Inc., Westport, Connecticut), 1960.

Price, J. F. and B. S. Schweigert, *The Science of Meat and Meat Products* 2nd Edition (W. H. Freeman and Company, San Francisco), 1971.

Saffle, R. L., "Meat Emulsions," in *Advances in Food Research*, C. O. Chichester, E. M. Mrak and G. F. Stewart, eds. (Academic Press, New York), Vol. 16, pp. 105–160, 1968.

CHAPTER 10

Microbiology, Deterioration and Contamination of Meat

Meat and meat products are extremely perishable, so special care and handling must be exercised during all operations. Deterioration begins soon after bleeding, as the result of microbial, chemical, and physical processes. If these processes were not curtailed they would soon make meat unfit for consumption. It is necessary to minimize deterioration in order to prolong the time during which an acceptable level of quality is maintained. Laxity in the exercise of quality control measures during any processing operations usually increases the rate and extent of the deteriorative changes that lead to spoilage and, ultimately, to putrefaction. The absolute necessity of maintaining preventive controls during all handling and processing of meat, in order to optimize its acceptability, and permit its storage is obvious. A number of methods are employed throughout the meat industry to retard deteriorative changes and extend the length of the acceptability period. These procedures constitute the various forms of meat preservation. The length of the storage time during which optimum acceptability is maintained depends upon the preservative method used, and the inherent properties of the specific meat item in question.

It is generally recognized that the postmortem changes associated with the conversion of muscle to meat, and the subsequent storage and handling are accompanied by some deterioration irrespective of the precautions taken during processing and handling. However, the extent to which these deteriorative changes occur is dependent upon the prevailing handling, processing, and storage conditions. These changes include those caused by microorganisms (bacteria, molds, and yeasts), insects, endogenous enzymes (naturally present in meat tissues), exogenous enzymes (secreted by microorganisms), chemical reactions other than enzymatic (oxidative rancidity), and physical effects (freezer burn, exudation (drip), light fading, and discoloration).

While each of these deteriorative changes may contribute to the unacceptability of meat, the major concern in the normal handling and storage of meat is microbiological contamination and activity. In most instances, meat spoilage is the result of the deteriorative action of microorganisms. However, insects may contribute to the spoilage of some products, such as "country-cured ham." Thus, all handling practices and storage methods are primarily concerned with minimizing microbial contamination and retarding microbial growth and activity. Fortunately, the proper application of methods that control microbial and insect activity also generally minimize other deteriorative changes. In other words, proper refrigeration for the control of microbial activity also retards enzymatic activity, as well as minimizing the deteriorative effects of the physical aspects such as the quantity of drip and freezer burn.

MICROBIOLOGICAL PRINCIPLES IN MEAT

Sources of Microbial Contamination

With the exception of the external surface (hair and skin) and the gastrointestinal and respiratory tracts, the tissues of living animals are essentially free of microorganisms. The animal's white blood cells, and the antibodies developed by the animal throughout its life, effectively control infectious agents in the living body. However, these internal defense mechanisms are lost with the removal of blood at slaughter. Thus, immediately following exsanguination (and at every stage thereafter) measures must be taken to minimize microbial contamination and reduce the growth and activity of the microorganisms that are present. We wish to emphasize here that not all microbial action is detrimental: the proper fermentation, curing, and aging of some meat products depends upon microorganisms for flavor enhancement or for producing other desirable effects.

The initial microbial contamination of meat results from the introduction of microorganisms into the vascular system when unsterile knives are used for exsanguination (e.g., in the processes of sticking and bleeding). Since blood continues to circulate for a short period of time following sticking, the microorganisms introduced by unsterile knives may be disseminated throughout much of the animal body. Subsequent contamination occurs by the introduction of microorganisms on the meat surfaces in almost every operation performed during the slaughtering, cutting, processing, storage, and distribution of meat. Meat can be contaminated by contact of the carcass with hides, feet, manure, dirt, and with the viscera (if it has been punctured) during the slaughtering process. Other potential sources of microbial contamination are: the equipment used for each operation that is performed until the final product is eaten, the clothing and hands of personnel, and the physical facilities themselves. Even the water used for washing carcasses and equipment, as well as the water used in brine solutions, can also be a source of contamination. Likewise, contamination might result from airborne microorganisms in the chilling, storage, and aging coolers, or in the processing and packaging rooms. Thus, it is obvious that everything, including other meat products, that comes in contact with meat is a potential source of microbial contamination. It should also be apparent that the use of proper sanitation is the best approach for the limitation of microbial contamination. We cannot overemphasize the fact that *there is no substitute for good sanitation in the meat industry.* Once microorganisms are present their activity can seldom be completely curtailed, no matter what control measures are subsequently applied. Thus, the "load" (amount of microbial contamination) is an important factor in determining the shelflife and acceptability of all meat products, whether fresh or processed.

Microorganisms in Meat

The microorganisms which are found on or in meat consist of *fungi* and *bacteria.* The fungi include molds and yeasts. *Molds* are multicellular organisms that are characterized by their mycelial (filamentous) morphology. They display a variety of colors, and are generally recognized by their mildewy, or fuzzy cottonlike appearance. Molds develop numerous tiny spores that are spread by air currents and other means. These spores produce new mold growths if they alight at a location where conditions are favorable for germination. In contrast to molds, *yeasts* are generally unicellular. Yeasts can be differentiated from common bacteria due to their larger cell sizes and morphology, and because they produce buds during the process of division. Yeasts, like mold spores, can be spread through the air, or by other means, and will contaminate meat surfaces wherever they settle. Most yeast colonies are moist (or slimy) in appearance, and they are generally creamy white in color. *Bacteria* are also unicellular, and they vary in morphology from elongated and short rods to spherical or ovoid forms. Some bacteria exist in clusters; in others the rods or spheroids are linked together to form chains. Other bacteria possess flagella and are motile. Colored pigments are produced by some bacteria, whose colors range from shades of yellow to brown or black. Pigmented bacteria having intermediate colors, such as orange, red, pink, blue, green, and purple are also found. These bacteria cause the discoloration of meat surfaces and greening in sausage. Some bacteria also produce spores, the properties of which vary considerably. Certain spores are extremely resistant to heat, chemicals, and other agents. The growth of bacteria on meat is usually characterized by slime formation.

The increase in number of microorganisms occurs in phases, as shown in the growth curve of Figure 10–1. When favorable conditions exist, the cells that are initially present increase in size during what is known as the *lag phase.* This is followed by a marked increase in the number of microorganisms. This is called the *logarithmic growth* (or *exponential*) *phase,* and it continues until some environmental factor becomes limiting. The rate of growth then slows down, reaches an equilibrium point, and becomes relatively constant, resulting in the *stationary phase.* When meat is subjected to certain preservation processes, and the destruction of the microorganisms occurs, the *death phase* is encountered. It is readily apparent from the growth curve that a prolonged lag phase will retard microbial proliferation. This phenomenon is the basic principle behind the use of refrigeration for the control of deteriorative changes or spoilage in meat.

The type, and numbers of each type of microorganism present are important factors that contribute to the rate of meat spoilage. However, a number of properties, inherent to the specific product and the sur-

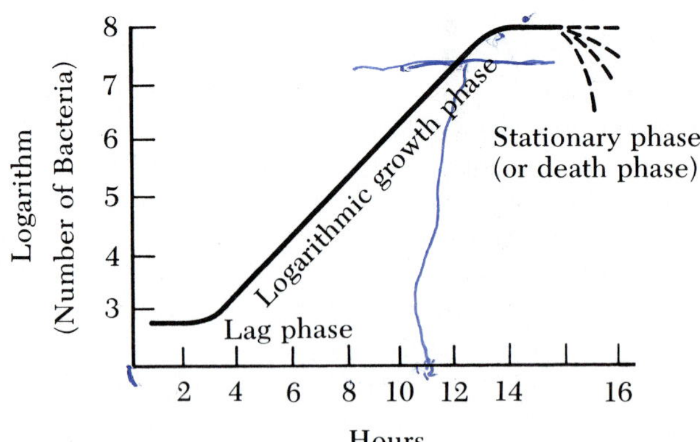

FIGURE 10-1
A typical growth curve for a pure culture of bacteria.

rounding environment, also markedly affect the kind, rate, and even the degree of such spoilage. Meat almost always contains populations of bacteria, molds, and yeasts in various proportions; the kind and number that are present depends upon local conditions. The organisms that eventually grow enough to cause spoilage will be those that have found the existing conditions most favorable. Even though many kinds of microorganisms may originally be present, usually only one, and seldom more than three, multiply rapidly enough to cause spoilage. The predominant microorganisms causing the spoilage may originally have been present in lower numbers than many others that did not grow, because of unfavorable conditions. Under similar conditions bacteria generally grow faster than yeasts, and yeasts outgrow molds. Thus, the growth of molds exceeds that of bacteria only if they were originally present in the largest numbers, or if conditions were more favorable for their growth than for the growth of bacteria or yeasts.

Factors Affecting Microbial Activity in Meat

Among the factors that affect microbial growth in meat are intrinsic properties, such as moisture content, pH, oxidation–reduction potential, nutritive value, and the presence or absence of barriers or inhibitory substances. In addition, extrinsic factors, such as temperature, relative humidity, the presence or absence of oxygen, and the physical form of the meat (carcass and wholesale or retail cuts *vs* the comminuted form) affect the growth of microorganisms. However, the factors having the greatest influence on the growth of microorganisms in meat and meat products are the storage temperature, and moisture and oxygen availability.

EXTRINSIC FACTORS. Each microorganism has an optimum, as well as a minimum and maximum, temperature for growth. Consequently, the temperature at which meat is stored markedly influences the kind, rate, and extent of microbial activity that occurs. A change of only a few degrees in temperature may favor the growth of entirely different organisms and result in a different type of spoilage. These characteristics provide the basis for the use of temperature as a means of controlling microbial activity.

The optimum temperature for the growth of most microorganisms is from 15–40°C. However, some organisms grow well at refrigeration temperatures; some even grow well at subzero temperatures, and other microorganisms can grow at temperatures exceeding 100°C. Microorganisms that have their optimum growth at temperatures lower than

20°C are called *psychrophiles.* Those that have growth optima at temperatures higher than 45°C are called *thermophiles.* Microorganisms with optima between the psychrophiles and thermophiles are called *mesophiles.* Bacteria, molds, and yeasts each have some genera with temperature optima in the range characteristic of thermophiles, mesophiles, and psychrophiles. However, molds generally are the most psychrophilic, followed by yeasts and bacteria, in that order. In keeping with this trend, molds and yeasts are the least thermophilic. Thus, although bacteria, molds, and yeasts might all be present in meat at the ordinary refrigeration temperatures of −1 to 3°C, if other conditions were favorable for the three types of microorganisms, mold and yeast growth would be more apt to occur than that of bacteria.

As the temperature approaches 0°C, fewer microorganisms can grow and their proliferation is slower. Temperatures below approximately 5°C greatly retard the growth of the more important spoilage microorganisms and also prevent the growth of nearly all pathogens. Consequently, 5°C is considered the *critical temperature* during meat handling and storage, and it cannot be exceeded for any length of time without a substantial reduction in the quality and appearance of meat.

Following slaughter, the surfaces of the carcass usually have a high proportion of bacteria of fecal and soil origin. During refrigerated storage these bacteria usually give way to the more psychrophilic bacteria. Among the important genera of bacteria found on meat during refrigerated storage are *Pseudomonas, Achromobacter, Micrococcus, Lactobacillus, Streptococcus, Leuconostoc, Pediococcus, Flavobacterium,* and *Proteus.* The particular species of these bacteria that are found will generally differ between fresh and cured meat. Some species of these psychrophiles can grow, although very slowly, at freezer temperatures. While the freezing process kills or damages many of the bacteria present on meat, and the number of organisms continues to decrease during subsequent freezer storage, species of the above genera can survive and will resume growth upon thawing.

Relative humidity is another of the extrinsic factors that affects microbial growth. It is usually considered at the same time as refrigeration temperatures because both characteristics contribute to the chilling and storage environment. The relative humidity level for maintaining optimal storage conditions varies with the temperature. In general, the higher the storage temperature, the lower the relative humidity should be. Thus, for the usual meat refrigeration temperatures of −1 to 3°C, the relative humidity should range between approximately 88–92 percent. When the relative humidity is too high, moisture will condense on the meat (sweating); if too low, moisture (primarily from the surface) will be lost to the atmosphere. If sweating occurs, the surfaces will become moist and very conducive for microbial growth and spoilage. On the other hand, microbial growth is inhibited by the dehydration and darkening of meat surfaces, but this results in economic losses due to shrinkage and a loss of eye appeal.

Of the various microorganisms bacteria require the highest relative humidity, usually in excess of 92 percent, for optimum growth. Yeasts are intermediate (90–94 percent), and molds have the lowest relative humidity requirement (85–90 percent). All microorganisms have high requirements for water to support their growth and activity. However, since most of this requirement is met by moisture that is available from the meat itself, the discussion of this factor, relative to microbial activity, will be considered with the intrinsic properties of meat.

The availability of *oxygen* is important because it determines what type of microorganism will be active. Some microorganisms have an absolute requirement for oxygen. Others grow in the complete absence of oxygen, and still others can grow either with or without available oxygen. Microorganisms that require free oxygen are called *aerobic organisms*, and those that grow in its absence are called *anaerobic organisms*. Microorganisms that can grow with or without the presence of free oxygen are called *facultative organisms*. All molds that grow in meat are aerobic, and the yeasts that are present also do best when aerobic conditions prevail. On the other hand, the bacteria found in meat may be aerobic, anaerobic, or facultative organisms.

Aerobic conditions prevail in meat stored in air, but only on or near the surfaces because oxygen diffusion into the tissues is strongly resisted. Thus, the microbial growth that occurs on the surface of meat is largely that of aerobes, perhaps with some facultative organisms, and the interior portions of meat contain primarily anaerobic and facultative bacteria. It is readily apparent that the atmosphere surrounding the meat will markedly affect the composition of its microbial population and their activity. The use of casings, packaging materials, vacuum packaging, and sealed containers, reduces, or entirely prevents, the activity of aerobic microorganisms.

The final extrinsic factor affecting microbial activity and the rate of meat spoilage is the *physical state of the meat*. This is determined by whether the meat is in carcass, wholesale cut, retail cut, or comminuted form, and on the processing treatment that is applied. Comminution results in a greater microbial load because of the larger amount of exposed surface area, more readily available water and nutrients, and greater oxygen penetration and availability. Hence, small cuts (and ground meat in particular) are conducive to the growth of microorganisms and more readily susceptible to spoilage. Factors that contribute to the increased microbial load that normally accompanies size reduction are the greater surface area that is exposed, combined with the additional time required for achieving size reduction, and contact with more sources of contamination such as saws, grinders, and choppers. In addition, the contaminating microorganisms are distributed throughout the meat product during comminution. Thus, rigid measures must be taken to minimize contamination by microorganisms, followed by steps to curtail their growth, in order to avoid spoilage.

INTRINSIC FACTORS. *Water* is required by all microorganisms, and a reduction in its availability constitutes a method of preservation. It is not the total amount of moisture present that determines the limit of microbial growth, but the amount of moisture which is readily available. The water requirement of microorganisms is usually expressed in terms of water activity. *Water activity* (a_w) is defined as the vapor pressure of the solution in question divided by the vapor pressure of the pure solvent, ($a_w = p/p_o$, where p equals the vapor pressure of the solution and p_o equals the vapor pressure of pure water). The a_w of fresh meat is usually 0.99 or higher, which is near optimum for the growth of many microorganisms. The relationship between relative humidity (*RH*) and a_w is as follows: $RH = a_w \times 100$. Thus, an a_w of 0.99 is equivalent to an *RH* of 99 percent. In general, bacteria have the highest water activity requirement and molds the lowest, with yeasts being intermediate. Most spoilage bacteria do not grow at an a_w below 0.91, but spoilage molds and yeasts can grow at an a_w of 0.80 or lower. Thus, molds and yeasts are most apt to grow on the surfaces of meat products that are partially dehydrated, while the growth of most bacteria is prevented.

Microorganisms have a *pH range* for optimum growth that is generally near neutrality (pH 7.0). Molds have the widest range of pH tolerance (2.0–8.0), although their growth is generally more favored by an acid pH. In fact, they can thrive in a medium that is too acid for either bacteria or yeasts. Although yeasts also can grow in an acid environment, they grow best in an intermediate acid (pH 4.0–4.5) range. On the other hand, bacterial growth is generally favored by near neutral pH values. At the normal ultimate pH of approximately 5.4–5.6 in meat, conditions favor the growth of molds, yeasts, and the *acidophilic* (acid loving) *bacteria*. In meat with low pH values (5.2 or lower), microbial growth is markedly reduced from that in the normal pH range. On the other hand, meat with a high ultimate pH (such as is found in dark cutters) is generally very susceptible to microbial growth, even under the best of management conditions and practices.

The *oxidation–reduction potential* of meat is an indication of its oxidizing and reducing power. In order to attain optimal growth, some microorganisms require reduced conditions and others, oxidized conditions. Thus, the importance of the oxidation–reduction potential of meat becomes readily apparent. Aerobic microorganisms are strongly favored by a high oxidation–reduction potential (oxidizing reactivity). A low potential (reducing reactivity) largely favors the growth of anaerobic organisms. Facultative microorganisms are capable of growth under either condition. Microorganisms are capable of altering the oxidation–reduction potential of meat to the extent that the activity of other organisms is restricted. Anaerobes, for instance, can decrease the oxidation–reduction potential to such a low level that the growth of aerobic organisms can be inhibited.

Following the elimination of the oxygen supply at exsanguination, and in the subsequent period during the conversion of muscle to meat, the oxidation–reduction potential falls. Reduced conditions generally prevail in postmortem muscle, oxygen penetration into the tissues is markedly inhibited, and many reducing groups are available. The oxidation–reduction potential in postmortem muscle, and the oxygen supply (from the air), are highest at the surface and lowest in the interior portions. Exposure of meat to oxygen from the air will increase the oxidation–reduction potential at the surface, and the interior will be affected in a manner that depends upon the rate of oxygen penetration. Grinding, or other comminution of meat, markedly increases its oxidation–reduction potential.

In addition to water and oxygen, aerobes have other *nutrient requirements.* Most microorganisms need external sources of nitrogen, energy, minerals, and B vitamins to support their growth. They generally obtain their nitrogen from amino acids and other nonprotein nitrogen sources, but some utilize peptides and proteins. Their usual source of energy is carbohydrate, but, since meat is low in this nutrient, proteolytic organisms use protein as an energy source, and a few utilize lipid for this purpose. Of the three groups of microorganisms, molds are the best equipped for the utilization of proteins, complex carbohydrates, and lipids because they contain enzymes capable of hydrolyzing these molecules into their simple components. Many bacteria have a similar capacity, but most yeasts require the simple forms of these compounds. Minerals are needed by all microorganisms, but their requirements for vitamins and other growth factors vary. Molds (and some bacteria) can synthesize enough B vitamins to meet their needs, but other microorganisms require a ready-made supply, even though some are needed only in very small amounts. Meat has an abundance of each of these nutrients and consequently, it is an excellent medium for microbial growth.

The final intrinsic factor affecting the growth of microorganisms in meat is the *presence or absence of inhibitory substances and protective tissues.* Substances or agents that inhibit microbial activity are referred to as *bacteriostatic,* and those which destroy the organisms are called *bactericidal* substances or agents. Some bacteriostatic substances are part of the normal ingredients added to meat products during processing. The bacteriostatic effects of these ingredients are described in Chapter 11 in the section discussing preservative agents.

Essentially no bacteriostatic substances are inherently present in meat. The absence, or low level, of simple carbohydrates in meat to which no sugar is added tends to dissuade the growth of the fermentative microorganisms. Fat, when and where present on carcasses, and wholesale and retail cuts, provides the lean tissues with a protective surface against microbial contamination. In the case of the fatter carcasses and wholesale cuts, the surface fat is usually trimmed away during retail cut fabrication, thereby eliminating much of the initial load of microorganisms. Similarly, a considerable amount of protection against microbial contamination of the muscles and associated tissues is given by the skin on poultry carcasses, and on pork carcasses and wholesale cuts, as well as by the scales and skin of fish.

SOME INTERRELATIONS BETWEEN THE PRECEDING FACTORS. The effects that such factors as temperature, oxygen, pH, and a_w have upon microbial activity are not entirely independent of one another. At temperatures near growth minima or maxima, microorganisms generally become more sensitive to a_w, oxygen availability, and pH. Under anaerobic conditions, for example, bacteria may require a higher pH, a_w, and minimum temperature for growth, than when aerobic conditions prevail. In fact, the microorganisms that grow at low temperatures are usually aerobic and they generally have a high minimum a_w requirement. Consequently, lowering a_w by adding salt or excluding oxygen from meat held at low temperatures markedly reduces the rate of microbial spoilage. Logically, if salt were to be added to meat held at low temperatures, it follows that if oxygen were also excluded at the same time, the preservative effect would be even greater. Usually, some microbial growth occurs when any one of the factors that controls the growth rate is at a limiting level. However, if two or more factors become limiting, growth is drastically curtailed or is even prevented entirely.

THE INTERRELATION BETWEEN TEMPERATURE, BACTERIAL NUMBER, AND LENGTH OF STORAGE. The effect of temperature on the generation interval (the time required for 1 bacterial cell to become 2) of a psychrophilic type bacterium in hamburger is shown in Figure 10–2. The marked increase in the generation interval below 5°C emphasizes the extreme importance of continually maintaining proper refrigerated temperatures during meat storage. Likewise, the hazards of leaving meat at ambient temperatures during transit and in the home, especially during the summer months, even for a few hours, are obvious. The effect of temperature on the shelf life or storage life of hamburger is illustrated in Figure 10–3. Freshly ground hamburger ordinarily contains about 1 million bacteria/gram. The number of bacteria required to cause spoilage in hamburger (as detected by abnormal odor and some slime development) is approximately 300 million/gram. From

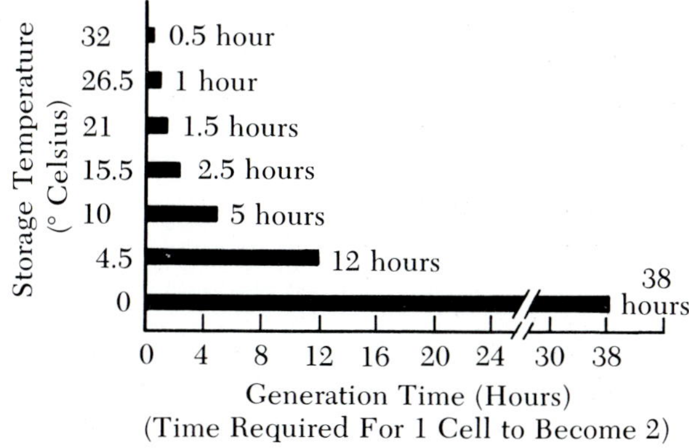

FIGURE 10-2
The generation intervals for one species of psychrophilic bacterium at different storage temperatures. [Adapted from E. A. Zottola, Introduction to Meat Microbiology, American Meat Institute.]

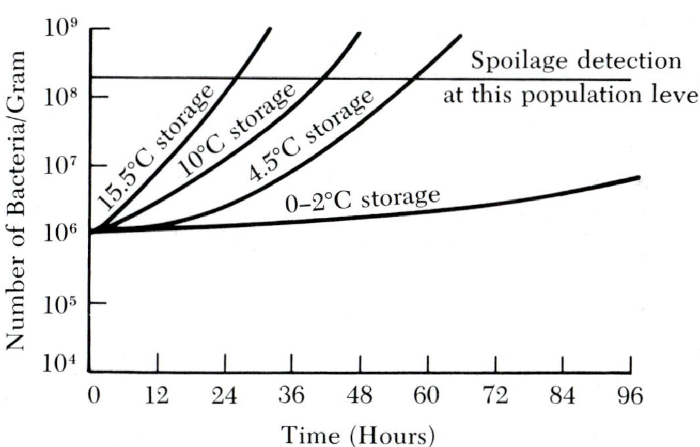

FIGURE 10-3
The time required at different storage temperatures for the spoilage of hamburger initially contaminated with 1 million psychrophilic bacteria/gram. The spoilage was detected by odor and slime formation at a population level of 300 million bacteria/gram. [From E. A. Zottola, Introduction to Meat Microbiology. American Meat Institute.]

Figure 10–3 it can be seen that the storage life of hamburger containing 1 million bacteria/gram is only approximately 28 hours at 15.5°C, at which time bacterial numbers reach 300 million/gram and detectable spoilage has developed. However, at a normal refrigerated storage temperature, about −1 to 3°C, the storage life greatly exceeds 96 hours.

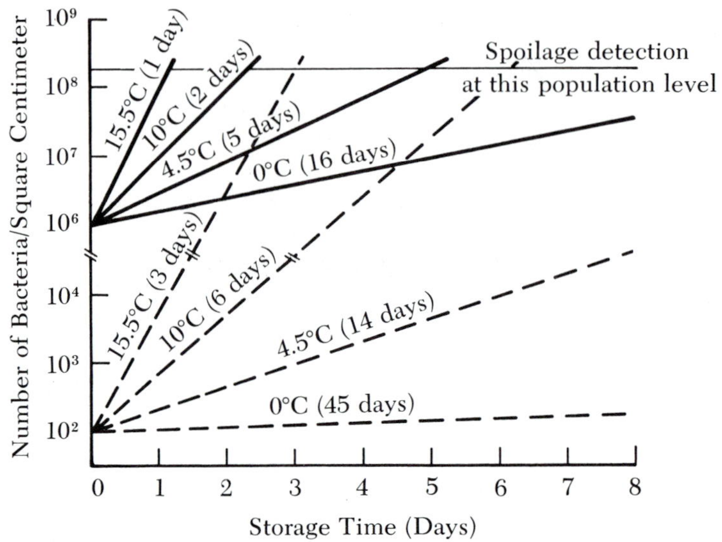

FIGURE 10-4
The time required for the spoilage of frankfurters that were contaminated with high and low levels of psychrophilic bacteria. Spoilage detection (by slime formation) was at a population level of 150 million bacteria/cm^2 of surface area. The high level of contamination was 1 million/cm^2, shown by the solid line; the low level was 100 bacteria/cm^2, shown by the broken line. [From E. A. Zottola, Introduction to Meat Microbiology, American Meat Institute.]

Figure 10–4 illustrates the influence of bacterial number (100 *vs* 1 million) on the storage life of pairs of frankfurters that were held at the same storage temperatures (0–2, 4.5, 10, and 15.5°C) as the hamburger meat of Figure 10–3. Spoilage of frankfurters as indicated by slime formation, occurred when bacterial numbers reached a level of 150 million/cm^2 on the product surface. The lower level of contamination markedly prolonged their storage life, especially at temperatures below 5°C. The information presented in these figures emphasizes the importance of low refrigerated storage temperatures, as well as the need for concerted efforts to minimize microbial load in meat products during all operations, in order to assure wholesomeness and a reasonable storage life.

Assessing Microbial Numbers, Growth, and Activity in Meat

A number of methods are available for the assessment of microbial growth and activity in meat and meat products. The method of choice is dependent upon the information needed, the specific product in question, and the nature of the microorganism(s). It should be emphasized that obtaining a representative sample of the product in question is probably the most important factor affecting the results. Additionally, the results of microbiological analysis are less objective than are chemical methods of analysis. Thus, considerable subjective judgment is usually applied in the interpretation of results of microbiological analysis. A good amount of experience and knowledge of the unique and specific problems encountered in meat products is essential for the selection of the most appropriate method and the eventual usefulness of the results that have been obtained.

Irrespective of these limitations, the application of microbiological analysis can provide the meat processor and handler with much useful information. It can provide an indication of sanitary conditions, microbiological load, spoilage problems, and give an estimate of expected shelf life. Two frequently used methods that provide this kind of information are the *total plate count*, and *indicator* and *dye reduction tests*. However, many other methods are used to assess microbiological status, but it is not our intent in this text to identify them.

TOTAL PLATE COUNT. This method provides an indication of the total population of microorganisms. It can be used to assess the load on or in a meat product, in the air and water of the plant, as well as on all equipment, related surfaces, and facilities. A swab of the equipment, walls, or product to be analyzed is applied to a culture medium that nonselectively supports the growth of microorganisms. The swab is applied to the medium in a sterile covered plate; or, a diluted sample of comminuted products can be added to the medium. The number of colonies that develop following an incubation period of several days gives an indication of the total microbial numbers, but provides little clue as to the specific organisms that are present. A typical plate, showing various microbial colonies is shown in Figure 10–5. Special methods, which permit the selective growth of specific microorganisms, are used to determine their presence and numbers.

INDICATOR AND DYE REDUCTION TESTS. Numerous microorganisms secrete enzymes as a normal metabolic function of their growth, and these enzymes are capable of inducing reduction reactions. Certain indicator substances (dyes) are used as the basis of these tests, and the rate of their reduction (as indicated by a color change) is proportional to the number of microorganisms present. The length of time required for the complete reduction of a standard amount of the indicator is used to assess microbial numbers in meat or in an extract of meat. The indicator and dye reduction tests are more rapid than a plate count incubation, and thus have become quite useful for microbiological assays

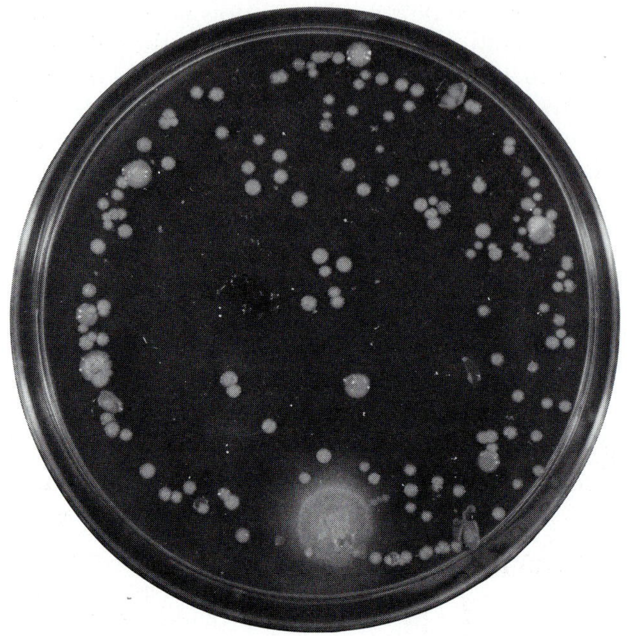

FIGURE 10-5
A typical culture plate, showing bacterial and mold colonies.

in meat processing and handling operations. Among the modifications of these methods is the application of a dye impregnated filter paper directly onto meat or a piece of equipment. The time required for the color changes to occur is used to evaluate the numbers of microorganisms that are present.

DETERIORATION OF MEAT

Deteriorative Changes Caused by Microorganisms

When is meat spoiled? This is a rather moot question, because what one individual describes as spoiled might well be considered edible by another. The usual characterization of spoiled meat, or of any food for that matter, is the point at which it becomes unfit for human consumption. However, since meat that is declared unfit for consumption during the normal inspection process is ordinarily not considered spoiled, this statement has to be qualified. In its usual connotation, spoilage is frequently equated with the decomposition and putrefaction that results from microorganisms. There is seldom any doubt about the fitness of meat for consumption when it shows evidence of decomposition or putrefaction, and as such it is definitely described as *spoiled.* But spoilage does not necessarily imply decomposition or putrefaction. This is particularly pertinent in the case of meat, because spoilage is not due solely to microbial action but to such factors as insects and intrinsic enzymatic and oxidative reactions as well. Some of the more obvious chemical and physical changes attributable to microorganisms will be described.

CHEMICAL CHANGES. The degradation of proteins, lipids, carbohydrates, and other complex molecules into simpler ones is accomplished by the action of endogenous hydrolytic enzymes that are present in the meat, as well as by the enzymes that microorganisms produce. Initially, the endogenous enzymes are responsible for the degradation of complex molecules; but, as the numbers of microorganisms and their activity increases they will contribute to (and may eventually account for) nearly all of the subsequent degradation reactions. These enzymes hydrolyze the complex molecules into simpler compounds that are then utilized as nutrient sources for supporting microbial growth and activity. The end products of microbial action depend upon the availability of oxygen. When oxygen is freely available, the end products of protein hydrolysis are simple peptides and amino acids. Under anaerobic conditions proteins are degraded to a variety of sulphur containing compounds, all of which are extremely odorous and generally obnoxious. The end products of nonprotein nitrogenous compounds usually include ammonia.

The *lipases* (lipid hydrolyzing enzymes) secreted by microorganisms hydrolyze triglycerides and phospholipids into glycerol and fatty acids, and, in the case of phospholipids, into nitrogenous bases and phosphorus as well. Extensive lypolysis can accelerate lipid oxidation and, if it does, the result is a tallowy or chalky flavor.

Meat is generally very low in carbohydrates, except when they are added during processing. However, microorganisms prefer carbohydrates to other compounds as a source of energy. Therefore, when they are made available, carbohydrates are readily utilized. Carbohydrate utilization by microorganisms results in a variety of end products, including alcohols and organic acids. In some sausage products, microbial fermentation of the sugar that has been added leads to the production of organic acids (primarily lactic acid) which contributes to their distinctive flavors.

PHYSICAL CHANGES. The physical changes that microorganisms produce usually are more apparent than their chemical changes. While microbial spoilage usually results in an obvious physical change in meat,

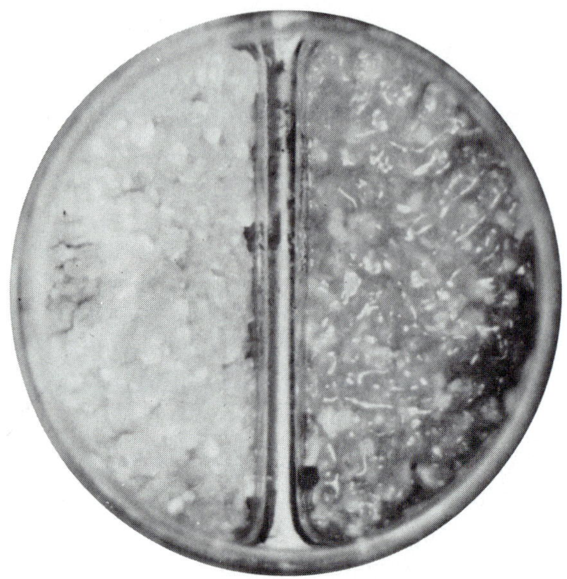

FIGURE 10-6
A photograph of ground pork muscle, showing slime formation in the right half of the bipetri dish. The ground pork on the right was innoculated with a psychrophilic species of bacteria while that on the left served as the control.

other less apparent changes appear in its color, odor, flavor, tenderness, and processing properties. Meat spoilage is usually classified as being either aerobic or anaerobic, depending upon the conditions under which it occurred as well as whether the principal organisms causing the spoilage were bacteria, molds, or yeasts.

Aerobic spoilage by bacteria and yeasts usually results in slime formation, undesirable odors and flavors (taints), color changes, and the previously mentioned changes in lipids. An example of slime formation is shown in Figure 10–6. The specific bacteria or yeast species causing the slime formation depends upon environmental conditions, particularly those of temperature and a_w. As a result of the production of oxidizing compounds by bacteria, myoglobin and oxymyoglobin may be changed to metmyoglobin and other oxidized forms of the pigment, resulting in a gray, brown, or green color. Some species of bacteria can cause "greening" in sausage, and pigmented bacteria and yeasts can cause various other surface colorations. The hydrolysis of lipids was referred to in the discussion of chemical changes brought about by microorganisms; however, the development of oxidative rancidity and undesirable odors and flavors is enhanced by some lipolytic bacteria and yeasts.

Aerobic spoilage by molds results in a sticky meat surface. The formation of "whiskers" by molds is also common. Meat that is aged for a long time almost always has some mold growth on it. Because of the colors associated with the growth of specific mold colonies, various surface colorations, such as creamy, black or green, will be seen. Like bacteria and yeast, spoilage that is caused by molds also results in lipolysis, and enhances oxidative rancidity as well as the production of musty odors and flavors.

Aerobic spoilage is essentially limited to the meat surface, where oxygen is readily available. Therefore, the affected area(s) can be trimmed off and the meat that remains is generally acceptable for consumption. This is certainly the case with aged meat; when the surface molds are trimmed off, the lean tissues underneath may have little or no microbial growth. However, if extensive bacterial growth occurs on the surface, it may have penetrated deeply into the tissues, particularly along bones and connective tissue septa. The increased tenderness and flavor development occurring in aged meat may partially be the result of microbial enzyme action, in addition to that of the endogenous enzymes present in the tissues.

Anaerobic spoilage occurs in the interior of meat products, or in sealed containers, where oxygen is either absent or present in limited quantities. This type of spoilage is caused by facultative and anaerobic bacteria, and is usually described by such terms as souring, taint, and putrefaction. *Souring* (the development of a sour odor or flavor) results mainly from the accumulation of organic acids during the bacterial enzymatic degradation of complex molecules. Proteolysis without putrefaction may also contribute to souring. Occasionally, souring is accompanied by the production of various gases. *Taint* is a nonspecific term used to describe undesirable odors and flavors. *Round sour* or *ham sour*, and *bone sour* are terms commonly used in the meat industry to describe the sour or putrid odor occasionally found in the tissues surrounding bones, particularly in rounds or hams. These sours, or taints, are caused by anaerobic bacteria that might originally have been present in lymph nodes or bone joints, or which might have gained entrance along the bones during storage and processing. If chill room temperatures are inadequate for the rapid dissipation of body heat, or if temperatures are improperly maintained during storage and processing, the growth of these microorganisms occurs, and souring may develop. The presence of one such ham in a vat during curing may contaminate an entire lot, making them unfit for consumption.

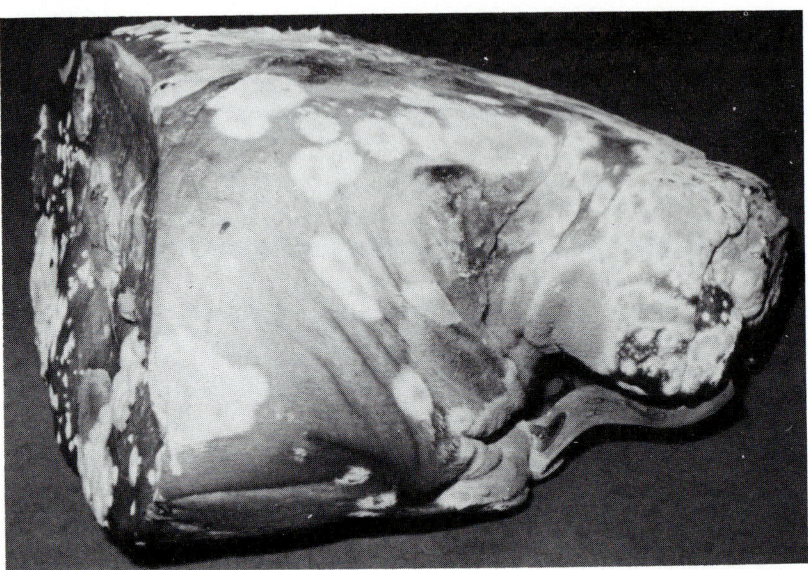

FIGURE 10-7
The shank portion of this cured and smoked ham shows considerable mold growth.

It should be recognized that all of the microbiological principles that we have discussed apply to meat in general, and to all meat products, including those of poultry and marine products. Fresh meat is more susceptible to spoilage than is most processed meat. Cured and smoked meat (as well as most types of sausage) either contain ingredients, or are processed in some manner, to restrict microbial growth and activity. They are also frequently placed in containers which protect them from further contamination. Hence, molds are more apt to be found on the surface of cured and smoked products, and sausages, than are either bacteria or yeasts. An example of mold growth on cured and smoked ham is shown in Figure 10-7. However, if sufficient moisture is available on the surface of sausages, bacteria and yeast spoilage can occur, resulting in slime formation and the other changes that are characteristic of such spoilage.

Deteriorative Changes Caused by Insects

Under the usual meat handling conditions in this country, insects seldom constitute a problem in meat deterioration, because the sanitation and insect control measures in general use are sufficiently stringent to prevent infestations. Insect infestation not only reduces product acceptability and the length of time before the onset of spoilage, but it constitutes a health hazard because of the disease transmission potential of the insects themselves.

Insect control is still a problem in certain aspects of meat production in this country, particularly in the processing of country cured ham and other similarly processed pork cuts. During the aging and storage phases of their preparation these pork cuts are held at relatively high temperatures (21–24°C), generally for many months. Thus, if the aging and storage rooms are not constructed so as to prevent entry of insects, or if the pork cuts are not protected by some other means, infestation by one or more species of insects will result. Among the commonly found insects in these products is the "skipper," a leaping larvae that feeds on the muscle and other soft tissues. Some other insects found on (or in) country cured pork products include the larder beetle, cheese or ham mite, and blow flies. In addition to the aesthetic effect of holes eaten in products, some discoloration and weight loss also occurs. This results in economic losses to the processor. Therefore, insect control can become an important aspect of the processing and storage of these products. The U.S. Department of Agriculture has approved methyl bromide for use as a fumigant. If it is properly used at the appropriate intervals, insect infestations can be effectively controlled.

Chemical Reactions Affecting Deteriorative Changes

OXIDATIVE RANCIDITY. Meat fats are susceptible to oxidation when they are exposed to the molecular oxygen present in air. This results in the production of a strong disagreeable odor and flavor in the cooked product. When these chemical reactions occur, they constitute a defect referred to as *oxidative rancidity.* The term *autoxidation* is used to describe the chemical reactions that cause oxidative rancidity. The characteristic flavor and odor of oxidized fat is caused by the presence of low molecular weight aldehydes, acids, and ketones that form during the oxidation and decomposition of the fatty acid molecules. Polyunsaturated fats are much more susceptible to autoxidation than are monounsaturated fats. Saturated fatty acids are the most resistant to oxidation and the development of rancidity. In the relatively short period of time that it usually takes for rancidity to develop, probably only the polyunsaturated fatty acids undergo any appreciable autoxidation. Thus, the polyunsaturated fatty acids are the focal point for understanding the autoxidation of fats.

The rate of autoxidation is enhanced by the presence of *prooxidants*, such as heat and exposure to light, especially ultraviolet light. Prooxidants favor oxidation by catalyzing the oxidation reactions, and *antioxidants* inhibit oxidation. Prooxidants include nitrite, sodium chloride, a number of metals (such as copper, iron, manganese, and cobalt), and numerous other substances and agents. Since heat and light increase the autoxidation rate, meat stored at refrigerator or freezer temperatures, should also be kept in the dark in order to retard rancidity development. Obviously, minimizing the amount of air in containers, or eliminating it completely during storage, especially during extended freezer storage, also will retard the development of rancidity.

The storage life of ground products is shortened because of the incorporation of oxygen during the grinding process. The addition of salt to processed meat products also shortens their storage life. However, some ingredients added during processing (such as ascorbate, sage, and the phenolic compounds present in wood smoke) possess antioxidant properties and tend to retard the development of rancidity.

In the case of edible fats such as lard, rancidity is inhibited by the addition of antioxidants during processing. Two widely used antioxidants are butylated hydroxytoluene (BHT) and butylated hydroxyanisole (BHA). The antioxidants used commercially generally contain various combinations of several antioxidants so as to take advantage of the desirable properties of each. Combinations of BHT, BHA, n-propyl gallate, and citric acid are commonly used in edible fats, and in foods with a moderate to high fat content.

DISCOLORATION. Any deviation from the bright red, light red, or grayish pink color of the muscle of beef, lamb, and pork or veal, respectively, may be described as *discoloration*. As was previously discussed in Chapters 8 and 9, such discolorations are usually associated with chemical changes in the muscle pigments, primarily in myoglobin. They result from separate and distinct conditions from those associated with abnormal color in the muscles of PSE and dark cutter carcasses.

Discoloration is usually due to the presence of metmyoglobin, which contributes the brown or grayish brown color of fresh and processed meat. The brown color of metmyoglobin is detectable when approximately 60 percent of the myoglobin is in the metmyoglobin form. As previously stated, metmyoglobin formation is accelerated by those conditions that cause denaturation of the protein portion of the myoglobin molecule, the absence of reducing compounds, and by low oxygen tension. Factors that cause denaturation of the protein portion of myoglobin include heat, salts, ultraviolet light, low pH, and surface dehydration. The latter effect concentrates salts which, in turn, promote metmyoglobin formation. Low temperatures, such as those encountered during refrigerated storage, suppress the activity of the oxygen utilizing enzymes surviving in the meat, which contributes to a higher oxygen tension at the meat surface. This, in turn, retards metmyoglobin formation. Cured meat products are also susceptible to discoloration. The most important discoloration problem is that of *light fading*, discussed in detail in Chapter 9.

Surface discoloration may become noticeable when meat is held at relative humidity levels low enough to result in dehydration. Moisture loss from the surface of carcasses and meat cuts reduces the amount of reflected light, due to the loss of intracellular water and, as a result, the meat appears dark in color. Such carcasses and cuts of meat are described as "having lost their bloom." Frozen meat generally possesses a darker color than fresh meat, but this depends somewhat on the rate of freezing. Upon thawing, the color of meat may be improved, but it usually does not correspond to the color of its fresh counterpart.

Physical Deteriorative Changes

DEHYDRATION. The loss of moisture from meat surfaces during storage produces a dried, stale, coarse textured appearance that adversely affects eye appeal and acceptability. Unless severe dehydration has occurred, the problem is confined largely to the surface. While a trimming of these surfaces will essentially restore consumer appeal, the removal of trimmings, and the labor involved, are serious economic losses. Severe dehydration usually results in a very dry product following cooking, and thus it can affect the palatability of meat as a food. Dehydration by freeze–drying usually results in some protein denaturation, so that these proteins will not bind water following rehydration. When cooked, such products would be relatively "dry." Since some flavor constituents may also be adversely affected, the product will taste different from, and is usually less acceptable than, its fresh counterpart.

FREEZING AND THAWING. During freezer storage an excessive loss of moisture from meat surfaces will result in localized areas of dehydration and discoloration. This harmless phenomenon is called *freezer burn.* It can result when the wrapping material has been punctured, or when moisture proof wrapping is not used. The improper maintenance of temperature, so that frequent cycles of partial defrosting are followed by refreezing during storage, also contributes to the development of freezer burn. Because of the unattractiveness resulting from dehydration, and the discolored grayish tan areas, eye appeal and consumer acceptability are adversely affected. During the development of freezer burn the proteins may become denatured, and rehydration may be poor. Meat with severe freezer burn is rather dry and tasteless. If oxidative rancidity has developed the quality of the meat becomes even more objectionable, because of the bitterness associated with the oxidation products accompanying rancidity.

If meat is frozen slowly, much of the water that is present in it collects extracellularly and freezes in large pools. The "spiked" edges of ice crystals forming in these pools may puncture the muscle fibers, releasing even more moisture. The physical damage caused by slow freezing results in a considerable loss of fluid from the meat when it is thawed. This fluid will collect in a package upon thawing, and is called *drip*. Excessive drip results in an unattractive package, a loss of nutrients, and in dryness in the cooked meat. Factors that influence the amount of drip are discussed in Chapter 11.

SHRINKAGE UNDER REFRIGERATION. During refrigerated storage meat loses moisture from its surfaces, resulting in a weight loss called *shrink*. Other than the economic loss that is associated with shrink, the mere loss of moisture during the first few days of refrigerated storage seldom has an adverse affect on meat acceptability. However, the physical changes that accompany shrink during prolonged refrigerated storage include surface dehydration and discoloration. These effects do contribute to the deteriorative changes that occur in meat.

CONTAMINATION OF MEAT

A contaminant is defined as any substance (or agent) present in food in an amount that renders it unacceptable or potentially harmful to the consumer. Contaminants include such diverse agents as residues from chemicals fed, absorbed, or inhaled by the animal that accumulate in its tissues. Contamination can occur by the incorporation of excess additives, dirt and refuse, excreta (and other evidence of rodents and insects), and finally, by the presence of parasites and microorganisms.

Chemical Residues

Chemical residues result primarily from materials that are consumed by the animal, inspired from the air, and absorbed through the skin. Residues derived from the feed include: antibiotics, hormones and hormone like compounds (e.g., progesterone and diethylstilbesterol), minerals (such as selenium from feeds grown in high selenium content soils), and residual chemicals in feeds resulting from the use of herbicides and fungicides for the control of diseases and insects in crop production. The occurrence of mercury residues in fish, and polychlorinated byphenyls in poultry, have received considerable publicity. Residues may also accumulate and cause tissue damage when animals are raised near industrial environments where gaseous pollutants, such as sulfur dioxide, contaminate the atmosphere. The absorption through the skin of chemical compounds, used to control external parasites, also contributes to tissue residues. Tolerance limits for residues vary, depending upon their effects on the tissues of the consumer.

Excess Additives

Several additives are approved for inclusion in meat products, including nitrite, ascorbic acid, and certain antioxidants. The principles involved in their use have been discussed in Chapter 9. The conditions under which additives may be used, and the permissible amounts, are clearly defined in meat inspection regulations. When additives are used in amounts greater than those specified by the regulations, the product is considered to be contaminated. For example, the maximum amount of nitrite permissible is 200 ppm. This limit is established because nitrites are toxic if ingested in very high levels, and also because of more recent concerns with the formation of nitrosamines (see Chapter 9).

Dirt and Refuse

The obvious contamination of meat and meat products with dirt, refuse, and insect and rodent excreta not only affects aesthetic properties, but also presents a potential health hazard from contamination with disease-causing microorganisms. Curtailment of such contamination involves good sanitation, as well as insect and rodent control in all facilities where meat is processed or handled.

Microbiological Contamination

The sources of microbial contamination have been discussed previously. Sanitation, proper refrigeration, and proper handling are paramount concerns, if such contamination is to be minimized and microbial activity curtailed. However, no meat product is completely sterile, and the potential for deteriorative changes and spoilage, as well as for food infections and poisoning, is a constant threat if meat is not properly handled.

The deteriorative effects and spoilage problem in meat products have been described, but the food infection and poisoning potential are discussed in detail in the following sections.

Food Poisoning and Infection

The development of gastrointestinal disturbances following the ingestion of food can result from any one of a number of causes. These include allergies, overeating, poisoning from chemical contaminants, toxic plants or animals, bacterial toxins, an infection by microorganisms, or an infestation of animal parasites. Although each of these factors is recognized as a potential source of illness in humans, the subsequent discussion is confined to those illnesses caused by microorganisms that are categorized as "food poisoning," and to the illness attributable to the parasite *Trichinella spiralis*.

Food poisoning is most appropriately defined as an illness caused by the ingestion of toxins. *Food infection* is defined as the ingestion of pathogenic (disease producing) organisms that grow and cause illness in the host. Toxins are produced by bacteria and fungi (molds and yeasts). The toxins produced by fungi are referred to collectively as *mycotoxins*. Since they are common in mold infested grains and legumes, mycotoxins often constitute a health problem for livestock. Meat (especially cured and smoked meat, and aged, dry sausages), frequently contains mold and yeast growth and associated mycotoxins. However, no evidence (to date) indicates that the ingestion of meat that had mold growth on it has caused disease or illness in humans. Also, meat from animals that have consumed mycotoxins in feeds does not contain any of these toxin residues. Thus, it appears that illnesses following the ingestion of foods containing toxins, or of microorganisms that produce toxins in the body, are of bacterial origin. These bacterial toxins are relatively odorless and tasteless. Thus, they are readily consumed by the unsuspecting victim.

BOTULISM. Botulism is a true food poisoning that results from the ingestion of a toxin produced by the bacterium *Clostridium botulinum*, during its growth in food. This bacterium is anaerobic, spore forming, gas forming, and is found primarily in the soil. The toxin produced by these organisms is extremely potent. It affects the central nervous system of the victim and death, which occurs in a large percentage of the cases, results from respiratory failure. (Some of the characteristics of botulism, as well as those of other common food poisonings and infections are presented in Table 10-1.)

Because of the presence of *Clostridium botulinum* in soil, it follows logically that it is also present in water. Hence, fish (and seafood,

TABLE 10-1
Characteristics of some common food poisonings and infections

Illness	Causative agent	Symptoms	Average time before onset of symptoms	Foods usually involved	Preventive measures
Botulism (food poisoning)	Toxins produced by *Clostridium botulinum*	Impaired swallowing, speaking, respiration, coordination. Dizziness, and double vision.	12–48 hours	Canned low acid foods including canned meat and seafood, smoked and processed fish.	Proper canning, smoking, and processing procedures. Cooking to destroy toxins, proper refrigeration and sanitation.
Staphylococcal (food poisoning)	Enterotoxin produced by *Staphylococcus aureus*	Nausea, vomiting, abdominal cramps due to gastro-enteritis (inflammation of the lining of the stomach and intestines).	3–6 hours	Custard and cream filled pastries, potato salad, dairy products, cooked ham, tongue, and poultry.	Pasteurization of susceptible foods, proper refrigeration and sanitation.
Clostridium perfringens (food poisoning)	Toxin produced by *Clostridium perfringens* (infection?)	Nausea, occasional vomiting, diarrhea and abdominal pain.	8–12 hours	Cooked meat, poultry and fish held at nonrefrigerated temperatures for long periods of time.	Prompt refrigeration of unconsumed cooked meat, poultry, or fish; maintain proper refrigeration and sanitation.
Salmonellosis (food infection)	Infection produced by ingestion of any of over 1200 species of *Salmonella* that can grow in the gastro-intestinal tract of the consumer.	Nausea, vomiting, diarrhea, fever, abdominal pain; may be preceded by chills and headache.	6–24 hours	Insufficiently cooked or warmed over meat, poultry, eggs, and dairy products; these products are especially susceptible when kept unrefrigerated for a long time.	Cleanliness and sanitation of handlers and equipment; pasteurization, proper refrigeration and packaging.
Trichinosis (infection)	*Trichinella spiralis* (nematode worm) found in pork	Nausea, vomiting, diarrhea, profuse sweating, fever, and muscle soreness.	2–28 days	Insufficiently cooked pork and products containing pork.	Thorough cooking of pork (to an internal temperature of 59°C or higher); freezing and storage of uncooked pork at minus 15°C, or lower, for a minimum 20 days; avoid feeding pigs raw garbage.

generally) are a greater potential source of botulism than are other flesh foods. However, improperly home canned vegetables and fruit, with a low to medium acid content, constitute the greatest potential source of botulism. Since the organism is anaerobic, canned and vacuum packaged foods including meat, poultry, and seafood are also potential sources. Fortunately, the distribution of botulinum spores in meat is usually very low, so botulism poisoning from canned or vacuum packaged meat products is rare.

In the prevention of botulism, there is no substitute for good sanitation. However, to protect people from possible outbreaks of botulism due to food that might be contaminated, proper refrigeration and thorough cooking are essential at all times. The toxin itself is relatively heat labile. However, the bacterial spores are very heat resistant, and severe heat treatment is required to destroy them. Thermal processing at 85°C for 15 minutes inactivates the toxin, but the following combinations of temperatures and times are required to completely destroy the spores.

Temperature (° Celsius)	Time (in minutes)
100°	360
105°	120
110°	36
115°	12
120°	4

Canned foods showing evidence of swelling (swelled cans result from the gas produced by the organism) should **never** be eaten. Smoked fish should be heated to at least 82°C for 30 minutes during processing. Processed, precooked–frozen, or raw–frozen foods should not be allowed to stand at room temperature for extended periods of time.

STAPHYLOCOCCAL FOOD POISONING. Staphylococcal food poisoning is caused by ingesting the *enterotoxin* produced by *Staphylococcus aureus*. The toxin produced by this facultative, nonspore forming organism is called an enterotoxin because it causes an inflammation of the lining of the stomach and intestines (gastroenteritis). However, it also affects the central nervous system. Death seldom results from staphylococcal food poisoning. When mortality does occur, it is usually due to an added stress in people already suffering from other illnesses. The organisms causing staphylococcal food poisoning are widely distributed in nature, and have been isolated from many healthy individuals. Thus, the handling of food by infected individuals is probably the greatest source of contamination. The most common foods associated with staphylococcal food poisoning are cream and custard filled pastries, potato salad, dairy products, cooked ham, tongue, and poultry. Under favorable conditions the organisms can multiply to extremely high levels in foods, including meat products, without significantly changing their color, flavor, or odor.

Staphylococcus aureus organisms are quite easily destroyed by heat (66°C for 12 minutes), but the destruction of the enterotoxin requires a severe heat treatment: for instance, autoclaving at temperatures of 121°C for 30 minutes. Ordinary cooking temperatures for most foods will not destroy the enterotoxin.

CLOSTRIDIUM PERFRINGENS FOOD POISONING. This food poisoning is referred to by the name of the causative bacterium, *Clostridium perfringens*. The organism is an anaerobic spore former that produces a variety of toxins. In addition, the organism also produces a considerable amount of gas during its growth. *C. perfringens* organisms, as well as their spores, have been found in many foods. These include fresh and processed beef, lamb, pork, poultry, veal, and fish. The fewest organisms have been found in sliced luncheon meats and loaves. On the other hand, the greatest number of these organisms have been observed in meat items which had been cooked, allowed to cool slowly, and then held for an extended period of time before serving. Large numbers of active organisms must be ingested for this type food poisoning to occur.

The spores from various strains of the organism differ in their resistance to heat; 100°C temperatures kill some in minutes, but others require from 1–4 hours at this temperature for their destruction. *Clostridium perfringens* food poisoning can be controlled by rapidly cooling cooked and heat processed foods. The proper refrigeration of foods at all times, especially of leftovers, and good sanitation are essential for preventing an outbreak. When foods are held on steam tables, the temperatures should exceed 60°C in order to prevent food poisoning by this organism. When leftover foods are reheated, they require a thorough heating to destroy the living organisms and their toxins.

SALMONELLOSIS. *Salmonella* "food poisoning" is a food infection, because it results from the ingestion of any one of numerous species of living *Salmonellae* organisms. These organisms grow in the consumer and produce an *endotoxin* (a toxin retained within the bacterial cell) that is the specific causative factor of the illness. The usual symptoms of salmonellosis consist of nausea, vomiting, and diarrhea. They are believed to result from the irritation of the intestinal wall caused by the endotoxins. It appears that quite a high number of organisms (about one million) must be ingested for an infection to occur. The time needed for the symptoms of salmonellosis to appear is generally longer than that of staphylococcal food poisoning symptoms, and this time to onset is used by doctors to distinguish between the two. Mortality from salmonellosis is generally low, with most deaths occurring in infants, the aged, or victims that are already debilitated from other illnesses.

Principally of intestinal origin, salmonellae are nonspore forming, facultative bacteria. Frequently, they are present in the intestine and other tissues of meat animals and poultry without producing any apparent symptoms of infection in the animal. While *Salmonellae* may be present in animal tissues, a major source of the infection results from the contamination of carcasses and meat during the slaughter operation. However, contamination of meat products by handlers during processing, and recontamination of meat and other foods are also potential sources of the infection. The thermal processing conditions necessary to destroy the *Staphylococcus aureus* organism (66°C for 12 minutes) will also destroy most species of *Salmonellae.*

OTHER BACTERIAL INFECTIONS. Infrequently, several other bacterial infections that occur in humans will cause illnesses having typical food poisoning symptoms. Perhaps the most commonly encountered of these infections is that caused by the *Streptococcus faecalis* bacterium. Many meat and poultry products have been the source of this infection, but dairy products, as well as other foods of animal origin, have been involved in some cases of this illness.

INFECTION FROM PARASITES. Among the food borne infections are several caused by parasites that were originally present in the animal body and are transmitted to man upon the ingestion of improperly cooked meat. Most notable among this group of infections is that caused by consuming the dormant *Trichinella spiralis* organisms found in insufficiently cooked pork. This nematode, a small wormlike organism, is present as encysted larvae in the muscles of infected animals. Among the meat producing animals, pigs are the greatest single source for human infection. The ingested dormant larvae that were not killed during cooking are released within the host because the digestive enzymes hydrolyze the cyst walls. The liberated larvae invade the walls of the upper portion of the intestine where they mature within 5–7 days. The mature worms reproduce in the intestine and the newly hatched larvae migrate into the circulatory system where they are carried to the muscles of the host. The larvae then complete the life cycle by becoming encysted in the muscles of the host, as shown in Figure 10-8. The symptoms of the infection occur as a result of the irritation of the intestinal lining caused by the burrowing of the immature larvae through the intestinal wall. These symptoms are followed by fever and a generalized weakness that accompanies the migration of the larvae into the muscles. During the encystment stage, muscle pains usually also occur.

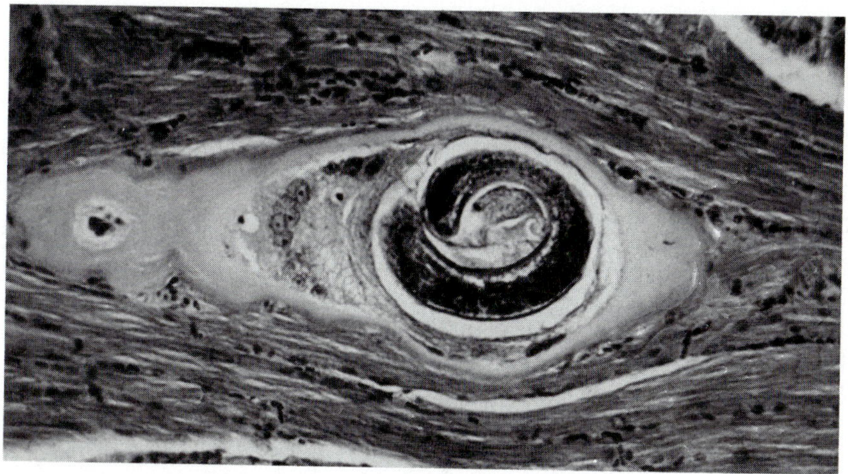

FIGURE 10-8
A photomicrograph showing an encysted trichinella spiralis larva in skeletal muscle (× 450). [Courtesy of Dr. W. J. Zimmerman, Iowa State University, Ames, Iowa.]

Trichinosis can be most easily prevented by cooking pork and products containing pork to an internal temperature of at least 58.5°C. The U.S. Department of Agriculture requires that pork products either be heated to this temperature, cured with salt and smoked, or frozen and held at specific temperatures for a set period of time. The length of this exposure depends upon the temperature employed and the thickness of the specific pork product (Table 10-2). Curing and smoking processes can also be used to destroy *Trichinella* larvae, but these differ for various pork products. (Those interested in this subject are referred to the appendix of the textbook *Meat Hygiene,* referred to at the end of this chapter.)

TABLE 10-2
The storage time required, at various freezer temperatures, to destroy *Trichinella* larvae in various thicknesses of pork*

Freezer temperature (in degrees Celsius)	Days of storage required	
	< 6 inches thick	> 6 inches thick
−15°	20	30
−23.3°	10	20
−20°	6	12

*Adapted from Brandley, P. J., G. Migaki and K. E. Taylor. *Meat Hygiene,* 3rd ed. Lea & Febiger, Philadelphia. 1966. p. 741, Appendix Table 1.

References

Brandly, Paul J., George Migaki and Kenneth E. Taylor, "Regulation Governing the Meat Inspection of the United States Department of Agriculture," in *Meat Hygiene* (Lea & Febiger, Philadelphia), 3rd ed., pp. 740–745, 1966.

Frazier, W. C., *Food Microbiology* (McGraw-Hill Book Company, New York), 2nd ed., 1967.

Jay, James J., *Modern Food Microbiology* (Van Nostrand Reinhold Company, New York), 1970.

Lechowich, R. V., "Microbiology of Meat," in *The Science of Meat and Meat Products*, J. F. Price and B. S. Schweigert, eds. (W. H. Freeman and Company, San Francisco), pp. 230–286, 1971.

Weiser, Harry H., George J. Mountney and Wilbur A. Gould, *Practical Food Microbiology and Technology* (The AVI Publishing Company, Inc., Westport, Conn.), 2nd ed., 1971.

Zattola, Edmund A., *Introduction to Meat Microbiology* (American Meat Institute, Chicago), 1972.

CHAPTER 11

Methods of Storing and Preserving Meat

Preservation by some means is absolutely essential for prolonging shelf life, and for the storage of all fresh meat and most processed meat products. The most common method of prolonging the shelf life of meat is the use of refrigeration. The term "refrigeration" in this textbook is confined to the use of temperatures between −2 and 5°C for the storage of meat. Almost all fresh meat is stored under such refrigeration. Refrigeration usually begins with the chilling of carcasses shortly after slaughter. It continues through their subsequent storage, breaking, transit, fabrication, retail cut display, and storage of these cuts in the consumer's refrigerator. Most processed meat products are also handled under refrigeration temperatures from the time of final processing until consumption.

Meat is also preserved by the processes of freezing, thermal processing, and dehydration. Preservation of meat by irradiation with gamma and x rays is still in the experimental stages. Some chemicals, such as curing and sausage ingredients, and some of the compounds present in wood smoke, provide a limited preservative effect to those meat products that are exposed to them. The preservative action of each of the preservation methods is accomplished by restricting, or in some instances completely inhibiting, microbial activity, as well as the enzymatic, chemical, and physical reactions that would otherwise cause deteriorative changes and spoilage.

REFRIGERATOR STORAGE

Initial Chill

At the completion of the slaughter process, the internal temperature of animal carcasses generally ranges between 30 and 39°C. This body heat must be removed during the initial chill period, and the internal temperature of the thickest portion of the carcass should be reduced to 5°C or less as rapidly as possible. Beef, pork, lamb, veal, and calf carcasses are chilled in *chill coolers* at temperatures ranging from −4 to 0°C; poultry and fish are generally chilled by immersion in ice water. The major factors affecting chilling rates include the specific heat of the carcass, its size, the amount of external fat, and the temperature of the chilling environment. Specific heat is directly related to the lean to fat ratio of the carcass. Fat reduces the efficiency of heat dissipation. For instance, in conventional chill coolers, beef carcasses usually require 48 hours or longer to reach an internal temperature of 5°C or lower (Figure 11-1). In coolers equipped for high velocity air movement, the chilling time can be reduced by up to 25–35 percent. The rela-

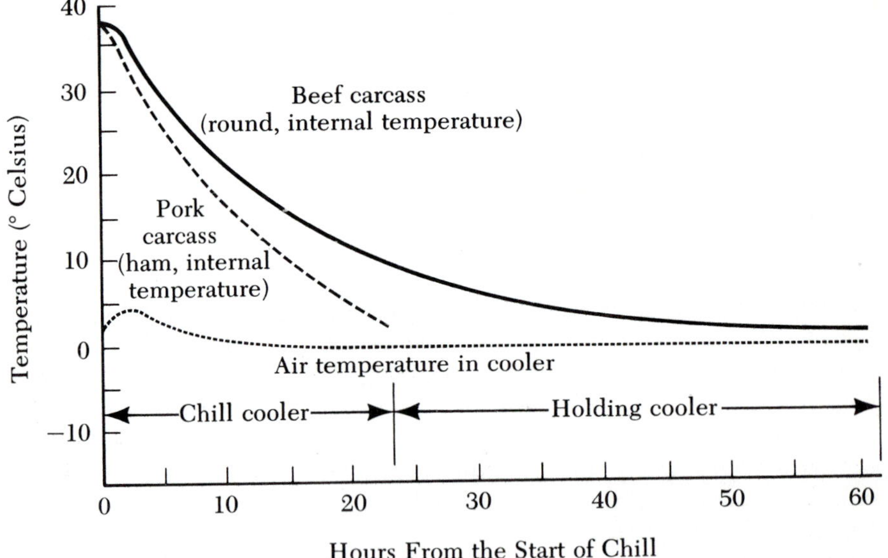

FIGURE 11-1
Typical cooling curves for beef (270 kg) and pork (75 kg) carcasses. Temperatures represent the interior of the round and ham, respectively. [After Retrum, R., Beef Carcass Chilling and Holding. *Am. Soc. Refrig. Eng.* Refrigeration and Meat Packing Conference. (1957)]

tive humidity in the chill cooler is usually kept high in order to prevent excessive carcass shrinkage due to moisture loss. Other factors affecting chilling rates include the number of carcasses placed in a chill cooler and the spacing between them. Sufficient spacing should be allowed for thorough air circulation in order to assure rapid heat dissipation. After approximately 12–24 hours of chilling, beef carcasses are removed from the chill cooler and placed in holding coolers at 0 to 3°C until they are fabricated or shipped. Lamb and veal carcasses are usually held until shipped in the same cooler as the one in which they were chilled. After approximately 24 hours in the chill cooler, pork carcasses are moved directly to the cutting line.

Immersion in ice water is a relatively rapid method of chilling. In the case of poultry, U.S. Department of Agriculture regulatory requirements demand that the internal muscle temperature must be decreased to 5°C or lower within 8 hours of slaughter. Although cold shortening is known to occur in chicken and turkey muscles at these low temperatures, the time that elapses between slaughter and chilling is sufficiently long to eliminate any measurable toughness problem. Likewise, the rapid chilling of a fish in cold water or ice water has no appreciable effect upon its tenderness. The more rapid chilling treatments described for beef, pork, lamb, and veal probably cause some toughening, particularly among beef and lamb carcasses. (Cold shortening of muscle is discussed at length in Chapter 7.)

Length of Refrigerator Storage

The refrigerated storage of meat and meat products is generally limited to relatively short periods of time, since deteriorative changes continue to occur and the rate of many of these changes actually accelerates with time. The major factors that influence the storage life of meat under refrigeration include the initial microbial load, the temperature and humidity conditions during storage, the presence or absence of protective coverings, the species of animal involved, and the type of product being stored. To achieve the maximum length of storage while maintaining an acceptable quality, all of the variables that influence storage life must be optimized. If only one of the variables is abused, then the effort expended in optimizing the other variables will probably have been wasted.

The initial microbial load has a profound effect upon the storage life of fresh and processed meat products, but the minimization of further contamination during all subsequent handling, processing, packaging, and storage is still essential in order to maintain optimum qualitative

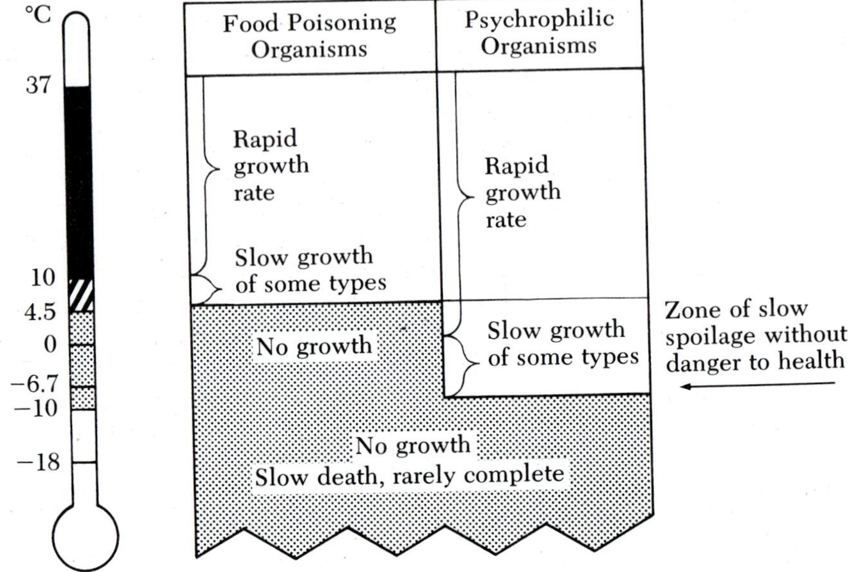

FIGURE 11-2
The food thermometer. [From McCoy, D. C., *Food Processing Operations.* Vol. 1, p. 364, 1963. AVI Publishing Company.]

properties and to prolong shelf-life. The maintenance of constant temperature storage conditions of 3°C or less is essential to the preservation of meat quality. The effect of temperature on the growth and activity of pathogenic and psychrophilic bacteria is shown in Figure 11-2. Since the comfort of workers is important, the temperature of most cutting, fabrication, and shipping rooms exceeds 5°C. The length of time meat is exposed to these elevated temperatures should be kept to an absolute minimum. The maintenance of proper refrigeration temperatures of 3°C or less is sometimes abused during transit, especially during loading at the point of origin and unloading at the destination. Maintenance of temperatures is especially important during hot, humid weather when the refrigeration capacity of trucks and railroad cars may be exceeded.

The presence of protective tissues (fat cover, skin, or scales) can prevent the microbial contamination, dehydration, and discoloration of meat surfaces. The covering of carcasses and wholesale cuts with a protective paper or film prior to shipment is a method that is widely used to prevent contamination and shrinkage losses during transit. Selecting the proper packaging materials for fresh retail cuts and processed meat products is important, in order to maintain a desirable meat color and to prevent dehydration and contamination during retail display. Under ordinary commercial conditions, the length of time that fresh meat maintains an acceptable appearance during retail display is generally about 3 days.

Pork, poultry, and fish possess more highly unsaturated fats than beef or lamb, and are therefore more susceptible to the development of oxidative rancidity. Thus, these meats are never aged for more than a few days. Also, the amount of contamination carcasses receive during slaughter varies among species. Poultry carcasses are subjected to more microbial contamination during the scalding and picking operations than are other species during their fabrication. Cured meat products are somewhat less perishable than fresh meat. Thus, the usual meat handling and storage conditions are sometimes abused when these latter products are being dealt with. This abuse can enhance the development of rancid flavors and bacterial discoloration.

The length of time that the consumer can keep meat under refrigerated storage in the home is also determined by the previous handling conditions. However, even under ideal home refrigeration conditions, fresh meat should be consumed within 4 days of its purchase. Fresh meat that is not consumed within this time, should be frozen. It is recognized that some deterioration will occur during the slow home freezing process, but such slight losses in quality are preferable to the bacterial spoilage and discoloration that could develop if fresh unfrozen meat were to be stored in the home refrigerator for extended periods of time.

FREEZER STORAGE

Freezing has long been recognized as an excellent method for the preservation of meat. It results in less undesirable changes in the qualitative and organoleptic properties of meat than the other methods of preservation. In addition, most of the nutritive value of meat is retained during freezing, and through the period of frozen storage. The only loss in nutritive value occurs when some of the water soluble nutrients are lost in the drip during thawing. The amount of drip varies with the freezing and thawing conditions, so the nutrient losses from frozen meat vary correspondingly. The nutrients found in the drip include salts, amino acids, some proteins and peptides, and water soluble vitamins. None of the nutrients present in meat is destroyed or rendered indigestible by the freezing process. When proper freezing and storage methods are used, there is little change in the color, flavor, odor, or juiciness of cooked meat products. The dark color of fresh frozen meat brightens upon exposure to oxygen or air. Thus, the qualitative properties of frozen meat approximate that of fresh meat.

In order to preserve optimum quality, meat that is going to be frozen must be handled in a manner similar to that used for refrigerated meat, especially if the product is going to be kept in freezer storage for several months or longer, because some deterioration continues to occur even at freezer storage temperatures. The length of time that meat is held in refrigerated storage prior to freezing also affects the ultimate qualitative properties of frozen meat. In addition, the quality of frozen meat is influenced by the freezing rate, length of freezer storage, and freezer storage conditions. These latter conditions include such important factors as the temperature, humidity, and the packaging material that is used. Included among the changes that can occur during frozen storage are the development of rancidity and discoloration, with the latter change being due to surface dehydration as well as to microbial activity. At temperatures below about −10°C, most deterioration due to microbial and enzymatic activity is essentially curtailed. On the other hand, in cases where meat has a relatively high microbial load, or is not thoroughly chilled prior to freezing, a slow freezing rate may allow a considerable amount of microbial growth before the temperature is lowered to a limiting point. Such meat products would undoubtedly undergo undesirable changes in qualitative characteristics.

Freezing Rate

It is a generally accepted fact that freezing rates affect the physical and chemical properties of meat. They are usually described as being "fast" or "slow." Freezing rates are influenced by the lean to fat ratio of the meat product. Because fat containing tissues have a lower thermal capacity than lean tissues, products with a high fat content freeze more rapidly than those containing low proportions of fat. Since rapid heat transfer is required for fast freezing to occur, this necessitates either extremely low temperatures (−40°C), rapid air movement, or direct contact with a freezing medium (direct immersion or spray application) at intermediately low temperatures (−20°C). Since air is a comparatively poor heat transfer medium, freezing in air, as is generally done in the meat industry, is a relatively slow process, even when high velocity air movement is used.

The most rapid freezing rates are obtained when condensed gases such as liquid nitrogen, dry ice (solid carbon dioxide), or liquified nitric oxide are used to freeze meat. Because of the extremely low temperatures of these condensed gases (liquid nitrogen, −195°C; dry ice,

–98°C; and liquid nitrous oxide, –78°C) freezing methods in which they are utilized are called *cryogenic.*

The undesirable physical and chemical effects occurring in meat during the freezing process appear to be associated with one or more of the following factors: (1) the nature and location of ice crystals that form within the muscle tissues, (2) mechanical damage to cellular structures resulting from volume changes, and (3) chemical damage caused by concentration of solutes, such as salts and sugars. (The presence of nonvolatile solutes such as these lowers the freezing point. This is why meat freezes at approximately −2° to −3°C instead of at 0°C.) The extent of the damage to meat tissues attributable to these three factors is influenced by the freezing rate.

SLOW FREEZING. During slow freezing, the temperature of the meat product being frozen remains near the initial freezing point for a long time. As a result, a continuous freezing boundary forms and proceeds slowly from the outside of the product inward. Extracellular water freezes more readily than intracellular water because it has a lower solute concentration. Slow freezing also favors the formation of pure ice crystals and the concentration of solutes in the unfrozen solution. Additionally, the intracellular compartment may be deficient in nucleation sites (suspended microscopic particles) that are necessary for the formation of many small ice crystals. These properties favor the gradual migration of water out of the muscle fibers, resulting both in the collection of large extracellular pools at the site of ice crystal formation, and in the intracellular concentration of solutes. Consequently, the intracellular freezing temperature is lowered even further. This process of the lowering of the freezing point by the concentration of solutes is called *eutectic formation.* The total contribution of this process to freezing damage is not completely understood, but it apparently causes chemical alterations, including such changes as protein insolubilization and a decrease in elasticity of the thawed muscle tissue.

During the slow freezing process, the long period of crystallization before freezing occurs, produces numerous large extracellular masses of ice crystals that are easily lost as drip during thawing. The freezing curves and periods of crystallization for various freezing temperatures are shown in Figure 11-3. Mechanical damage due to volume changes is more likely to occur during slow freezing because of the expansion associated with the formation of large ice masses, as well as the concomitant shrinkage of muscle fibers that have lost water to extracellular pools. Such muscle tissue has a distorted appearance in the frozen state (Figure 11-4b) that completely obliterates the normal striated appearance (Figure 11-4a).

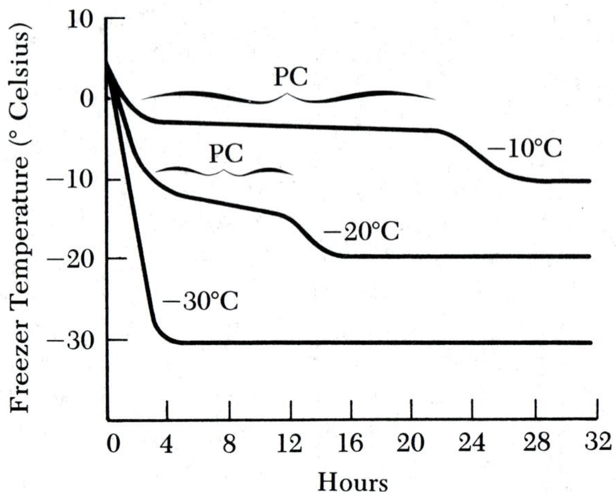

FIGURE 11-3
Freezing curves, showing the relative rates of freezing at various freezer temperatures. (Key: PC = period of crystallization.)

FAST FREEZING. During cryogenic fast freezing the temperature of the meat product being frozen rapidly falls below the initial freezing point. Numerous small ice crystals tend to form uniformly throughout all of the meat tissues. These small ice crystals have a filamentlike appearance, and they are formed with approximately the same speed, both intra- and extracellularly. Because of the rapid temperature drop due to the rapid rate of heat transfer, the numerous small ice crystals that form have little opportunity to grow in size. Thus, fast freezing causes the spontaneous formation of many individual small ice crystals resulting in a discontinuous freezing boundary and very little translocation of water. Since most of the water inside the muscle fibers freezes intracellularly, drip losses during thawing are considerably lower than in the thawing of slow frozen meat. In addition, muscle fiber shrinkage and distortion effects are minimized during fast freezing, resulting in in a near normal ultrastructural and striated appearance in the frozen state (Figure 11-4c). Volume changes are less, and the periods of crystallization are shorter (Figure 11-3) than in slow frozen muscle and, as a consequence, mechanical damage is correspondingly less. The filamentlike ice crystal formation which occurs during fast freezing entraps solutes, and thus minimizes the concentration effect. Thus, the fast freezing of meat generally causes less deleterious effects than slow freezing. In addition, the smaller and more numerous ice crystals in rapidly frozen meat reflect more light from the surface, resulting in a lighter color than that of slow frozen meat.

home freezer units also fall within this range. Because of the slow freezing rate in home freezers, large quantities of meat should not be frozen simultaneously.

PLATE FREEZER. The heat transferring medium in this freezing method is metal, rather than air. Trays containing the products, or the flat surfaces of meat products are placed directly in contact with the metal freezer plates or shelves. Plate freezer temperatures usually range from about −10°C to −30°C in commercial practice, and the method is generally limited to thin pieces of meat (such as steaks, chops, fillets, and patties). Conduction rather than convection is the important factor in heat transfer in this method and, as a consequence, the freezing rate is slightly faster than it would be in still air (Figure 11-5). Although plate freezing is still slow, it can be speeded up by circulating cold air over the product. Several modifications of the conventional plate freezer principle are used in the meat industry to provide for the use of batch, semicontinuous, or continuous production operations. Other modifications also include a double plate system in which one plate is in direct contact with each of the two flat surfaces of the meat product.

BLAST FREEZING. The most commonly used method for freezing meat products is cold air blast freezing in rooms or tunnels that are equipped with fans to provide rapid air movement. Air is the medium of heat transfer, but because of its rapid movement the rate of heat transfer is greatly increased over that in still air, and thus the rate of freezing is markedly increased (Figure 11-5). High air velocity increases both the cost of freezing and the severity of freezer burn in unpackaged meat products. Because of the speed at which meat products freeze when this method is used, it is identified by any one of several names: quick frozen, fast frozen, sharp frozen, or blast frozen. Commercially, air velocities range from approximately 30–1070 meters/minute (mpm) and temperatures range from about −10°C to −40°C in the blast freezer. However, an air velocity of about 760 mpm and a temperature of −30°C are probably the most practical and economical now being used in the meat industry.

The proper spacing and stacking of meat products on pallets, or on shelved racks, in the blast freezer room is important for rapid and efficient freezing. In blast tunnels, the meat products that are to be frozen are placed on moving metal mesh belts or some similar conveyor system and passed through the tunnel. The speed of the conveyor will depend on the time necessary to freeze the product. Sometimes meat products are frozen or partially frozen in blast freezer tunnels prior to wrapping, and are then packaged at a later time. In another procedure meat products are passed through the blast freezer tunnel just long enough to produce surface freezing and hardening. They are then wrapped, and the partially frozen packaged products are transferred to a still air freezer for the completion of freezing, and for subsequent storage.

LIQUID IMMERSION AND LIQUID SPRAYS. Liquid immersion or spray is the most widely used commercial method for freezing poultry. However, some red meat products, and fish, are also frozen by this method. Because of the rapid heat transfer, higher temperatures are generally used than in blast freezing. However, the freezing rates are comparable.

The products to be frozen are placed in plastic bags, stacked on pallets or in shelved racks, and then either immersed into the freezing liquid by fork lift trucks, or moved through the cold liquid by a conveyor. In another application, the product is conveyed through an enclosed freezing cabinet while the cold liquid is continuously sprayed on its surface. After the product is removed from the immersion tank or freezing cabinet, the freezing liquid is rinsed from its surfaces with cold water. The length of time that the product is immersed or sprayed determines the extent of freezing. When the surface of the product is frozen (crusted) the product is generally transferred to a freezer room for completion of the freezing process and subsequent storage.

The liquids used for liquid freezing must be nontoxic, relatively inexpensive, and should have a low viscosity, low freezing point, and high heat conductivity. Sodium chloride brine has been commonly used, but glycerol, and glycols (such as propylene glycol), are currently achieving wide usage. One problem with salt brines in particular, in contrast to other liquids, is the corrosion of metal tanks and equipment. Another important factor in immersion freezing is the presence of holes in the protective package. No matter how small these breaks are, a seepage of liquid into the package necessitates a washing and repackaging of the product.

CRYOGENIC FREEZING. Any one of three systems may be used for cryogenic freezing. These are direct immersion, liquid spray, or the circulation of the cryogenic agent vapor over the product to be frozen. The most commonly used cryogenic agents are nitrogen, either as a liquid or vapor, and carbon dioxide which is stored either as a liquid under high pressure, or as a vapor, or snow. Occasionally liquified nitrous oxide is also used. Large pieces of meat are rarely immersed directly into liquid nitrogen because of the extensive shattering or cracking that might occur. Therefore, present systems generally evaporate liquid nitrogen in the freezing chamber and utilize its tremendous cooling capacity as it changes into nitrogen gas in order to freeze the meat products. Liquid nitrogen spray or liquid carbon dioxide spray (released as a snow), combined with a conveyor belt system, are used to rapidly freeze meat products of relatively small size, such as patties, and diced meat, fish, and shell fish. Because of the excellent qualitative properties of meat products frozen by these methods, they are gaining in acceptance.

Length and Conditions of Frozen Storage

The previous discussion has emphasized the effects of freezing rates on the retention of quality in frozen meat. However, the conditions under which frozen meat is stored may be even more important for maintaining quality. The length of time that frozen meat can be successfully stored varies with the species and the type of product, and is influenced by freezer temperature, temperature fluctuations, and the quality of the wrapping materials that are used in packaging.

In general, the storage time of all types of frozen meat can be extended by decreasing the storage temperature. The rate of all chemical deteriorative changes is greatly reduced by freezing, but reactions such as oxidative rancidity continue at a slow rate, even in the frozen state. Most chemical changes could essentially be eliminated by reducing the temperature to −80°C, but such temperatures are not economically feasible in most storage facilities. The growth of putrefactive and spoilage microorganisms, and most enzymatic reactions, are greatly reduced (if not entirely curtailed) at temperatures below −10°C. In general, storage temperatures of less than −18°C are recommended for both commercial and home freezer units. Most of these units operate at temperatures between −18 and −30°C. Although it is more expensive to maintain the lower temperatures in this range, the length of storage can be significantly extended.

Temperature fluctuations during frozen storage should be avoided as much as possible, in order to minimize ice crystal growth, formation of large ice crystals, and drip losses associated with large ice crystals in the frozen meat. Almost all of the water in meat is frozen at about −18°C (Figure 11-6). However, as the temperature increases the unfrozen percentage increases, and becomes especially marked above −10°C. Small ice crystals are thermodynamically less stable than large crystals. Water molecules tend to migrate from the smaller crystals through unfrozen pools of water to recrystalize and form large ice crystals. Migration, recrystallization, and ice crystal growth are enhanced by higher storage temperatures and by temperature fluctuations. Fluctuating storage temperatures can also cause excessive frost accumulation inside packages, much of which will be lost as drip upon thawing.

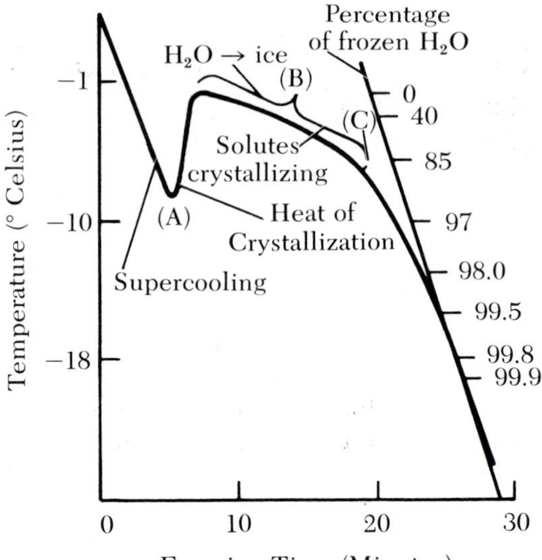

FIGURE 11-6
Freezing curve of a thin section of beef showing the period of crystallization and percentage of frozen water. [From N. Desrosier, *The Technology of Food Preservation*, p. 37, 1959. AVI Publishing Company.]

Acceptable quality can be maintained in frozen meat products for a number of months, only if certain critical packaging requirements are met. These requirements include the use of a moisture vapor proof packaging material in order to keep the moisture in the package, and to keep oxygen from the air out. Moisture losses, due to improper packaging materials or techniques, result in dehydration and freezer burn, and oxygen from the air in the package causes oxidative changes, including rancidity. Another requirement in order to maintain quality for long periods of time, is to eliminate as much air as possible from the package. Tight fitting bags, such as those usually used for freezing poultry, are ideal. Other requirements include the use of odorless, grease proof packaging materials that are strong when wet, and resist scuffing, tearing, and puncturing under normal handling conditions.

The amount of time that meat can be held in frozen storage, while still maintaining an acceptable quality, also depends on the nature of the meat fats (degree of saturation). Since poultry and pork fats are more unsaturated than beef and lamb fats, they are more susceptible to oxidative changes, including the development of rancid flavors and

TABLE 11-1
Maximum recommended length of storage of certain meat items at various temperatures for the preservation of optimum quality

Item	−12°C	−18°C	−24°C	−30°C
	Months			
Beef	4	6	12	12
Lamb	3	6	12	12
Veal	3	4	8	10
Pork (fresh)	2	4	6	8
(cured, unsliced)*	0.5	1.5	2	2
Variety meats (liver, heart, tongue)**	2	3	4	4
Poultry	2	4	8	10
Ground beef and lamb	3	6	8	10
Seasoned sausage*** (pork, bulk)	0.5	2	3	4

*It is not recommended that sliced bacon and luncheon meat products be frozen.
**It is not recommended that brains and sweetbreads be frozen.
***It is not recommended that pork sausage links and patties, and other sausages (such as bologna, franks, and braunschweiger) be frozen.

odors. Hence, the recommended storage time for pork and poultry is less than that for beef and lamb (Table 11-1). The gradual decrease in flavor and odor acceptability during frozen storage is primarily due to oxidation of the lipids.

In addition to the items already mentioned, the length of frozen storage is influenced by such factors as whether the meat product is fresh, seasoned, cured, smoked, precooked, sliced or diced, contains preservatives, or is contained in sauce or gravy. Salt enhances the development of rancidity, and processed meat products containing salt have a rather limited frozen storage life. It is generally recommended that cured and smoked products not be stored frozen but, if they are frozen, they should be stored only for very limited periods of time. Sliced meat products that contain salt (such as bacon and luncheon meat) should never be frozen because the air incorporated during slicing, together with the salt effect, leads to the development of rancid flavors in a matter of several weeks. Thus, if frozen bacon is to be stored at all, it should not be sliced prior to freezing. Precooked frozen meat and poultry products lose their "fresh cooked" flavor during frozen storage and develop a "warmed over" flavor, and eventually a rancid flavor, due to fat oxidation. These oxidative changes are directly related to the degree of cooking; they increase with cooking time. The treatment of meat products with alkaline polyphosphates during precooking will

TABLE 11-2
Maximum recommended length of storage of some precooked meat products stored at −18°C for the preservation of optimum quality

Product	Packaging material	Recommended maximum storage time
Fried chicken, breaded	aluminum pan	3 months
Steam cooked breaded chicken parts	aluminum pan	9 months
Steam cooked breaded chicken parts (chicken soaked in a tripolyphosphate† solution prior to breading)	aluminum pan	12 months
Sliced turkey and gravy	aluminum pan	18 months
Turkey rolls with a tripolyphosphate added†	fibrous casing	2 years
Broiled beef patties, seasoned	layer packed in polyethylene	6 months
Broiled beef patties, seasoned, extended with textured soy flour*	layer packed in polyethylene	6 months
Breaded deep fried beef patties (patty extended with textured soy flour)	layer packed in polyethylene	6 months
Roast beef, unsliced	vacuum packed in cryovac	18 months
Corned beef sandwiches	overwrapped with saran	60 days
Ham sandwiches	overwrapped with saran	60 days
Roast beef sandwiches	overwrapped with saran	60 days
Frozen entrees containing adequate sauce or gravy (beef burgundy, chicken fricassee, salisbury steak and gravy, meatballs and spaghetti sauce)	aluminum pan mylar pouches	12 months 18 months

The data in this table supplied courtesy of Armour and Company.
*Textured soy flour tends to have an antioxidant effect.
†An alkaline polyphosphate.

partially inhibit these oxidative changes. When precooked meat products are covered by a sauce or gravy that includes flours (such as soy flour), the frozen storage life is greatly extended. These flours apparently have a marked antioxidant effect. The frozen storage life of precooked meat products can also be extended considerably by packaging them in an inert gas, such as nitrogen. The elimination of oxygen from the packages is responsible for extending the storage life.

The recommended length of frozen storage for some fresh and processed meat items, at various storage temperatures, is shown in Table 11-1. The maximum recommended length of frozen storage at −18°C or lower, in specific packaging materials, is presented for a number of precooked meat products in Table 11-2. The recommended storage time indicated for each product is applicable only when near optimum handling, packaging and storage conditions have been followed. These storage times are recommendations for the preservation of near optimum quality. If these recommended maximum times are exceeded, the meat products will still be edible but probably will have decreased in their qualitative characteristics such as flavor, odor, and juiciness as compared to the fresh products.

Thawing and Refreezing

The thawing process probably does greater damage to meat than freezing. Several factors are largely responsible for the damaging effects that occur during thawing. First of all, thawing occurs more slowly than freezing, even when the temperature differential is the same. However, the temperature differential in thawing is generally much less in practice than that encountered in freezing. Secondly, during thawing the temperature rises rapidly to the freezing point, but then remains there throughout the entire course of thawing (Figure 11-7). This situation further increases the length of the thawing process, compared to freezing. The thawing process thus provides a greater opportunity for the formation of new, large ice crystals (recrystallization), for increased microbial growth, and for chemical changes. Thus, the time–temperature pattern of thawing is more detrimental to meat quality than that of freezing. An additional potentially serious problem is that most thawing is done at the point of consumption by individuals who generally have little knowledge of the problems associated with thawing meat products.

Meat products may be thawed in any of several ways: (1) in cold air, such as in the meat cooler or home refrigerator, (2) in warm air, (3) in water, or (4) during cooking, without prior thawing. Unless meat products are being cooked directly from the frozen state, they should be

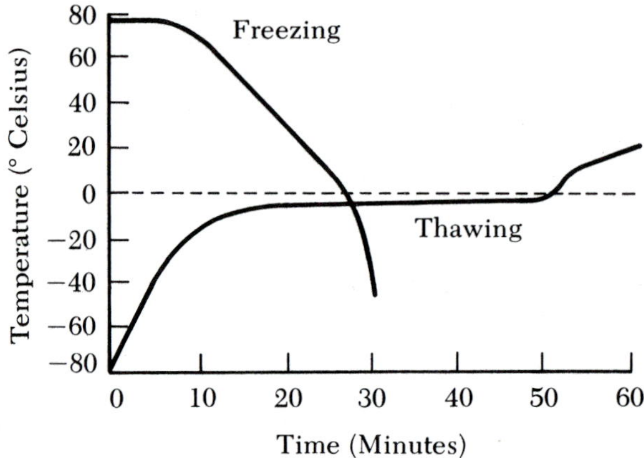

FIGURE 11-7
Graphs of the average temperature at the geometric center of cylindrical specimens of a starch gel during freezing and thawing under equal temperature differentials. Can size 303 × 407 (3-3/16″ diameter by 4-7/16″ tall). [From Fennema, D. and W. D. Powrie, *Advances in Food Research*, 13:219, 1964. Academic Press, Inc.]

thawed with the packaging material left on them, in order to prevent dehydration. It is generally believed that thawing methods affect the amount of drip in direct proportion to the rate of thawing. The time required for thawing frozen meat products depends upon a number of factors. Among the more important of these factors are: (1) the temperature of the meat product, (2) the thermal capacity of the meat product (lean meat products have higher thermal capacities and therefore thaw more slowly than do fat meat products), (3) the size of the meat product, (4) the nature of the thawing medium (water provides faster heat transfer than air), (5) the temperature of the thawing medium, and (6) whether or not the thawing medium is circulated.

The recommended procedure is to thaw meat products at refrigerated temperatures. Admittedly, this is a slow process and might require several days for items as large as whole turkeys. Consequently, this process is frequently abused by the thawing of meat products for hours at room temperature or (even worse) in warm water. These latter procedures not only accelerate the thawing process, but they also increase the opportunity for microbial growth and excessive drip losses. A thawed meat product is no more susceptible to microbial growth than is fresh meat. However, once the surface reaches temperatures above 0°C, microbial growth can readily occur. Thus, frozen meat products should not be thawed too long before being cooked. In fact, cooking from the frozen or partially thawed state largely eliminates the opportunity for any appreciable microbial growth to take place.

Refreezing frozen meat that has been thawed is another problem area that is generally beset with considerable confusion and misunderstanding, particularly at the household level. There is basically nothing wrong with refreezing meat, and the practice undoubtedly occurs in normal commercial practice. However, there are two important considerations for deciding whether meat products should be refrozen. The temperature of the defrosted product must be considered, as well as the length of time that it has been defrosted. Neither should have reached the point where any appreciable microbial growth has been allowed. Never *refreeze a product as a last resort to save it,* since at this point it is already too late. It should be recognized that all the undesirable physical and chemical effects associated with freezing meat products will occur again during refreezing. Therefore, repeated freezing is not recommended. Microbial growth is the key factor in determining whether meat should be refrozen. If the thawing time and temperature relationships are such that no appreciable contamination or growth has been permitted to take place meat products can be refrozen more than once, and still be safe for human consumption. However, it should be recognized that meat that has been frozen and thawed several times decreases in qualitative properties, especially in flavor and juiciness.

Bone Darkening

Bone discoloration or *bone darkening* is a condition induced by freezing and thawing young chickens, particularly fryers. It rarely occurs in old chickens, and has not been observed in turkeys, ducks, or geese. After freezing and thawing, muscle areas adjacent to bones exhibit a bloody appearance in the uncooked state. During cooking the red color changes to a dark gray or brown, and in severe cases to black. The red color is due to hemoglobin that has leached from the marrow of the relatively porous bones of the young chickens. During cooking the leached hemoglobin is oxidized and converted to methemoglobin, thus causing the dark discoloration. In general, the discoloration occurs around the bones of the leg, thigh, knee joint, and second joint of the wing, but it is sometimes also seen in the back and breast. Bone darkening does not affect the flavor, odor, or texture of the cooked meat, but it can be a serious problem because of the undesirable appearance. The problem of bone darkening is greatly reduced by precooking prior to freezing.

Factors Affecting Quality of Frozen Meat

The various factors affecting the quality of frozen meat can be ranked, in order of importance, as follows: (1) frozen storage conditions, (2) thawing conditions, (3) freezing rate, and (4) prefreezing treatment and handling. Frozen storage, as conducted under commercial and household conditions, is considered to be the most damaging phase in the handling and processing of frozen meat. Even though damage due to recrystallization occurs during the thawing process, it is generally less severe and extensive than that which occurs during storage because of the shorter time interval involved. The original difference in product quality, caused by the use of a fast, rather than a slow freezing process, gradually diminishes during frozen storage over a period of several months. In addition, the cooking process that is necessary for the evaluation of frozen meat quality, further reduces those differences in quality caused by the original freezing rate.

In summary, the specific factor that exerts the most damaging effect on meat products during frozen storage is temperature, particularly if it fluctuates. The deleterious effects produced by temperature conditions are primarily the result of the changes associated with recrystallization. The damage that occurs during frozen storage can be minimized, but not entirely prevented, by maintaining a low, constant storage temperature, and by not keeping the product in frozen storage for a long time.

THERMAL PROCESSING

Heat processing, as a method of preservation, is used to *kill* spoilage and potentially toxic microorganisms in meat and meat products, and to inactivate endogenous enzymes that could cause deteriorative changes. This is in contrast to the refrigeration and freezing processes, which *slow* or *stop* microbial growth but generally do not kill microorganisms. Two general levels of heat processing are employed in meat preservation. A *moderate heating* process, where products reach temperatures of 58°–75°C, is employed in the cooking of most processed meat items. This heat treatment kills part, but not all, of the microorganisms present, and will inactivate other organisms. This process is referred to as *pasteurization.* The shelf life of meat products is extended by this pasteurization process, but they must still be stored under refrigeration after being processed. A more *severe heating*, generally to temperatures above 100°C, is used to prepare "commercially sterile" meat products that are stable at room temperature for one or more years. This process is called *sterilization.* This treatment either kills all potential spoilage organisms, or causes sufficient damage to microbial cells so that growth is impossible. The palatability of meat generally decreases as it is heated to temperatures above those used for pasteurization. Shelf stable canned meat products have a typical strong sulhydryl flavor due to extensive protein denaturation during heat processing. In addition, the texture of the product is modified, due to a breakdown of connective tissues.

The discussion in the following sections of this chapter is directed more toward thermal processing for commercial sterilization. (The effects of heating meat products to moderate temperatures were discussed in Chapter 9, and Chapter 12 includes an additional discussion of meat cookery.)

Principles of Thermal Processing

HEAT TRANSFER. All conventional methods of thermal processing involve heat transfer by conduction, convection, and/or radiation. Heating by conduction involves the direct transfer of heat from particle to particle without the use of a medium other than the product itself. For example, the transfer of heat from the surface to the center of a solid piece of meat is almost entirely by conduction. Convection heating involves heat transfer by the mass movement of heated particles in a "fluid" such as air, steam, or water. Heating by radiation is the transfer of heat energy through space. One or more of these heat transfer mechanisms are involved in all of the procedures used to thermally process meat products.

Several variables affect the rate of heat transfer to a product, and the extent of heating. Of course, the rate and extent of the temperature increase are directly proportional to the *temperature differential* between the product and the heat source, and to the *length of time* that heat is applied. Products differ in *specific heat* and *thermal conductivity*. The specific heat of a substance is a measure of the amount of heat energy required to change the temperature of 1 gram of product by 1°C. The thermal conductivity of a substance determines how fast heat moves through it by conduction. A block of metal will heat to a given temperature more rapidly than a piece of meat of the same size, because the metal has both a higher thermal conductivity, and a lower specific heat. Meat with a high fat content has a lower specific heat than does very lean meat, so that less heat energy is required for a given temperature increase. However, the *rate* of temperature increase can be quite similar in the two types of meat tissue, because fat has a lower thermal conductivity. Product *consistency*, and *agitation* of the container affects (to a certain extent) the amount of heat transfer accomplished by conduction *vs* convection. Fluid or semifluid products heat faster than solid products, because heat transfer in the former occurs by both conduction and convection. Agitation during thermal processing further increases convective heat transfer by fluids in the container. Finally, products with a large *surface area per unit weight* heat faster than those with a smaller surface area.

It is important to consider the factors that affect the rate and extent of heating, as well as fundamental heating principles, when specific thermal processing schedules are developed for a meat product. However, meat products vary widely in amounts of fat, water content, consistency, and homogeneity, so that a calculation of exact processing schedules is virtually impossible. Thermal processing schedules usually must be developed on an individual product basis, and sometimes even on an individual plant basis, due to variations in specific equipment.

HEAT RESISTANCE OF MICROORGANISMS. Cells and spores of microorganisms differ widely in their resistance to high temperatures. Some of these differences are the result of factors that can be controlled, and others are due to characteristics of the organism that cannot be controlled. Factors that are known to affect the heat resistance of cells and spores must be considered when thermal processes for the destruction of microorganisms are developed for, or applied to, the production of meat products.

The concentration of cells (or spores) that is present in the product has an important effect on the heat treatment that is needed. The more spores or cells that are present, the greater will be the heat treatment required to kill all of them. The previous history of cells or spores also affects their heat resistance. For example, bacterial cells are most resistant to thermal processing during the logarithmic phase of their growth. Immature spores are less heat resistant than mature ones. The composition of the medium in which cells or spores are heated has a great effect on their heat resistance. Moist heat is much more effective for the killing of microorganisms and spores than is dry heat, thus products with a low level of moisture require more heat for their sterilization than do foods with a high moisture content. In general, microorganisms are most heat resistant at (or near) neutral pH. An increase in either acidity or alkalinity hastens killing by heat, but any change in pH toward increased acidity is more effective than a similar change toward alkalinity.

The heat resistance of microorganisms is usually expressed in terms of their *thermal death time* (TDT). This is defined as the time required to kill a certain number of cells or spores under certain specific conditions. Those conditions which must be specified, in order for TDT values to have any meaning, include the temperature, numbers and type of organisms, and the characteristics of the medium in which the organism is heated. The TDT at 121°C is arbitrarily taken as a reference point, and is designated F_o.

In order to establish thermal processes that will yield a commercially sterile meat product, the following data must be known: the thermal death time curve for the most heat resistant organism likely to be present in the product, the heat penetration and cooling curves for each product, and the size and type of container to be used. When all of these factors are known, methods are available for the calculation of a processing schedule that will theoretically be adequate for the product. However, it must be remembered that the calculated processing schedule applies only to the specific product and type of container in question, and not to any other. In commercial practice, an additional safety margin, beyond that which has been calculated, is allowed to insure the destruction of all potential spoilage and toxigenic cells and spores.

Methods of Thermal Processing

The heating of sausage and smoked meats has been discussed as one of the steps in their manufacture in Chapter 9. Therefore, this section is restricted to a discussion of methods that are used to thermally process meat after it has been packed into metal cans or glass jars. The commercial heat processing of canned products (such as canned hams and luncheon meat) to pasteurization temperatures is generally done by immersing the sealed cans in water that is heated in open kettles or vats. The water temperature is usually less than 100°C under these conditions, but higher temperatures can be reached by adding salts, such as calcium or sodium chlorides, to increase the boiling point. Heating in open kettles is not recommended for the production of commercially sterile products because of the extremely long processing times required, and the danger of inadequate heat treatment in these meat products.

Thermal processing to achieve sterility is usually done in large metal drums, known as *retorts*, which are capable of withstanding pressures of up to 32 kg/cm^2. The sealed cans are placed in the retort, which is then closed, sealed, and heated. Heat is applied by heating water under pressure, or by injecting superheated steam, or a mixture of steam and air. Temperatures of 120°C or higher are routinely achieved in the retort, which greatly reduces processing time over the open kettle method. In addition, the cans are often agitated, in order to further shorten the processing time.

Techniques have been developed for the thermal processing of fluid products outside the container, followed by an aseptic transfer of the sterile product into sterilized containers. However, these processes have not been adapted to solid products.

DEHYDRATION

The drying of meat in the sun or over a fire dates back to prehistoric times. The preservative effects of dehydration are due to the reduction of water activity (a_w) to such a low level that microbial growth is inhibited, and these meat products are stable without refrigeration. Some products, such as dry and semidry sausages, are partially dried, usually in air, during the aging process. Since they are also fermented sausages, the acidity developed during aging provides some additional preservative action.

Dehydration Methods

HOT AIR DRYING. Acceptable quality comminuted, cooked meat products can be produced by hot air drying. However, such factors as temperature, particle size, and the rate of air movement must be carefully controlled. Meat products dehydrated in this manner have a residual moisture content of about 5 percent. As a consequence, certain deteriorative changes can develop during prolonged storage. The fat, especially that of pork, tends to become rancid following hot air drying. This can be retarded either by the addition of antioxidants, or packaging the product in a manner that eliminates oxygen. Hot air drying is a slow process that is not applicable to uncooked meat or to large pieces of cooked meat (such as roasts, steaks, or chops) because the resultant surface hardening yields a product with poor consumer acceptability. Meat products dried by hot air also shrivel considerably, and have poor rehydration properties due to the protein denaturation that occurs during the drying operation.

FREEZE DRYING. In conventional freeze drying, the meat product remains frozen throughout the drying cycle while its internal frozen water (ice) is removed by the application of sufficient heat to transform it directly to water vapor without going through the liquid state. The transformation of a substance directly from a solid to a vapor, without passing through an intermediate liquid state, is called *sublimation.* A rapid sublimation of water from the meat product cools it sufficiently to prevent thawing and, as the drying process proceeds from the outside inward, the low heat exchange coefficient of the dehydrated outer portion prevents the frozen inner portion from reaching a temperature high enough to cause thawing. The freeze drying process is carried out in a vacuum chamber, and if vacuum pressures of 1.0–1.5 mm of mercury are used, the drying chamber temperature can be as high as 43°C without causing any thawing. The rate of freeze drying is limited by the rate that heat can be transferred into the meat product in order to continue the process of sublimation. The rates of heat transfer and drying can be increased by placing the meat directly on heated shelves, or by using microwave or infrared heating. Infrared is the most commonly used heating method in commercial use at present. The residual moisture content of freeze dried meat products is generally below 2 percent.

Meat that is cooked prior to freeze drying is generally more stable than uncooked freeze dried meat, and has a shelf life that is 2–4 times longer. The shelf life of cooked freeze dried beef is approximately 24–28 months. The stability of freeze dried meat products depends upon the method of drying, the residual moisture content, packaging method, storage temperature, and the quality of the product prior to drying. The principle requirement for maintaining product stability is the exclusion of oxygen and moisture. This requirement necessitates the use of impermeable packaging materials. The most prominent physical and chemical factors affecting the stability, rehydration, and palatability characteristics of freeze dried meat products are rancidity development, nonenzymatic browning, and protein denaturation. Most nutrients that are present in meat, with the exception of thiamin, are relatively stable to freeze drying. The protein denaturation that occurs has little effect on the biological value of the protein.

Freeze dried meat products retain essentially their original shape and size. Consequently they are very porous, and are much more readily rehydrated than hot air dried meat products. However, the texture and flavor of freeze dried raw and precooked meat is greatly affected by the method of rehydration. If 1–2 percent sodium chloride and 0.1–0.15 percent alkaline pyrophosphate are added to the water used for rehydrating freeze dried products, their palatability characteristics can be greatly improved. The extent of rehydration is greatest, if the rehydrating solution is maintained at a temperature just below the boiling point. However, even thinly sliced freeze dried meat generally does not rehydrate to the original moisture content. This characteristic is responsible for a reduction in flavor, texture, and juiciness. The "dehydrated" flavor that is a characteristic of freeze dried meat products can largely be masked by the addition of seasonings, such as those used in freeze dried chili.

The most common freeze dried meat products at the present time are dehydrated soup mixes. Because of the problem of fat rancidity, meat items which are low in fat are preferred for dehydrated soups. Other dehydrated meat products include those used by the military, and a limited number of products available for campers.

IRRADIATION

Radiation as a method of meat preservation generally involves the application of *ionizing radiation* to products. *Ionizing radiation* is defined as radiation having energy sufficient to cause the loss of electrons from atoms to produce ions. This includes the high speed electrons produced by a variety of electron generators, x-rays generated by electrons when they strike a heavy metal target, and electrons and gamma radiation emitted from radioactive isotopes such as $^{60}_{27}Co$ (radioactive cobalt) or $^{137}_{55}Cs$ (radioactive cesium). Ionizing radiation kills microorganisms in and on meat without raising the temperature of the product. It is therefore referred to as *cold sterilization.*

The amount of radiation energy absorbed by the meat product being irradiated is expressed in units of *rads.* One million rads (1 megarad) are approximately equal to 2 calories. In order to sterilize a product so that it will be stable during subsequent unrefrigerated storage, a radiation dose of about 4.5 megarads is required. This amount of irradiation insures the destruction of the most resistant spoilage and potentially toxigenic microorganisms. The use of less than sterilizing dosages to extend the shelf life of fresh meat has been termed *radiation pasteurization.* A radiation level of about 100,000 rads eliminates many of the spoilage microorganisms and makes possible significant extensions in storage time under refrigeration. However, such products will eventually spoil, due to the growth of those organisms which were resistant to the radiation.

Ionizing radiation causes a number of undesirable chemical and physical changes in meat products, including discoloration and the production of very objectionable odors and flavors. Some objectionable odor and flavor occurs at all levels of irradiation, but the flavor problem is significant only at the higher dosages used for sterilization. Irradiation levels up to about 4 megarads produces an odor characteristic of sulfur compounds. At higher levels (4–10 megarads) the odor is best described as smelling like wet dog hair. The extent of objectionable flavor development varies with the meat from different species. Pork and chicken develop little "irradiated flavor", even at relatively high doses. Beef develops a strong irradiated flavor, and veal, lamb, and mutton are intermediate.

At present the use of ionizing radiation as a method of meat preservation is not approved by the Food and Drug Administration. It is being used, in some countries, for the sterilization of foods other than meat.

Nonionizing radiations also can have a lethal effect on microorganisms, and are useful in meat preservation. Microwaves (high frequency radio radiation) and infrared radiation are capable of generating heat in the irradiated object, and are used in some thermal processing and cooking applications. These radiations do not have sufficient energy to cause ionization, and any preservative effect is entirely attributable to the heat that they produce. Ultraviolet radiation can cause death when absorbed by microorganism; thus, they possess a germicidal effect. However, the practical value of ultraviolet light is limited since this radiation possesses a very low penetrating power. Thus, it can be used to sterilize only the surfaces of carcasses and meat products. Ultraviolet irradiation also accelerates the discoloration of meat and the development of rancidity in fat. The only significant use of ultraviolet radiation in the meat industry has been to control surface microbial growth on beef carcasses during accelerated, high temperature aging.

PRESERVATIVE EFFECTS OF CERTAIN CHEMICALS AS MEAT INGREDIENTS

A number of ingredients that are added to meat products during processing, including the application of smoke, impart varying degrees of preservative properties. The preservative action of sodium chloride in curing meat has long been known, and probably has the longest history of usefulness. Preservation by salt is achieved by its effectiveness in lowering water activity. However, lowering the water activity to a level necessary for effective preservation requires a salt content in the finished product of approximately 9–11 percent. This is considerably higher than the 2–3 percent commonly found in commercially cured meat products. While some microorganisms are inhibited by these latter salt concentrations, the water activity is usually high enough to support the growth of molds, yeasts, and halophilic (salt loving) bacteria. Thus, the salt in commercial processed meat products today only provides a limited preservative effect, and other methods of preservation are necessary in order to prolong their shelf life.

The addition of nitrite to processed meat products provides them with marked bacteriostatic properties. Nitrite effectively inhibits the growth of a number of bacteria including pathogens, most notable among which is *Clostridium botulinum.* The nitrite in cured canned meat products, that are thermally processed, aids in destroying the spores of anaerobic bacteria and inhibits germination of the surviving spores. In fact, the addition of 150–200 ppm of nitrite to canned or vacuum packaged processed meat products prevents the formation of botulinum toxin that may occur at lower nitrite levels, or in its absence. (See Chapter 9 for a more detailed discussion.) Even at low concentrations, nitrite functions synergistically with salt to provide certain cured meat products, such as canned hams and canned luncheon meats, with effective preservative and bacteriostatic properties.

The use of sugar as a preservative agent would require levels well above those normally used in cured meat or other processed meat products. However, the sugar that is added to fermented sausage products indirectly serves as a preservative because of the lactic acid that is formed. In addition to their lowered pH, these sausage products are also partially dried during the aging process and, as a result, develop a relatively high salt concentration. All these factors add up to a high degree of stability and a longer shelf life.

A number of the seasonings and spices, especially those containing essential oils (such as mustard and garlic), contribute some preservative effects to processed meat products through their bacteriostatic action. Some also have antioxidant properties. However, seasonings and spices are now added solely for their flavoring effects, and not because of their preservative potential.

Some of the aldehydes, ketones, phenols, and organic acids in wood smoke impart bacteriostatic and bacteriocidal effects to smoked meat products, as was mentioned previously in the discussion of meat processing. Even though most smoked meat products have greater stabilities and shelf lives than their fresh meat counterparts, they still require some other means of preservation in order to prevent spoilage.

The addition of acids to meat products, such as in pickled pigs feet and pickled sausage products, provides them with limited bacteriostatic properties. A decrease in pH by one unit increases the bacteriostatic effects about tenfold. Acetic acid (vinegar) is the most commonly used acid in meat preservation by pickling. The pH of meat products preserved by pickling is generally somewhat below the ultimate postmortem pH, and this excess degree of acidity essentially inhibits microbial activity.

Carbon dioxide and ozone gases have received a certain limited amount of use for the preservation of meat. The use of these gases in the holds of ships with prolonged shipping schedules selectively curtails aerobic growth, and favors the proliferation of anaerobic microorganisms. However, the use of these gases can have deleterious effects on meat quality. Carbon dioxide can cause discoloration in concentrations above 25 percent. The prolonged storage of sausages in carbon dioxide gas, at concentrations above 50 percent causes souring to occur, because the gas dissolves in the product, resulting in the formation of carbonic acid. Ozone enhances the development of rancidity and, at bacteriostatic levels, it causes loss of bloom in meat.

Antibiotics are potentially excellent preservative agents for enhancing shelf life, but their use in meat is not permitted in the United States. These restrictions by the Food and Drug Administration are based on the possibility of allergic sensitivity development by consumers, as well as the possible development of bacterial resistance to the antibiotic, causing it to become ineffective when used therapeutically.

PACKAGING REQUIREMENTS AND MATERIALS

The principal functions of meat product packages are to provide protection against damage, physical and chemical changes, further microbial contamination, and to attractively display the product to the consumer. Thus, packages are designed to maintain the quality of the product that is placed in it, but they can not improve that quality in any way.

The packaging requirements for fresh and cured meat differ primarily due to the chemical nature of the pigments that are present. However, they are also affected to some degree by the processing and merchandising methods used, and the nature of the microorganisms that limit shelf life. One of the most important considerations in meat merchandising is that of maintaining an optimum color. In fresh meat, the optimum color occurs when oxymyoglobin is the predominant form of the pigment. Thus, the packaging materials used for fresh meat display must allow a sufficient amount of oxygen to pass through them, in order to maintain a predominantly oxymyoglobin pigment. On the other hand, the packaging materials must be moisture proof in order to prevent product dehydration and surface discoloration. In contrast to fresh meat color, the retention of a stable color in cured meat depends upon the absence of oxygen. The exposure of cured meat products to light and oxygen causes the familiar color change known as light fading. Therefore, the packaging requirements for cured meat products necessitate the use of materials that are oxygen impermeable. In fact, some cured meat products are vacuum packaged, and others are placed in opaque packages with only a limited portion of the product visible through a "window" in the package, in order to reduce light fading. As in fresh meat, the packaging materials for cured meat products must be moisture proof.

The packaging materials for both fresh and cured meat must protect the product from further microbial contamination during all subsequent storage, handling, and merchandising. Packaging materials must not impart any odor or flavor to the product, and they should retain the natural flavors and odors that are inherent to the product. They should have sufficient tensile strength and resistance to tearing and scuffing to withstand normal handling. Packaging materials should also be grease proof so that meat fat will not be absorbed and reduce their strength or moisture vapor transmission properties.

The packaging requirements for frozen meat include low moisture vapor transmission, pliability, strength, and grease resistance. The materials used for heat shrink packaging of meat products should take the shape of the product and yet retain their strength and moisture vapor transmission characteristics. The overwrap films used in packaging meat for retail display should be strong, have good stretch and heat sealing properties, and retain the seal under normal storage and handling conditions. The packaging materials used for "boil in the bag" meat products must withstand the freezing conditions, freezer storage temperatures, and the rigorous temperature changes encountered during cooking.

A large number of materials are available for the packaging of meat products. These include glass and metal containers, aluminum foil, paper and paper board, cellophane, a large number of films manufactured from polyethylene, polypropylene, polyesters, nylon, polystyrene, polyvinyl chloride (PVC), saran, and chemically treated rubber (Pliofilm). Combinations of these materials, called laminates, are prepared by bonding together two or more papers, films, or foils to produce a packaging material with a wide variety of functional properties that depend upon the materials used in the laminate. In addition, many materials are available that can be applied as coatings to the basic packaging materials to improve their functional properties. It is not the intent of this discussion to describe all of the available packaging materials, and their application to meat products. However, the asterisked references at the end of this chapter will provide additional information.

REFERENCES

American Chemical Society, "Radiation Preservation of Foods," 1965 Symposium Advances in Chemistry, Washington, D.C., No. 65 (1967).

Brandly, P. J., G. Migaki and K. E. Taylor, *Meat Hygiene* (Lea Febiger, Philadelphia), 3rd ed., 1966, pp. 740–745.

Burke, R. F. and R. V. Decareau, "Recent Advances in the Freeze-Drying of Food Products," in *Advances in Food Research,* C. O. Chichester, E. M. Mrak and G. F. Stewart, eds. (Academic Press, New York), 1964, Vol. 13, pp. 1–88.

Fennema, D., "General Principles of Cryogenic Processing," Proceedings of the Meat Industry Research Conference, American Meat Institute Foundation, Chicago, pp. 109–118 (1968).

Fennema, D. and W. D. Powrie, "Fundamentals of Low Temperature Food Preservation," in *Advances in Food Research,* C. O. Chichester, E. M. Mrak and G. F. Stewart, eds. (Academic Press, New York), 1964, pp. 219–347.

Goldblith, S. A., M. A. Joslyn and J. T. R. Nickerson, *The Thermal Processing of Foods* (AVI Publishing Company, Westport, Conn.), 1961.

Luyet, B. J., "Physical Changes in Muscle During Freezing and Thawing," Proceedings of the Meat Industry Research Conference, American Meat Institute Foundation, Chicago, pp. 138–156 (1968).

Modern Packaging Encyclopedia (McGraw-Hill Book Company, Inc., New York), 1967, Vol. 40, No. 13A.

*Ramsbottom, John M., "Packaging," in *The Science of Meat and Meat Products*, J. F. Price and B. S. Schweigert, eds. (W. H. Freeman and Company, San Francisco), 2nd ed., 1971, pp. 513–537.

*Tressler, D. K., W. B. Van Arsdel, M. J. Copley, eds., *The Freezing Preservation of Foods* (AVI Publishing Company, Westport, Conn.), 4th ed., 1968, 4 Vols.

Urbain, W. M., "Meat Preservation," in *The Science of Meat and Meat Products*, J. F. Price and B. S. Schweigert, eds. (W. H. Freeman and Company, San Francisco), 2nd ed., 1971, pp. 402–451.

CHAPTER 12

Palatability and Cookery of Meat

People eat meat for reasons that include tradition, nutritive value, availability, wholesomeness, variety, satiety value, and social or religious custom. Of even greater consequence is the fact that meat is a delectable food that has become the central item of most meals in many countries.

The ultimate test of the value of meat is its degree of acceptability to the consumer. The extent to which satisfaction is derived from meat consumption depends on psychological and sensory responses that are unique to each individual. A consideration of such factors as appearance, purchase price, aroma during cooking, cooking losses, ease of preparation for serving, edible portion, tenderness, juiciness, flavor, and accepted nutritive value can govern the composite reaction of an individual, as measured by the *hedonic* (like or dislike) *scale.* Among individuals, there is a wide variation in the importance attributed to such factors.

PALATABILITY

The term *palatability* is only slightly more precise than the broad spectrum of acceptability factors listed above. The characteristics of meat that contribute to its palatability are those which are agreeable to the eyes, nose, and palate. Consequently, assessing palatability, or its absence, begins with the appearance of the meat item.

Appearance

Consumers expect raw meat to have an attractive color. With the exception of fish and some poultry, meat is normally some shade of red (Chapter 8). A dark color is often associated by the consumer with a lack of freshness, even though it usually indicates an old animal, or one that was slaughtered under stress. Such an impression reduces the expectation for (or prejudices the perception of) flavor when the meat is consumed. With respect to fat, the most desired color is a creamy white. Yellow fat, as is found in some grass fed cattle, is less appealing to most consumers, although it does not affect the palatability of the cooked product.

The color of the cooked product has impact on the meat consumer's enjoyment. The brown surfaces of meat cooked with dry heat are associated with crispness and a unique flavor. Golden brown pork roasts, or thick juicy broiled T-bone steaks, readily stimulate the salivary glands and thus are usually considered as being good even before they are tasted. The interior color of many cuts also influences palatability reactions depending on the consumer's preference for a rare (pink), medium (light pink to grey) or well done (uniform grey) product.

The textural properties of cooked meat affect its appearance and impart sensory impressions related to its adhesion, mealiness, or fragmentation. Overcooked meat that is stringy in appearance is associated, by previous experience, with dryness and lack of flavor. The texture of the cooked fat associated with meat cuts is a major appearance and palatability factor. Most consumers prefer that the external fat on meat cuts be browned and cooked to a degree of dryness. Many consumers who do not eat the adhering fat object to its texture, rather than to its flavor.

The ratio of muscle to bone and fat has a pronounced effect on the appearance and, consequently, on the palatability of meat. Less pleasure is derived from consuming a product that is considered to be excessively fat, both because of the dissection that is required and the reduction of the edible portion.

Tenderness

No palatability factor has received more research study than tenderness. Many experiments have been conducted to identify the constituents of muscle that are responsible for tenderness, yet much of the variation in this attribute cannot be explained. The problem is complicated by the facts that the sensation of tenderness itself has several components of varying importance, and the perception of tenderness by humans is very difficult to duplicate by scientific instrumentation. Therefore, it is difficult to describe the experience of eating tender meat in a few simple terms.

PERCEPTION OF TENDERNESS. The perception of tenderness has been described in terms of the following conditions of the meat during mastication.

(1) *Softness to tongue and cheek* is the tactile sensation resulting from contact of the meat with tongue and cheek. There is a wide variation in the softness of meat, ranging from a mushy to a woody consistency.

(2) *Resistance to tooth pressure* relates to the force needed to sink the teeth into the meat. Some meat samples can be so hard that little indentation can be made with the teeth, whereas other samples are so soft that they offer practically no resistance to such force.

(3) *Ease of fragmentation* is an expression of the ability of the teeth to cut across the fibers. Rupture of the sarcolemmae is required in order to accomplish fragmentation.

(4) *Mealiness* is an exaggerated type of fragmentation. Small particles cling to the tongue, gums, and cheeks and give the sensation of dryness. This condition apparently exists when the fibers fragment too easily.

(5) *Adhesion* denotes the degree to which the fibers are held together. The strength of the connective tissues surrounding the fibers and bundles influences this characteristic.

(6) *Residue after chewing* is detected as that connective tissue remaining after most of the sample has been masticated. The coarse strands of connective tissue in the perimysium or epimysium are responsible for this component.

The major components of meat that contribute to tenderness, or the lack of it, can be conveniently divided into three groups. These are the connective tissues, the muscle fibers, and the lipids associated with muscle tissue. However, meat lipids are of only minor importance in this respect.

CONNECTIVE TISSUE. Much of the muscle to muscle variation in tenderness that exists within an animal is due to differences in the amount and nature of the connective tissues. Estimates of the amount of collagen (white, fibrous connective tissue protein) in meat cuts generally indicate that a low connective tissue content is associated with tenderness. Although most of the connective tissue fibers in the muscles are collagenous, elastin and reticulin fibers are also present and may contribute to toughness.

The connective tissue content of a muscle is probably a reflection of the functional demands placed on it during life. For example, the powerful muscles of the legs must develop more strength than some of the muscles surrounding the backbone. The supporting and connecting tissues must therefore undergo extensive development. The quantity of collagenous connective tissue fibers is great in such muscle, and the chemical crosslinking (Chapter 3) in these fibers is extensive. The result is a tough weblike structure surrounding the muscle fibers (endomysium), bundles (perimysium), and the muscle (epimysium) itself.

The decreases in tenderness that accompanies an animal's aging are believed to result largely from connective tissue changes. It is likely that the additional exercise that old animals have experienced causes a strenthening of the connective tissue fiber structure. Although the quantity of connective tissue apparently changes very little after maturity, the number of intermolecular crosslinks in the collagen fibrils probably increases. This results in a decreased solubility of the collagen, and an increased resistance to shearing or chewing action.

Although the muscles of the very young animal are much more tender than those of the aged animal, the changes that occur as the animal ages are not linear with increasing age. During the rapid phase of growth, tenderness seems to actually increase with time, in some animals. It is likely that the rapid development of muscle fiber size "dilutes" the existing connective tissue. Thus, market weight beef animals (12–18 months of age) often have more tender meat than the growing calf (6 months of age).

A substantial muscle toughening in beef animals becomes evident at about 30 months of age. Beyond this time there is a further gradual toughening, but the rate of change becomes progressively slower with advanced age. Similar patterns of meat toughening occur in animals of other species during their growth and development, but the various stages occur at different ages.

Although the above discussion of age associated changes in tenderness has included reference to chronological age, it should be emphasized that the changes that occur reflect the physiological environment within the animal. Since all animals of a species, and even all the muscles within an individual animal, do not age at the same rate, the indicators of physiological condition in the animal body are generally recognized as being better predictors of tenderness than is the chronological age. These indicators, located principally in the skeleton, are described in Chapter 15.

The action of many methods of meat tenderization is on the connective tissues. The use of weak acids, such as vinegar or lemon juice, is a traditional method for overcoming connective tissue toughness. These *marinades* promote the swelling of collagen, which requires some disruption of hydrogen bonds within the collagen fibril.

Enzyme tenderizers of plant origin are also used on meat to reduce connective tissue toughness. These are proteolytic enzymes, the most commonly used of which is *papain*. In general, they degrade several tissue proteins, including collagen and elastin. They are sometimes introduced into the tissue by infusion into the circulatory system, either before or after slaughter, or by a direct application to the surface of meat cuts. As the temperature rises during cooking, the enzymes are activated. However, when high temperatures are reached, they become denatured and lose their activity.

The tenderizing effects of mechanical treatments (such as pounding, chopping, grinding, or cubing) depend for their effectiveness on the mechanical destruction of the structure of connective tissues, and of the muscle fibers as well.

MUSCLE FIBER CHARACTERISTICS. Another factor that determines the degree of tenderness of a muscle is its postrigor contraction state, a condition that is controlled in part by the degree of tension on the muscle during rigor onset. This relationship is illustrated by differences in tenderness that exist within some muscles. For instance, there are marked differences in the tenderness of various parts of the longissimus muscle. Generally speaking, the anterior and posterior ends of this muscle are more tender than the center portion; and, in transverse section, the medial and lateral portions are more tender than the center portion. These tenderness differences are probably related to the degree of local tension, but the relative importance of contraction state and connective tissue stretching is unknown. Figure 12-1 shows the relation of sarcomere length and tenderness (resistance to shear) in the longissimus of beef. The inverse relationship between sarcomere length and resistance to shear holds closely at all cooking temperatures. However, as the discussion later in this chapter indicates, further toughening of the contractile protein occurs only at high internal meat temperatures. Consequently, the close association of sarcomere length and tenderness at relatively low cooking temperatures could indicate that other proteins are also involved in the muscle tension/tenderness relationship.

Certain muscles in the carcass that are free to shorten during rigor mortis onset often are lacking in tenderness. Table 12-1 shows the tenderness of some beef muscles as determined both by their resistance to mechanical shearing, and by palatability tests. These values are representative of the muscle to muscle differences that would be expected in beef carcasses handled in a conventional way after slaughter. However, it is possible to change these relationships by hanging or placing the carcass in a different position during the rigor process (Chapter 7). The degree of tension placed on the individual muscles by the skeletal system, and the temperature achieved prerigor, influences

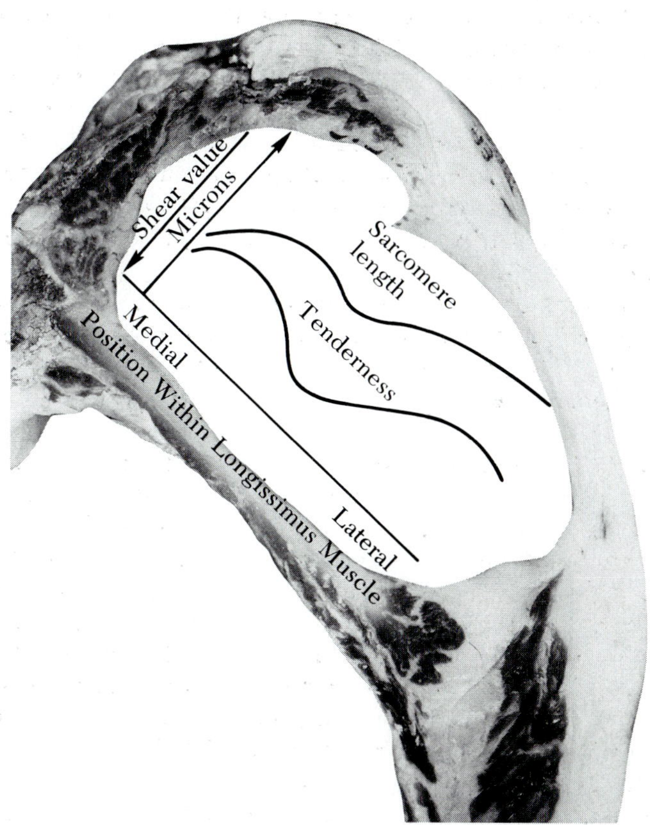

FIGURE 12-1
Some of the variation in tenderness across the *longissimus* muscle can be explained by differences in sarcomere length. Long sarcomeres are associated with tender meat (low shear values). [Summarized from Howard, R. D. and M. D. Judge, "Comparison of Sarcomere Length to Other Predictors of Beef Tenderness," J. Food Sci., 33, 456 (1968).]

their final sarcomere length and ultimate tenderness. (Nevertheless, some of the differences in tenderness between muscles would persist under any postmortem handling conditions because of their differing quantities of connective tissue.) These muscle to muscle differences are the basis for the variations in tenderness often noted within a single cut of meat. The normal differences in contraction state, quantity of connective tissue, and cutting angle through the difference muscles contained within any single cut all contribute to variability in eating quality. Consequently, a consumer's first impression of a meat cut can depend greatly on the particular muscle originally tested.

TABLE 12-1
Resistance to shear, and the tenderness rating of selected muscles of cooked beef

Muscle	Shear*, kg	Tenderness
Rectus femoris	4.3	Slightly tough
Semimembranosus	5.4	Slightly tough
Semitendinosus	5.0	Slightly tough
Biceps femoris	4.1	Medium
Gluteus medius	3.7	Slightly tender
Psoas major	3.2	Very tender
Longissimus	3.8	Slightly tender
Infraspinatus	3.6	Slightly tender
Supraspinatus	4.2	Medium
Triceps brachii	3.9	Medium
Trapezius	6.4	Very tough
Sternocephalicus	7.1	Very tough

SOURCE: Summarized from the data of Ramsbottom and Strandine (1948).
*Warner-Bratzler shear, 1.27 cm diameter cylinders. Low values indicate tender meat.

In spite of the fact that some muscle toughening can be prevented by maintaining tension during rigor onset, the formation of crossbridges is a toughening process that occurs even in the absence of shortening. The loss of tenderness that occurs in the first few hours postmortem has been referred to as *actomyosin toughening.*

Water holding capacity and tenderness are minimal after the muscles have undergone the rigor associated changes just described. Thus, tenderness and water holding capacity are well correlated. The characteristic toughness found immediately after the completion of rigor mortis may be partially due to the fact that the water is free to escape from the cell during cooking. According to this theory, cells shrunken in this way do not rupture easily when masticated, and an impression of toughness is given.

The integrity of the structural components of the muscle fiber influences tenderness but the relative importance of natural cellular disruption to tenderness is uncertain. The subtle changes that occur naturally within the tissues can have profound effects on tenderness. The release of cathepsins, or other proteolytic enzymes, enables the degradation of muscle proteins, including the contractile elements of the fiber to occur. Although it is unlikely that this enzyme degradation is extensive, the changes that result, either in membrane permeability or in its general structural integrity, could permit the redistribution of ions, and shifts in water binding activity. If these shifts were to occur, much of the tenderization accompanying the *aging* process might be attributable to cationic shifts and the resultant improvements in the ability of the proteins to retain water (Chapter 8).

Among the structural changes that occur during aging, which probably relate directly to tenderness, are those taking place within the myofibril. Progressive degradation of the Z line, and some dissociation of the actinmyosin complex occur in muscles with increasing time postmortem. These changes are accompanied by a fragmentation of the filamentous structure, and lengthening of the sarcomeres. However, it is possible that these structural alterations are caused by the enzyme activity discussed above.

The tenderizing process that accompanies aging may be speeded up by raising the air temperature in carcass holding rooms to 18–20°C. The carcasses warm to the room temperature in less than 24 hours, and there are resultant increases in the rate of protein hydration, and associated reactions.

INTRAMUSCULAR LIPIDS. The intramuscular lipids have been credited by meat industry personnel with making meat tender. However, little research can be cited to show a strong positive influence of the lipid component of the tissue (marbling) on tenderness. It is likely that some lipid acts as a lubricant in the mastication of less tender meat, thus improving the apparent tenderness, and easing the process of swallowing. However, the animals that are normally slaughtered for retail meat cuts are youthful and tender per se, making any tenderness due to marbling relatively less important.

Juiciness

The meat juices play an important role in conveying the overall impression of palatability to the consumer. They contain many of the important flavor components, and assist in the process of fragmenting and softening the meat during chewing. Regardless of the other virtues of meat, the absence of juiciness severely limits its acceptability, and destroys its unique palatability characteristics.

The principal sources of juiciness in meat, as detected by the consumer, are the intramuscular lipids and water content. In combination with water, the melted lipid constitutes a broth that, when retained in the meat, is released upon chewing. This broth may also stimulate the flow of saliva, and thus improve the meat's apparent juiciness.

The marbling that is present also serves to enhance juiciness in an indirect way. During the cooking process the melted fat apparently becomes *translocated* along the bands of perimysial connective tissue.

This uniform distribution of lipid throughout the muscle may act as a barrier to moisture loss during cooking. Consequently, meat with some marbling shrinks less during cooking and remains juicier. Subcutaneous fat also minimizes drying and moisture loss during dry heat roasting.

In spite of the influences just described, the major contributor to the sensation of juiciness is the water remaining in the cooked product. Since the fat free water content of meat is relatively uniform, differences in juiciness must relate to the ability of the muscle to retain its water during cooking.

Flavor and Aroma

Many of the psychological and physiological responses experienced when meat is being eaten are elicited by the flavor and aroma of the product. The meaty flavor and aroma stimulate the flow of saliva and gastric juices, thus aiding in the process of digestion.

The flavor and aroma sensations that are produced result from a combination of factors that are difficult to separate. Physiologically, the perception of flavor involves the detection of four basic sensations (salty, sweet, sour, and bitter) by the nerve endings on the surface of the tongue. Aroma is detected when numerous volatile materials stimulate the nerve endings in the lining of the nasal passages. The total sensation is a combination of gustatory (taste) and olfactory (smell) stimulations.

The components of meat that are responsible for flavor and aroma have not been completely identified. It is likely that many constituents of the tissue become flavor compounds upon being heated. Some evidence shows that inosine monophosphate (IMP) and hypoxanthine enhance flavor or aroma. Since IMP and hypoxanthine are breakdown products of ATP, it is obvious that muscles with large energy stores would have a more pronounced flavor. This may be partially responsible for the intensity of flavor found in the frequently used muscles of the carcass, and the strong flavor of some game animals.

The most generally accepted view is that most of the constituents of meat responsible for the "meaty" flavor are water soluble components of the muscle tissue. Species flavor and aroma are thought to arise from some materials in the fat that become volatilized upon heating. Further research will be required to determine the validity of these concepts.

Changes in flavor and aroma may take place in meat as a result of several factors. Of particular importance is the length and conditions of storage. Flavor changes occur after extended periods of storage due to chemical breakdown of certain constituents, escape of volatile materials, oxidation of certain components, and microbial growth.

Some flavor changes that occur during storage are considered desirable and others are undesirable. The aging of meat results in a combination of changes that is appreciated by many consumers. Such changes probably include the destruction of the mononucleotides adenosine monophosphate (AMP) and IMP, and the production of some flavor compounds by microbes, and most particularly by yeasts and (in long term aging) molds.

Undesirable flavors can develop during storage, due to oxidative changes in the fat. *Rancidity* of fat will occur when the fatty acid chains are broken at points of unsaturation (double bonds) by the chemical addition of oxygen. The formation of carbonyls, particularly low molecular weight volatile aldehydes, is directly responsible for the rancid flavor and sharp aroma.

The end products of microbial growth can produce undesirable flavor if they are present in large quantities. An undesirable flavor usually results from the growth of aerobic organisms that are confined to the free surfaces of the meat. Putrid odors develop in the deep portions of the meat where anaerobic organisms can grow. The condition known as *bone sour* occurs when incomplete chilling or inadequate cooking rates permit such organisms to grow and break down the muscle proteins.

An objectionable onionlike or perspiratory odor is sometimes noted in pork, and is referred to as a *sex odor*. This odor becomes evident upon heating, and although it is noted with special frequency in the meat of boars, it is sometimes identifiable by some consumers in pork of all sex conditions. Other consumers are relatively insensitive to it. Tentative identification of the material responsible indicates that it is 5-α-androst-16-ene-3-one, a metabolite of the hormone testosterone.

Several other types of undesirable flavor have been noted. Examples are the undesirable flavors produced by consumption of certain weeds, such as wild onion, by the animal prior to slaughter, the previously described flavor of some irradiated meats, the strong mutton flavor of old sheep, and the absorbed flavors of meat products that have been stored with other volatile materials. However, it is important to recognize that flavors objectionable to one consumer might be enjoyed and appreciated by another. The variation in the individual preferences of meat consumers precludes the formulation of universal flavor standards for meat.

Palatability Interrelationships

The experience of eating meat does not consist of separate impressions of the individual palatability factors of tenderness, juiciness, and flavor. Rather, the consumer gets an overall impression of satisfaction or dissatisfaction, unless there is one particularly outstanding characteristic. Consequently, a product that is juicy and flavorful might seem more tender than shearing tests would verify. Palatability tests of meat by laboratory or consumer taste panels usually show high correlations among the various factors that are evaluated. These close relationships are partially due to the psychological factors mentioned previously. However, some tissue properties, such as water binding capacity or the level of energy metabolites, can affect more than one palatability attribute, and can thus provide a basis for palatability interrelationships.

COOKERY

The art of meat cookery exists because of the sharing of culinary experience from generation to generation. Apprentice cooks learn that certain principles of meat preparation must be observed, in order to increase the likelihood of high palatability in the finished product. These principles include knowledge of the time and temperature combinations that will assure the preparation of meat having the maximum eating satisfaction. Knowledge about various characteristics of a cut of meat and its probable response to heat is also vital to successful meat cookery.

Effects of Heat on Meat Constituents

When the proteins of muscle are exposed to heat, they lose their native structure and undergo several changes in configuration. In general, even though specific changes are unique to each protein, *denaturation* (alteration of structure due to non-proteolytic changes) of the protein occurs. This may be accompanied or followed by an aggregation, or clumping, of the protein molecules (*coagulation*), the presence of which indicates a loss in protein solubility.

The coagulation of the myofibrilar proteins is associated with easily observed changes in the gross physical character of the tissues. Even though the microscopic striated appearance of the muscle fibers persists after heating, in well cooked meat, an increased rigidity typically occurs. This phenomenon is sometimes referred to as *protein hardening*. The temperatures associated with various phases of protein hardening are incompletely identified, but some research indicates that it does not occur below approximately 64°C. Heating the myofibrillar proteins above this temperature renders them less tender.

The amount of protein solubility that is lost depends on the time and temperature of heating. These changes are measurable in terms of water holding capacity. Figure 12-2 illustrates that high cooking temperatures reduce water holding capacity, and that the duration of heating is important at temperatures between 30–70°C.

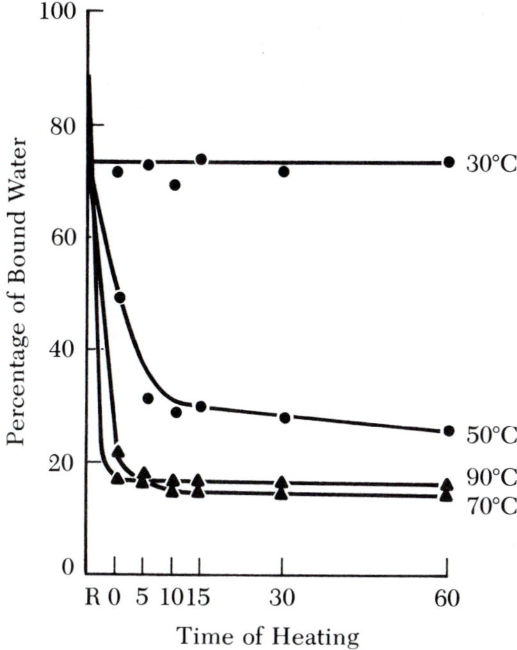

FIGURE 12-2
Influence of time and temperature of heating on the waterholding capacity of beef muscle. R—0 is the decrease of waterholding capacity during heating from 20°C up to the given temperature. [From Hamm, R., Z. Lebensm. Untersuch. Forsch., 116, p. 120 (1962)]

These changes occur in most of the proteins of the muscle fiber, upon heating. On the other hand, when the collagen of the surrounding connective tissue is heated, it undergoes some physical changes that cause increased solubility. The first change that takes place is a physical shortening of the collagen fibril to one-third of its original length. This can occur at a temperature as low as 56°C in some fibrils, and is usually complete in half the muscle collagen fibrils at 61–62°C. Called *collagen shrinkage,* this change is accompanied by an increased collagen solubility. Upon heating it further for an extended time in the presence of moisture, the collagen is further hydrated and hydrolyzed, and forms *gelatin.* Thus, collagen becomes more tender with heating, and its water binding capacity increases.

The connective tissue protein elastin is not susceptible to the effects of heat. Even though properly cooked, some meat will have a persistent toughness because of a high elastin content. The only feasible method of tenderizing such products is by use of the plant proteolytic enzymes that were previously discussed.

Several other alterations occur when the muscle proteins are subjected to heat. These include changes in pH, reducing activity, ion binding properties, and enzyme activity. A slight upward shift in pH (approximately 0.3 unit) is believed to result from the exposure of a reactive group on the amino acid histidine. The increased reducing activity develops as a result of the unfolding of protein chains and the exposure of sulfhydryl groups. Likewise, the conformational changes in the proteins alter and usually reduce their power to bind various ions, such as Mg^{2+} and Ca^{2+}. With regard to enzyme activity, heat is an inactivator, but varying degrees of heat resistance are shown by muscle enzymes.

Certain volatile materials are driven off during the heating of meat that contribute to the unique flavor and aroma of cooked meat. These include various sulfur and nitrogen containing compounds, as well as certain hydrocarbons, aldehydes, ketones, alcohols, and acids.

The amine groups of muscle proteins react with any available reducing sugars, such as free glucose, in the tissue. This sugar–amine browning starts occurring at high (approximately 90°C) temperatures, such as those found at the surface of a cut during roasting or broiling.

The process of *fat translocation* in cooked meat was discussed earlier in this chapter in relation to palatability. Although the extent of this migration of fat is uncertain, the action is brought about by raising the muscle temperature. The solubilization of collagenous connective tissue provides channels through which the melted fat may diffuse. Thus, the cooking action results in a movement, and possibly an emulsification, of the fat with soluble protein.

Effects of Heat that Are Associated With Palatability

Heat can cause both the tenderization *and* the toughening of meat. In general, those heat induced changes in proteins that result in coagulation and hardening, reduce tenderness. Conversely, those changes that result in greater solubilization increase tenderness. Specifically, the myofibrilar proteins react to heat by toughening, and the collagenous connective tissue proteins break down upon being heated.

Figure 12-3 shows the degree of tenderness that might be observed after heating beef at three cooking temperatures for various periods of time. At all three temperatures, an initial reduction in shear resistance occurs because of the previously discussed *collagen shrinkage*. At the

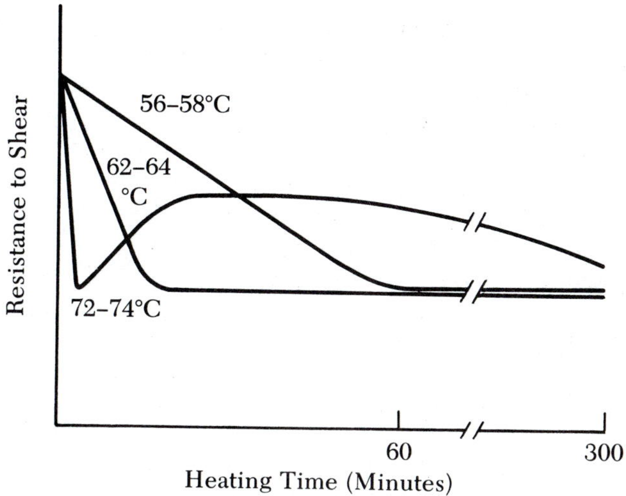

FIGURE 12-3
Effect of heating temperature and time on the tenderness of 1.27 cm diameter cylinders of beef (*semitendinosus*) muscle. [Summarized from Machlik, S. M. and H. N. Draudt, "The Effect of Heating Time and Temperature on the Shear of Beef *Semitendinosus* Muscle," J. Food Sci., 28, 711 (1963).]

very low cooking temperature of 56–58°C, the tenderization is relatively slow. Upon raising the temperature slightly to 62–64°C, the shrinkage reaction occurs more quickly, and some improvement in tenderness occurs upon prolonged heating. At the higher temperature of 72–74°C, the rapid shrinkage of collagen is followed by *protein hardening* and toughening. However, continued heating at this temperature results in a substantial amount of *gelatin formation* and eventually, this change causes meat tenderization.

Practical guides to meat cookery can be developed on the basis of time and temperature influences on tenderness. For example, the unnecessary toughening of most cuts of meat can be avoided by preventing the internal temperature from rising to a level that causes protein hardening. However, it is necessary to cook some meats in this temperature range, in order to develop other desirable palatability characteristics. To achieve the best flavor development, and a complete conversion of pigment to the denatured (brownish-grey) form, the recommended end point for most fresh pork is 77°C, and for poultry is 77–82°C. Other meats, notably beef, are cooked to a degree that is consistent with the preferences of the consumer. Meat heated to an internal temperature of 58–60°C is considered rare; 66–68°C, medium rare; 73–75°C, medium; and 80–82°C, well done.

The end color of the cooked meat is a function of the combination of cooking times and temperatures. The sugar–amine browning discussed previously is confined to the surface of dry heated cuts. Yet, the interior of the cut may vary in color from red to grey. This is due to the fact that the internal temperatures in the "rare" cooking range are not sufficient to denature the myoglobin, whereas those coinciding with "medium" and "well done" cause progressive increases in pigment denaturation. Consequently the extent of heating is an important influence on consumer acceptability through its effect on the chemical state of the muscle pigments.

Obviously, severe cooking procedures that cause extensive dehydration of the meat also render it less juicy to the consumer. Meat that contains 68–75 percent moisture in the raw state, will contain about 70, 65, and 60 percent moisture after being dry-heat roasted to 60, 70, and 80°C, respectively. The losses that occur are due to evaporation and drip losses. Of course, the extent of these losses also depends on the water holding capacity of the tissues, and the degree of protection afforded by lipid translocation (as discussed earlier in this chapter) and subcutaneous fat.

Flavor development is an important result of the meat cooking process. The changes in the quantity and type of volatile materials that are present are extensive, but poorly understood. Yet, the effects of heat are so unique that the type and conditions of cookery often may be identified solely by the flavor and aroma of the product. Dry heat cookery imparts certain flavors, particularly at the exposed surfaces where temperatures become very high. On the other hand, cooking with moisture under pressure causes the development of pronounced and unique flavor changes in the deep tissues of the cut.

Methods of Cookery

There are several systems by which temperature increases are accomplished in meat. Although the major objective in meat cookery is to achieve a particular internal temperature, the rate of heating, the equipment used, and many other factors influence the characteristics of the final product.

Of particular importance in determining the cooked character of meat is the amount of moisture present during heating. Since all meat contains water, there will be some moisture effects. Water is a good conductor of heat, and its presence aids in the penetration of heat into the deepest parts of the cut. On the other hand, the moist surfaces of meat can also delay the heating process, due to the evaporative cooling that takes place. Water is also necessary for developing the tenderness and final texture of cooked meat because of the hydrolysis of connective tissue that occurs.

One of the essentials for success in meat cookery is a knowledge of the proper duration of heating. For thin cuts, such as steaks or chops, the criteria are subjective. The experienced cook can determine the proper end point from the color and rigidity of the cut. However, the proper cooking of very thick cuts, such as roasts, requires the use of a meat thermometer. The thermometer is inserted into the thickest part of the cut, avoiding pockets of fat and bones, so that the temperature in the coolest region will be detected. For thick cuts, this is the only method that will completely assure attaining the desired degree of doneness.

DRY HEAT COOKERY. Cooking with dry heat can be accomplished by any method that surrounds the cut of meat with hot dry air. Broiling and roasting are the best examples of dry heat cookery.

Broiling is appropriate only for tender cuts, such as steaks or chops, because the heating period is usually of rather short duration and there is inadequate time to achieve connective tissue breakdown. The high surface temperature results in the development of a unique flavor in the cut, and of an extensive browning on it. The duration of the heating is extremely critical, since the cuts are relatively thin and the temperatures used are quite high.

Charcoal broiling is a very popular method of cooking that also imparts a unique flavor. Many products, such as chops, steaks, chicken, ribs, kabobs, sausage, or roasts, are cooked by this method. The temperatures used are usually lower than those of oven broiling. The products acquire a pronounced smoked flavor from the combustion of the charcoal and the melted fat that drips on it.

Roasting with dry heat is appropriate for tender roasts. It is usually accomplished in an oven at temperatures of 150–175°C. It imparts unique flavor by the sugar–amine browning action mentioned earlier. However, the cut should be protected during roasting by a layer of external fat to prevent excessive moisture losses. If the cut is a large one, such as the intact rounds that are sometimes cooked in restaurants, it is possible to use relatively low (120°C) oven temperatures for an extended period of cooking. The limited surface area per unit of meat in these large cuts prevents extensive moisture losses.

MOIST HEAT COOKERY. If a meat cut contains a relatively large amount of connective tissue, it is desirable to provide extra water during cooking. This provides all the water that is necessary for the complete hydrolysis of collagen into gelatin. Low temperatures are prescribed, over a long cooking time, to allow this conversion to occur without hardening the myofibrillar proteins. It is emphasized that this action occurs in collagen, but does not occur in elastin. Thus, a cut that does not tenderize, in spite of extensive cooking, is probably high in elastin connective tissue fibers.

Braising, cooking in water, or pot roasting are examples of moist heat cookery. The heating is accomplished in a closed container with added water. Seasoning, sauces, or flour may also be added to enhance the development of a desired flavor and texture in the final product.

The tenderizing action in moist heat cookery may be achieved by wrapping cuts in moisture proof materials and heating them in a dry oven. The natural juices are trapped and moisture loss is minimized. Consequently, the heating may be extended for long periods of time to allow the collagen to change into gelatin. The heating temperatures are usually in the range of 95–100°C for this method.

MICROWAVE COOKERY. The use of microwaves for heating meat represents a modern method of cookery that is extremely rapid. The heating results from the conversion of microwave energy to heat by friction from internal molecular rotations caused by the interactions of molecules with the rapidly fluctuating electromagnetic field. Frequencies of 915 and 2450 megahertz (millions of cycles per second) have been approved in the United States for such purposes. This method permits meat cookery that is many times faster than conventional methods, although some variation exists in doneness and textural properties. Some of these are alleviated when external heating coils are used to brown the outer surface of the meat, while the interior is being cooked by the microwave radiation. Its utilization by the food service industry represents the most widespread application to meat cookery. However, it is also gaining popularity as a cooking method in the home.

REFERENCES

Cover, S. and R. L. Hostetler, "Beef Tenderness by New Methods," Texas Agr. Expt. Sta. Bull. 947 (1960).

Hamm, R., "Heating of Muscle Systems," in *The Physiology and Biochemistry of Muscle as a Food*, E. J. Briskey et al., eds. (University of Wisconsin Press, Madison), 1966, p. 363.

Hornstein, I., P. F. Crowe and W. L. Sulzbacher, "Constituents of Meat Flavor: Beef," J. Agric. Food Chem., 8, 65 (1960).

Howard, R. D. and M. D. Judge, "Comparison of Sarcomere Length to Other Predictors of Beef Tenderness," J. Food Sci., 33, 456 (1968).

Jones, N. R., "Meat and Fish Flavors. Significance of Ribomononucleotides and Their Metabolites," Agr. Food Chem., 17, 712 (1969).

Machlik, S. M. and H. N. Draudt, "The Effect of Heating Time and Temperature on the Shear of Beef Semitendinosus Muscle," J. Food Sci., 28, 711 (1963).

Martin, A. H., "The Problem of Sex Taint in Pork in Relation to the Growth and Carcass Characteristics of Boars and Barrows: A Review," Can. J. Animal Sci., 49, 1 (1969).

Paul, P. C. and H. H. Palmer, *Food Theory and Applications* (John Wiley & Sons, Inc., New York), 1972.

Ramsbottom, J. M. and E. J. Strandine, "Comparative Tenderness and Identification of Muscles in Wholesale Beef Cuts," Food Res., 13, 315 (1948).

Sanderson, M. and G. E. Vail, "Fluid Content and Tenderness of Three Muscles of Beef Cooked to Three Internal Temperatures," J. Food Sci., 28, 590 (1963).

Wang, H., E. Rasch, V. Bates, F. J. Beard, J. C. Pierce and O. G. Hankins, "Histological Observations on Fat Loci and Distribution in Cooked Beef." Food Res., 19, 314 (1954).

CHAPTER 13

Nutritive Value

The nutritive value of meat is attributed to its proteins, fats, carbohydrates, vitamins, and minerals. Whereas meat does provide calories from the proteins, fats, and the limited quantities of carbohydrates that are present, its more vital contributions to the diet are derived from the high quantity and quality of its protein, the available supply of B vitamins and certain minerals, and the presence of essential fatty acids.

PROTEINS

Meat proteins are largely those of the muscle and connective tissues. The largest proportion of total muscle proteins are those of the myofibrils. The sarcoplasmic proteins, consisting of muscle enzymes and myoglobin, make up the next largest fraction. This is followed in abundance by the connective tissue proteins, consisting largely of collagen and some elastin. Although raw muscle contains approximately 18–22 percent protein, this content is quite variable in many meat products, and varies inversely with the amount of fat that is present. Nevertheless, meat products generally supply a major portion of the recommended dietary allowance (RDA) of protein. As prescribed by the Food and Nutrition Board of the National Research Council, the RDA for a grown man, for example, is 56 grams per day. Large amounts of protein cannot be stored in the body, so it is essential that protein be consumed every day. Since a normal sized serving of cooked meat is approximately 100 grams (3.5 ounces), and its protein content is 25–30 percent, this serving furnishes 25–30 grams of protein, or about 45–55 percent of the RDA.

On a per capita basis, Americans eat approximately 4 ounces of cooked, edible meat, poultry, or fish each day.

In addition to its protein content, skeletal meat provides a high quality protein with a high biological value. A *high quality protein* is one that contains all of the essential amino acids, in amounts that are equivalent to the requirements of the human body, is highly digestible, and is easily absorbable. *Amino acids* are the basic building blocks of which all proteins are composed, and the *essential amino acids* are those that cannot be synthesized by the body in amounts sufficient to meet its requirements. Adult humans need eight essential amino acids. These are phenylalanine, valine, tryptophan, threonine, methionine, leucine, isoleucine, and lysine. Most animal products such as meat, milk, and eggs have high biological value. On the other hand, most individual vegetable proteins and the connective tissues of animals are somewhat low in biological value. The high amounts of the nonessential amino acids glycine, proline, and hydroxyproline in collagen are responsible for the lower biological value of some processed meats that contain connective tissue as the primary source of protein.

In addition to proteins, meat also contains some nonprotein nitrogenous compounds, such as the free amino acids, simple peptides, amines, amides, and creatine. Although these compounds do not contribute significantly to nutritive value, they are a potential source of nitrogen that can be used for amino acid and protein synthesis.

LIPIDS

The lipid content of meat is generally the most variable component. The amount of lipid depends upon the cut of meat, and the amount of fat that is left after cutting and trimming. The lipid components that are a major concern, from a nutritional standpoint, are the triglycerides, phospholipids and cholesterol, and limited quantities of fat soluble vitamins. The caloric value of the lipids in meat is derived from the fatty acids in triglycerides and phospholipids, of which the triglycerides constitute the bulk.

The fatty acids constituting the triglycerides of meat are relatively saturated, particularly when compared to the vegetable fats. Many of the latter are oils, since they contain a high proportion of unsaturated fatty acids. In meat fat, the most abundant fatty acid is unsaturated oleic acid, with one double bond. However, the other fatty acids that are present in high proportions are saturated, and include palmitic and stearic acids (Table 3-3). Thus, meat fats (or fats of animal origin) are generally described as saturated fats, and vegetable fats are labeled as either unsaturated, or polyunsaturated fats (more than one double bond per fatty acid molecule).

The American Medical Association and American Heart Association have publicly linked a high rate of saturated fat and cholesterol consumption as contributing factors in cardiovascular diseases. The processors of unsaturated and polyunsaturated fats (such as margarines and cooking oils) have promoted these allegations, and the sales volume of vegetable fats has increased at the expense of that of animal fats. The publicity has adversely affected the consumption of animal products, especially those that are considered to be high in saturated fats. Among the products so affected is pork, which almost always is removed from the diet by physicians when a patient demonstrates symptoms that will eventually lead to cardiovascular disease. Recent nutrition studies have shown that, except for bacon, cooked pork has no more calories derived from fat per serving than do corresponding cuts of beef or lamb.

The results of several research studies have also indicated that, in general, diet *per se* is not related *directly* to the incidence of cardiovascular disease. However, it does follow logically that high caloric intake is related to obesity. In turn, obesity, stress, and relative inactivity are factors which have been shown to be directly related to the occurrence of cardiovascular diseases. Thus, discretion should be exercised in the amount of fat eaten, whether of animal or vegetable origin, in order to curtail caloric intake. If, while a person is eating meat, large fat deposits are trimmed away and discarded, the number of calories derived from the meat will be drastically reduced.

Meat fats contain variable quantities of cholesterol. Even though blood cholesterol values rise following the ingestion of cholesterol in foods, it should be pointed out that the body is capable of synthesizing more cholesterol than is normally ingested. The cholesterol controversy, like the saturated *vs* unsaturated fat question, is unresolved at the time of writing.

Meat fats contain ample quantities of the fatty acids that are essential in the diet of humans. Since the daily need of these essential fatty acids is relatively small, the RDA is easily met from intramuscular fat, even when most of the external fat is cut away. The fatty acids that are known to be essential are linoleic and arachidonic. It is not definitely known yet whether linolenic acid is also essential in the diet.

CARBOHYDRATES

Carbohydrates constitute less than 1 percent of the weight of meat, most of which is present in the form of glycogen and lactic acid. Since the liver is the principal storage site for glycogen, most of the carbohydrate in the animal body is present in the liver. Thus, most meats are poor sources of carbohydrates, except those products (such as cured meats) to which sugars or carbohydrate materials have been added.

MINERALS

Meat is generally a good source of all minerals except for calcium. Most of the calcium in the body is present in bones and teeth, and the small quantity present in muscle and the other edible tissues is considerably below the RDA. The minerals in meat are associated with the lean tissues. (The approximate mineral composition of muscle is presented in Table 3-4.)

Meat is an especially good source of iron, a nutrient that is essential for maintaining good health. Iron is required for the synthesis of hemoglobin, myoglobin, and certain enzymes. Since little iron is stored in the body, a regular intake of dietary iron is important, and meat provides it in a form that is easily absorbed.

VITAMINS

Meat is generally an excellent source of the water soluble B complex group but, it is a poor source of the water soluble vitamin C, and of the fat soluble vitamins A, D, E, and K that are found primarily in the body fat. All of the B complex vitamins are present in meat, but thiamine, riboflavin, and niacin are present in the highest quantities. Pork contains higher levels of the B complex vitamins than beef, veal, lamb, fish, or poultry. In fact, the lean portion of pork is 8–10 times higher in thiamine than other meats, as well as being slightly higher in riboflavin, pyridoxine, pantothenic acid, and biotin content. Pork contains approximately the same amount of niacin and slightly less vitamin B_{12} than other meats.

All meat is a very poor source of the water soluble vitamin C, except when ascorbate has been added to processed meat products.

MEAT IN THE DIET

From the preceding discussion, it is apparent that meat is a very nutritious food and supplies much or all of the daily human needs for many nutrients. The approximate composition and caloric content of the lean portion of retail cuts of beef, pork, lamb, veal, chicken, and halibut are shown in Table 13-1. The composition includes protein, moisture, fat, and ash (minerals). Collectively, the carbohydrates, vitamins, and non-protein nitrogenous compounds constitute less than one percent of the total composition, so they are not presented in the table.

The caloric value of meat depends upon the amount of fat in and on meat cuts that is actually eaten. The number of calories from lean meat is frequently less than that derived from equal weights of many other

TABLE 13-1
Approximate composition and caloric content of lean portions of raw and cooked retail cuts of beef, chicken, halibut, lamb, pork, and veal

Meat source* and physical state	Percent				Calories per 100 grams
	Protein	Moisture	Fat	Ash	
Beef					
Raw	21.5	69.5	8.0	1.0	160
Cooked	30.0	58.0	10.0	1.4	230
Chicken, dark meat					
Raw	20.6	73.7	4.7	1.0	130
Cooked	28.0	64.4	6.3	1.2	176
Chicken, light meat					
Raw	23.4	73.7	1.9	1.0	117
Cooked	31.6	63.8	3.4	1.2	166
Halibut					
Raw	20.9	76.5	1.2	1.4	100
Cooked	25.2	66.6	7.0	1.7	171
Lamb					
Raw	19.5	71.5	7.0	1.5	145
Cooked	27.0	61.5	8.5	2.0	200
Pork					
Raw	19.5	69.5	9.5	1.0	170
Cooked	29.0	57.0	12.0	1.3	230
Veal					
Raw	20.0	75.0	3.5	1.0	130
Cooked	29.0	63.0	5.5	1.6	175

*The table values are weighted to represent the entire carcass of each species.

foods. The amount of visible fat that is present in meat as marbling, is an excellent indication of its caloric value.

While beef, fish, lamb, pork, poultry, and veal differ in the relative proportions of each of the nutrients, a 100 gram serving (a normal size portion in institutional meal planning) of cooked meat from these species will, on the average, supply approximately the following nutrients for an adult male:

10 percent of the recommended daily allowance of calories;
50 percent of the protein recommendation;
35 percent of the iron recommendation (100 percent if the serving is liver);
25–60 percent of the B complex vitamins (higher, if the serving is pork or liver).

Several of the variety meats have slighly less protein and fat than skeletal meats (Table 13-2). Yet these variety items are often more economical sources of protein and vitamins than the conventional retail

TABLE 13-2
Approximate protein and fat composition percentages for cooked variety meats

	Beef		Pork		Veal		Lamb	
	Protein	Fat	Protein	Fat	Protein	Fat	Protein	Fat
Brain	11.5	9.0	12.2	8.7	10.5	7.4	12.7	9.2
Heart	28.9	6.0	23.6	4.8	26.3	4.5	21.7	5.2
Kidney	24.7	3.6	25.4	4.7	26.3	5.9	23.1	3.4
Liver	22.9	4.3	21.6	4.7	21.5	7.6	23.7	10.9
Lung	20.3	3.7	16.6	3.1	18.8	2.6	20.9	3.0
Pancreas	27.1	17.2	28.5	10.8	29.1	14.6	23.3	9.6
Spleen	25.1	4.2	28.2	3.2	23.9	2.6	27.3	3.8
Thymus	20.5	24.9	–	–	18.4	2.9	–	–
Tongue	22.2	21.5	24.1	18.6	26.2	8.3	21.5	20.5

SOURCE: National Live Stock and Meat Board (1966).

TABLE 13-3
Approximate protein, moisture, and fat composition percentages for selected sausages and smoked meats

	Percent		
	Protein	Moisture	Fat
Sausage			
Bologna	12	56	29
Braunschweiger	15	53	27
Dutch loaf	15	51	26
Frankfurters	12	56	29
Headcheese	15	59	22
Salami, cooked	17	51	26
Salami, dry	24	30	38
Pork sausage, cooked	18	35	44
Pork sausage, fresh	10	38	50
Pork sausage, smoked	15	50	31
Thuringer	18	49	25
Smoked meats			
Ham, cured, cooked lean portion	28	63	5
Bacon, regular sliced, cooked	33	3	60

cuts of meat. In addition, liver provides the richest source of iron, vitamin A, niacin, and riboflavin of any ordinary food. Nutritionists recommend that liver be included in the diet on a regular basis.

Processed meat contains less protein and water and more fat and minerals than the consumed portions of fresh meat, the percentage of minerals being increased by the added salt and seasonings (Table 13-3).

TABLE 13-4
Cost of a 3-ounce (84 gram) serving of cooked lean meat from selected kinds and cuts of meat, at specified retail prices per pound*

Kind and cut of meat	Price of retail cuts (in cents/pound)													
	50	55	60	65	70	75	80	85	90	95	100	105	110	115
BEEF														
Roasts														
Brisket, bone in	26	29	31	34	36	39	42	44	47	49	52	55	57	60
Chuck, bone in	22	25	27	29	31	33	36	38	40	42	45	47	49	51
Chuck, bone out	17	19	21	23	24	26	28	30	31	33	35	36	38	40
Ribs-7th, bone in	22	25	27	29	31	33	36	38	40	42	45	47	49	51
Round, bone in	17	18	20	22	23	25	27	28	30	32	33	35	37	38
Round, bone out	16	17	19	20	22	23	25	27	28	30	31	33	34	36
Rump, bone in	22	24	26	28	31	33	35	37	39	41	44	46	48	50
Rump, bone out	17	19	20	22	24	26	27	29	31	32	34	36	38	39
Steaks														
Chuck, bone in	22	25	27	29	31	33	36	38	40	42	45	47	49	51
Chuck, bone out	17	19	21	23	24	26	28	30	31	33	35	36	38	40
Club, bone in	28	31	34	37	40	43	45	48	51	54	57	60	62	65
Porterhouse, bone in	26	29	31	34	36	39	42	44	47	49	52	55	57	60
Round, bone in	17	18	20	22	23	25	27	28	30	32	33	35	37	38
Round, bone out	16	17	19	20	22	23	25	27	28	30	31	33	34	36
Sirloin, bone in	21	23	26	28	30	32	34	36	38	40	43	45	47	49
Sirloin, bone out	20	21	23	25	27	29	31	33	35	37	39	41	43	45
T-bone, bone in	28	30	33	36	39	41	44	47	50	52	55	58	61	64
Ground beef, lean	13	14	16	17	18	20	21	22	23	25	26	27	29	30
Short ribs	29	32	35	38	41	44	47	50	53	56	58	61	64	67
PORK, FRESH														
Roasts														
Loin, bone in	25	28	30	33	36	38	41	43	46	48	51	53	56	58
Loin, bone out	17	19	21	23	24	26	28	30	31	33	35	36	38	40
Picnic, bone in	27	29	32	35	37	40	43	45	48	51	53	56	59	62
Chops														
Loin	22	25	27	29	31	33	36	38	40	42	45	47	49	51
Rib	25	28	30	33	36	38	41	43	46	48	51	53	56	58
PORK, CURED														
Roasts														
Butt, bone in	18	20	22	23	25	27	29	31	32	34	36	38	40	42
Ham, bone in	17	19	21	23	24	26	28	30	31	33	35	36	38	40
Ham, bone out	13	14	16	17	18	20	21	22	23	25	26	27	29	30
Picnic, bone in	23	25	27	30	32	34	37	39	41	43	46	48	50	53
Picnic, bone out	18	19	21	23	25	27	28	30	32	34	35	37	39	41
Ham slices	16	17	19	20	22	23	25	27	28	30	31	33	34	36
LAMB														
Roasts														
Leg, bone in	21	23	25	27	29	31	33	35	38	40	42	44	46	48
Shoulder, bone in	23	25	27	30	32	34	37	39	41	43	46	48	50	53
Chops														
Loin	23	25	27	30	32	34	37	39	41	43	46	48	50	53
Rib	28	30	33	36	39	41	44	47	50	52	55	58	61	64

*After U.S. Department of Agriculture Bulletin No. 183

TABLE 13-4 (cont'd)
Cost of a 3-ounce (84 gram) serving of cooked lean meat from selected kinds and cuts of meat, at specified retail prices per pound

Price of retail cuts (in cents/pound)																
120	125	130	135	140	145	150	155	160	165	170	175	180	185	190	195	200
62	65	68	70	73	76	78	81	83	86	89	91	94	97	99	102	105
54	56	58	60	62	65	67	69	71	74	76	78	80	83	85	87	89
42	43	45	47	49	50	52	54	56	57	59	61	62	64	66	68	69
54	56	58	60	62	65	67	69	71	74	76	78	80	83	85	87	89
40	42	43	45	47	48	50	52	54	55	57	59	60	62	64	66	67
38	39	41	42	44	45	47	48	50	52	53	55	56	58	59	61	63
52	55	57	59	61	63	66	68	70	72	74	76	79	81	83	85	87
41	43	44	46	48	49	51	53	55	56	58	60	61	63	65	67	68
54	56	58	60	62	65	67	69	71	74	76	78	80	83	85	87	89
42	43	45	47	49	50	52	54	56	57	59	61	62	64	66	68	69
68	71	74	77	80	82	85	88	91	94	97	99	102	105	107	110	113
62	65	68	70	73	76	78	81	83	86	89	91	94	96	99	102	104
40	42	43	45	47	48	50	52	54	55	57	59	60	62	64	66	67
38	39	41	42	44	45	47	48	50	52	53	55	56	58	59	61	63
51	53	55	57	60	62	64	66	68	70	72	74	77	79	81	83	85
47	49	51	53	55	57	59	61	62	64	66	68	70	71	73	75	77
66	69	72	75	77	80	83	86	88	91	94	97	99	102	105	108	110
31	33	34	35	36	38	39	40	42	43	44	46	47	48	50	51	52
70	73	76	79	82	85	88	91	94	96	99	102	105	108	111	113	116
61	63	66	69	71	74	76	79	81	84	86	89	91	94	96	99	101
42	43	45	47	49	50	52	54	56	57	59	61	62	64	66	68	69
64	67	70	72	75	78	80	83	86	88	91	94	96	99	102	104	107
54	56	58	60	62	65	67	69	71	74	76	78	80	83	85	87	89
61	63	66	69	71	74	76	79	81	84	86	89	91	94	96	99	101
43	45	47	49	51	52	54	56	58	60	61	63	65	67	69	71	73
42	43	45	47	49	50	52	54	56	57	59	61	62	64	66	68	69
31	33	34	35	36	38	39	40	42	43	44	46	47	48	50	51	52
55	57	59	62	64	66	68	71	73	75	78	80	82	85	87	89	92
42	44	46	48	49	51	53	55	57	58	60	62	64	66	68	69	71
38	39	41	42	44	45	47	48	50	52	53	55	56	58	59	61	63
50	52	54	56	58	60	62	65	67	69	71	73	75	77	79	81	83
55	57	59	62	64	66	68	71	73	75	78	80	82	84	87	89	91
55	57	59	62	64	66	68	71	73	75	78	80	82	85	87	89	91
66	69	72	75	77	80	83	86	88	91	94	97	99	102	105	108	110

TABLE 13-5
Cost of chicken, whole and parts*

If the price/pound (in cents) of whole fryers, ready to cook, is:	Chicken parts are an equally good buy if the price/pound (in cents) is:				
	Breast half	Drumstick and thigh	Drumstick	Thigh	Wing
29	41	37	36	39	23
31	44	40	38	41	25
33	47	42	41	44	26
35	49	45	43	47	28
37	52	47	46	49	29
39	55	50	48	52	31
41	58	53	50	55	33
43	61	55	53	57	34
45	63	58	55	60	36
47	66	60	58	63	37
49	69	63	60	65	39
51	72	65	63	68	41
53	75	68	65	71	42
55	78	71	68	73	44
57	81	73	70	76	46
59	84	76	73	79	48

*After U.S. Department of Agriculture Bulletin No. 183.

TABLE 13-6
Approximate percentage composition, and caloric content, of the nonskeletal portion of different U.S. Department of Agriculture grades of beef carcasses*

Carcass Grade	Percent				
	Protein	Moisture	Fat	Ash	Calories/100 grams
Prime	13.6	44.8	41.0	0.6	428
Choice	14.9	49.4	35.0	0.7	379
Good	16.5	54.7	28.0	0.8	323
Standard	18.0	60.1	21.0	0.9	266
Commercial	15.8	52.4	31.0	0.8	347
Utility	18.6	62.5	18.0	0.9	242

*After U.S. Department of Agriculture Home Economics Research Report No. 31.

Processed meat also contains more calories than fresh meat, because of the added cereals, flours, or skimmed milk (carbohydrates and proteins), and it also frequently has a higher fat content.

To properly utilize the nutritional attributes of meat, it is usually necessary for the dietician or meal planner to consider the cost of the cuts and products that are available. Since retail cuts of meat are variable in their bone and fat content, the number of servings they provide per pound is also variable. Consequently, the price per pound is not a good guide to the cost per serving.

Table 13-4 lists the costs, for several retail cuts, for a three ounce (84 gram) serving of cooked lean meat at various prices per pound. This information emphasizes the fact that the lowest priced cut of meat is not always the cheapest source of standard sized servings. Similar comparisons must be made when purchasing poultry; but, in this case, the options to buy a whole bird or only the carcass parts must be considered. Table 13-5 lists the prices for chicken carcass parts that are consistent with various prices for whole birds.

Knowledge of the composition of intact carcasses is useful to the consumer who buys on this basis for a home freezer. Table 13-6 lists the composition of the nonskeletal portion of several grades of beef carcasses. These figures include all the muscle and fat in the carcass.

REFERENCES

Leverton, R. M. and G. V. Odell, "The Nutritive Value of Cooked Meat," Oklahoma Agr. Exp. Sta., Misc. Publ. MP-49 (Oklahoma State University, Stillwater), 1959.

Mann, G. V., "Saturated *vs* Unsaturated Fat Controversy," Proc. Meat Industry Research Conf., American Meat Institute Foundation, Chicago, Illinois, p. 73 (1972).

National Live Stock and Meat Board, *Lessons on Meat* (Chicago), 1966.

U.S. Department of Agriculture, *Composition of Foods,* Handbook 8, 1963.

U.S. Department of Agriculture, "Nutritive Value of Foods," Home and Garden Bulletin No. 72 (1960).

U.S. Department of Agriculture, "Proximate Composition of Beef from Carcass to Cooked Meat," Home Economics Research Report No. 31 (1965).

U.S. Department of Agriculture, "Your Money's Worth in Foods," Home and Garden Bulletin No. 183 (1970).

CHAPTER 14

Meat Inspection

HISTORY

Since earliest recorded history, people have recognized the importance of the source and the proper processing of their meat supply. In the interest of public health, the earliest civilizations of the Mediterranean area regulated and supervised the slaughter and handling of meat animals. For instance, the Mosaic laws proclaimed which animals were suitable and unsuitable for human food. These food laws (Leviticus, 11, and Deuteronomy, 14) forbidding the consumption of pork and many other types of meat, are still strictly adhered to by Orthodox Jews.

Several noted Greek and Roman writers, including Aristotle, Hippocrates and Virgil, noted the similarity between diseases in animals and humans. However, the medieval Roman Catholic Church voiced displeasure toward any speculation about a possible relation between them. As a consequence, meat inspection in medieval Europe was often carried out in opposition to the Church, and in a sporadic and superficial manner. Nevertheless, meat inspection was practiced in France as early as 1162, in England by 1319, and in Germany by 1385.

Meat inspection in the United States was carried out in a rudimentary manner prior to the passage of the Meat Inspection Act in 1906. In early colonial times the production and slaughter of livestock was entirely a local enterprise. Animals were slaughtered by local butchers and sold to customers who were generally able to identify the product closely with the butcher and even, in many cases, with the farmer who produced the animal. As a rule, the local butcher was a prominent member of the community and, in an attempt to eliminate diseased animals, usually scrutinized every animal that was slaughtered. But, because of a general lack of training in detecting diseased conditions, much unwholesome meat was inevitably processed. Later, as the population increased and transportation systems developed, the livestock and meat industry expanded from a local to a national enterprise. More animals were slaughtered and processed by a few large slaughter plants and fewer animals by the local butcher.

With the concentration of livestock marketing and slaughtering at central locations, there developed a need for improved sanitation and better inspection procedures, in order to safeguard the consumer from the large scale sale of unwholesome meat and meat products. In the early 1880s the press focused public attention upon the problem of quality and purity of food products being sold to the public. Meat was the most suspected product, and consequently attention was focused on the growing packing industry. Chicago newspapers published charges of unsanitary conditions in the packing industry, and of diseased animals being slaughtered for human consumption. These reports adversely affected meat exports, and created a growing suspicion of the packers and their products in the mind of the public. The Chicago packers and processors of meat products were aroused to action and, in an effort to correct the deplorable situation that existed, they cooperated with the health authorities to create an improved inspection system that would detect diseased animals. However, this improved system did not entirely correct the existing situation.

In 1890, in response to the growing demand for action, Congress passed a law that provided for the inspection of meat "in the piece," but only when it was intended for export. Obviously, this initial move by Congress was designed to assist foreign trade, rather than to protect the American public. The provisions of this law provided for an inspection that would determine the character and manner in which products (such as salted pork and bacon) were packed, and the condition of these products at the time of shipment. Since the law did not authorize the inspection of animals at the time of slaughter, diseased animals were still not detected. Foreign governments consequently refused to remove the prohibition they had placed against American pork, because the inspection required by the law was deemed inadequate.

In 1891 Congress passed a law that extended the inspection system. This law provided that the Secretary of Agriculture should establish a preslaughter inspection system for all cattle, hogs, and sheep that were intended for interstate commerce. The success of this law led to a further extension of the inspection system, with the passage of an amended law, in 1895. This act conferred power upon the Secretary of Agriculture to make such rules and provisions as he deemed necessary to prevent the transportation, from one state or territory, or to any foreign country, of the condemned carcasses of cattle, hogs, or sheep. However, these laws and their administration did not satisfy the demands of the American public for an adequate national system of meat inspection. No provision had been made to control the sanitation of the surroundings in which animals were slaughtered and their meat processed. The press continually focused public attention on the unsanitary manner in which packing houses were being operated. Upton Sinclair's book *The Jungle* depicted existing conditions in the packing industry and aroused public anger against the industry. These reports led to the appointment of committees by the Secretary of Agriculture and the President to investigate conditions of the packing industry in the Chicago area. A message to Congress, entitled "Conditions in Chicago Stockyards," from President Theodore Roosevelt brought about the enactment and passage of a comprehensive Meat Inspection Act on June 30, 1906. This act applied not only to inspection of live cattle, hogs, sheep, and goats before slaughter, and their carcasses following slaughter, but to the meat and meat products in all stages of processing as well. It also applied to the surroundings in which livestock were slaughtered and the meat was processed. The Horse Meat Act of 1919 extended the provisions of the Meat Inspection Act to cover the slaughter of horses and the processing of horse meat products.

The inspection of poultry was not covered in the Meat Inspection Act of 1906. Until recently, the bulk of poultry was bought by the consumer either alive or "New York dressed" (only blood and feathers removed). The housewife dressed and prepared the bird and determined whether the meat was wholesome. By the 1920s, with the development of the poultry processing industry, the need for legislation covering the inspection of poultry had become more apparent. In 1926, a voluntary Poultry Inspection Program evolved, patterned on an agreement among the United States Department of Agriculture, the New York Live Poultry Commission Association, and the Greater New York Live Poultry Chamber of Commerce. The voluntary inspection was paid for by the poultry slaughter and processing plants. Subsequently, other cities passed ordinances requiring that poultry and poultry products shipped into these cities be inspected.

A Poultry Products Inspection Act was signed into law by President Eisenhower in 1957. The provisions of this law were similar to those contained in the Meat Inspection Act of 1906. In 1968 this act was amended by the Wholesome Poultry Products Act. This amended law provides for inspection of poultry and poultry products by the USDA.

The Humane Slaughter Act of 1958 required plants desiring to sell meat products to Federal Agencies to slaughter animals in a humane manner, as described by the Secretary of Agriculture. However, an inspected plant may still continue slaughtering in a manner not described as humane by the Secretary of Agriculture. The only penalty for not complying with the recommended humane slaughter methods is removal from an approved list of plants eligible to sell products to government agencies, such as the army, navy, and veterans administration, among others. Those methods of immobilizing animals prior to bleeding that are approved include chemical, mechanical, and electrical methods. The exposure of animals to carbon dioxide gas, in a way that will accomplish anesthesia with a minimum of excitement and discomfort, is an approved chemical method. The use of captive bolt stunners, compressed air concussion devices, and firearms are examples of approved mechanical devices. The administration of electrical current to produce surgical anesthesia is approved as an electrical method. Animals that are kosher slaughtered are exempted from the provisions of this act.

The Meat Inspection Act of 1906 and the Poultry Products Inspection Act of 1957 applied only to the animals slaughtered and those products that were processed for interstate and foreign commerce, but did not apply to intrastate commerce. Small slaughterhouses and processors doing a strictly local business were not required to comply with the provisions of these acts. They only needed to comply with nonuniform and often poorly enforced city or state laws. To improve upon this situation, the Wholesome Meat Act was enacted by congress and signed into law by President Johnson on December 15, 1967. This law provided for a State–Federal cooperative program, under which all intrastate slaughter and processing plants were to be placed under State supervision. The act provided for federal financial, technical, and laboratory assistance in setting up state meat inspection programs that were equivalent to the federal standards. Individual states were given two years in which to bring their meat inspection programs up to these standards. Some states failed to act, and the federal government has taken over all meat inspection services within them.

When the Wholesome Meat Act was passed in 1967, the provisions of this act and those of the Federal Meat Inspection Act of 1906, and its amendments, were consolidated into one Federal Meat Inspection Act. Under the provisions of the current act, antemortem inspection is now mandatory for all cattle, sheep, swine, goats, horses, mules, and other equine animals. Under the provisions of the act of 1906, antemortem inspection was not mandatory, but the Secretary of Agriculture, at his discretion, could require an antemortem inspection of all cattle, sheep, swine, and goats.

Application and Enforcement of Inspection Laws

The Secretary of the United States Department of Agriculture is given the responsibility of administering federal meat inspection laws, and is also authorized to promulgate regulations that will implement these laws and enable him to effectively administer their provisions. The federal inspection program, as conducted by the Animal and Plant Health Inspection Service, consists of (1) the examination of animals and of their carcasses at the time of slaughter, (2) inspection at all stages of the preparation of meat and meat food products to assure sanitary handling, the sanitation of equipment, buildings, and grounds, (3) the destruction of condemned product to prevent its use for human food, (4) the examination of all ingredients used in the preparation of meat food products to assure their fitness for food, (5) the prescription and application of identification standards for inspected meat food products, (6) the enforcement of measures that insure informative labeling, (7) the prohibition of the use of false and deceptive labeling, (8) the inspection of foreign meat and meat food products that are offered for importation, and (9) the administration of a system of certification to assure acceptance of domestic meats and meat food products in foreign commerce.

Federal and state governments hire, pay, and assign meat inspectors. The cost of the meat inspection program to the federal government amounts to about 50 cents per person per year. The packer or processor is required to compensate the government only for the cost of overtime inspection.

MEAT INSPECTORS. Federal inspection personnel are classified into three groups: food inspectors, laboratory inspectors, and veterinary medical officers. Food inspectors are experienced laymen who assist in antemortem and postmortem inspections and report to the veterinary medical officers, who are graduate veterinarians. These veterinary medical officers are in charge of inspection, wherever slaughtering is conducted. Laboratory inspectors are technically trained in the bacteriological and chemical examination of meat and meat food products. All permanent employees engaged in federal meat inspection are Civil Service employees.

IDENTIFICATION OF INSPECTED PRODUCTS. The presence of the federal or state meat inspection stamp on large cuts of meat (Figure 14-1) or meat products (Figure 14-2) assures consumers the product was wholesome when it was shipped from the plant where it was inspected. There are definite legal provisions that control the use of the inspection

FIGURE 14-1
A Federal inspection stamp similar to those used on inspected fresh meat cuts.

FIGURE 14-2
A Federal inspection stamp similar to those used on inspected meat products.

stamp. The circular stamp inscribed with the legend "U.S. INSP'D & P'S'D" must also have a number on it to identify the official establishment. For example, the number 38 is reserved by the U.S. Department of Agriculture for illustrative purposes. Each plant under federal inspection has a different number. In some cases, letters will appear with the number to indicate a specific plant in a multiple plant operation. Special inks that have been approved for use by the Food and Drug Administration are now used to apply the inspection legend on fresh meats.

Meat that is imported into the United States from foreign countries is subjected to the same controls as federally inspected meat. A country that desires to export meat to the United States requests that its system of inspection be accepted as equivalent to that maintained by the United States, and asks that its certificates be accepted by the Animal and Plant Health Inspection Service as evidence of such inspection. Investigations of approved foreign plants are then made to determine whether the inspection program in the foreign country is comparable to that maintained in the United States. If the foreign inspection is satisfactory, arrangements are made to provide a system of certification whereby each shipment of meat, upon arrival at a port of entry in the United States, is identified as to its inspectional background. If, upon inspection by the Animal and Plant Health Inspection Service, the meat is found to be wholesome, free from adulteration, and properly labeled, it is stamped (Figure 14-3), permitted to enter the United States, and can move freely in interstate commerce.

FIGURE 14-3
Imported meat bears the inspection stamp of the country exporting the product and that of the United States. The latter stamp includes the name of the port of entry.

Meat exported from the United States carries certificates of inspection that are recognized in foreign countries. A consignment of meat for a foreign country is not permitted to leave the United States unless the meat is identified by export stamps or an export certificate indicating that the product was prepared under Animal and Plant Health Inspection Service supervision.

Requirements for Granting Inspection Service

The owner or operator of any meat processing plant who plans to engage in the interstate or foreign meat trade or to furnish meat to federal agencies, must send detailed information relative to the nature and volume of the proposed operation to the U.S. Department of Agriculture. The operator will then be informed whether the proposed plant requires or is entitled to federal inspection. If so, a formal application form will be furnished. Along with this application, the operator will be required to furnish detailed plans and specifications of the proposed plant. The application must be accompanied by drawings that show the character of plant construction, the arrangement of all essential equipment, and the location of the establishment. Included in the federal requirements covering the granting of inspection are a potable and ample water supply, an approved sewage system, adequate natural or artificial light, adequate ventilation, adequate refrigeration, and an ample supply of pressurized hot water for cleanup purposes. If these requirements are met, inspectors are assigned to the plant and the system of inspection is applied in accordance with the inspection provisions of the law.

Federal meat inspection is conducted only at establishments that meet sanitary standards, and are equipped with facilities that will assure

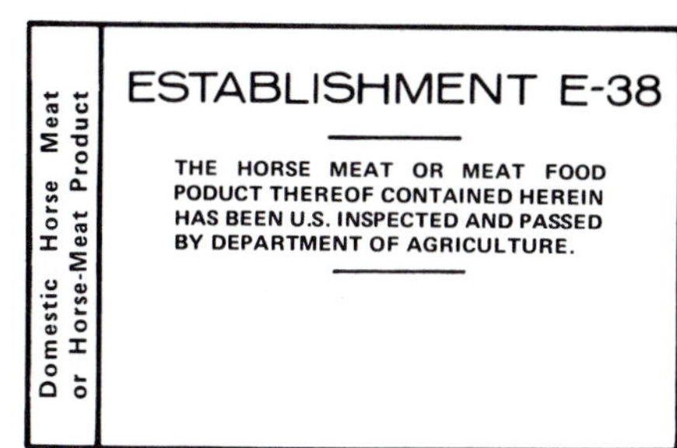

FIGURE 14-4
A Federal inspection stamp similar to those used on inspected fresh horse meat, and a label similar to those used on inspected horse meat products.

the sanitary preparation, handling, and storing of meat and meat food products. If a plant does not comply with these standards inspection is withdrawn, and the plant cannot sell meat. Any anticipated change in the facilities and equipment or expansion of a plant under federal inspection must be in compliance with federal regulations and be approved by the U.S. Department of Agriculture. States that have established a cooperative meat inspection program in compliance with the Wholesome Meat Act of 1967, have requirements and procedures that are similar to the federal system for the granting of inspection service.

U.S. Department of Agriculture regulations require that the slaughter of horses, and the processing of horse meat products, must be conducted in a plant that is separate from those that process other meat animals. Any meat or food products from horses must be plainly and conspicuously labeled, marked, branded, or tagged "Horse meat" or "Horse meat products" as the case may be. A hexagonal brand in green ink is used for horse meat (Figure 14-4). Labels on horse meat must be printed with black ink on light green paper.

U.S. Department of Agriculture regulations also require that poultry be slaughtered and processed in facilities that are separate from those used for other meat animals.

Elements of Inspection

The elements of inspection consist of the following essential parts: sanitation, antemortem, and postmortem inspections, the control and disposition of condemned materials, product inspection, laboratory inspection, and marking and labeling. Although the discussion of these essential parts of the meat inspection process that follows is directed toward inspection of cattle, sheep, swine, and goats, the same points are applicable to the slaughter and processing of poultry and equine animals.

SANITATION. Inspection is conducted only at establishments that meet the legal requirements for sanitation, and have those facilities that are required for the inspectors and the inspection process. The requirements for and enforcement of sanitary standards begin with the livestock pens, and extend to every subsequent operation. Attention is given to every detail in the environment where the animals are slaughtered and their meat is handled. This includes structural aspects of the premises, water supply, all equipment used for processing and handling of edible product, sanitary facilities, sewage disposal, and rest room and locker room facilities for personnel.

No establishment is allowed to employ, in any department where meat products are handled or prepared, any person affected with a serious communicable disease, such as tuberculosis.

When it is decided by the inspector that any equipment, utensil, room, or compartment of an establishment is unclean, or that its use would be in violation of sanitary requirements, a "U.S. Rejected" tag is attached to it. No equipment, utensil, room, or compartment so tagged can be used again until it has been made acceptable. Failure by any establishment to maintain the required sanitary conditions will result in the immediate discontinuance of slaughter or processing activities in the particular area in question.

ANTEMORTEM INSPECTION. An antemortem examination and inspection is made of all cattle, sheep, swine, goats, poultry, and horses, for the purpose of eliminating those animals that are unfit for the preparation of food. Such inspection is made on the day of slaughter (Figure 14-5). Animals plainly showing symptoms of a disease or condition that would cause condemnation of their carcasses on postmortem inspection are marked "U.S. Condemned" and are killed by the official establishment and "tanked" in a manner prescribed for condemned carcasses. These include any dying animal, and those showing symptoms of such diseases as rabies, tetanus, hog cholera, anaplasmosis, parturient paresis, and anthrax. Such animals can not be taken into any establishment to be slaughtered or dressed, nor conveyed into any department of the establishment that is connected in any way to the production of edible products. Any animal that does not yet plainly show symptoms of such diseases, but which is suspected of having a disease or condition that may cause condemnation of part or all of the carcass upon postmortem inspection, is marked "U.S. Suspect". Its identity is maintained during the slaughter process until a final postmortem inspection is performed.

FIGURE 14-5
All animals are examined by inspectors before slaughter.

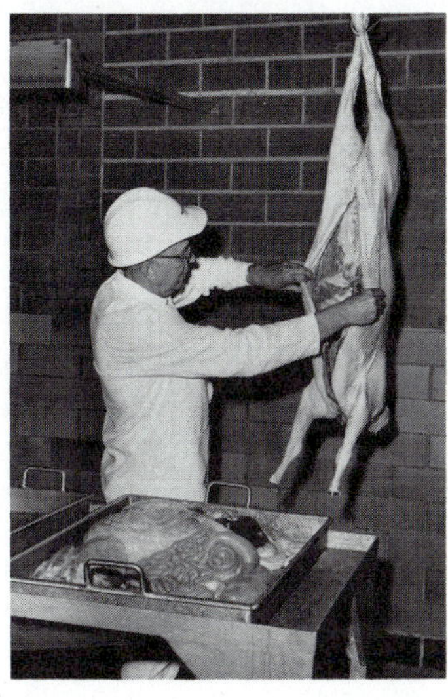

FIGURE 14-6
All parts of the carcass receive a thorough postmortem examination by an inspector.

POSTMORTEM INSPECTION. A careful inspection and examination is made simultaneously with the slaughtering and dressing operations (Figure 14-6). The postmortem inspection is conducted under the supervision of a veterinary medical officer. Inspectors are located at specific stations along the dressing line so that the carcass, head, and viscera of each animal can be inspected. The lungs, liver, lymph glands, spleen, and heart receive particular attention, because symptoms of disease are readily detectable in these organs. Positive identity of all parts of each animal is maintained through the use of duplicate numbered tags, or by synchronized chains and moving tables containing the carcass and the corresponding viscera.

When an inspector observes a lesion (or other condition) in any carcass, internal organ, or detached part that might render it unfit for food purposes, and will therefore require a subsequent inspection, the inspector retains the carcass and parts. "U.S. Retained" tags are then affixed by the inspector to the carcass, detached parts, and organs and they are removed from the production line. They are completely isolated from other carcasses until after a complete final examination is made by a veterinarian. In some instances a laboratory examination is necessary before a final decision can be made. If the carcass or part is finally found to be unwholesome, or unfit for human food, the veterinarian conspicuously marks the surface of the carcass or part "U.S. Inspected and Condemned". If they are found to be wholesome and fit for human food, then the inspector allows them to reenter the production line and marks them "U.S. Inspected and Passed". In cases where the lesion or condition is localized, the inspector condemns the affected part, and may pass the remainder of the carcass. Bruises are trimmed out, and the remainder of the carcass passed, provided that only a limited area is involved.

CONTROL AND RESTRICTION OF CONDEMNED MATERIAL. When a carcass, part of a carcass, or an organ is condemned, it must be plainly marked "U.S. Inspected and Condemned". All condemned material remains in the custody of the inspector until it is finally disposed of by an approved method. The most commonly used method for disposal is called "tanking." The condemned materials are cooked at temperatures that are sufficiently high to destroy any disease producing organisms, and also to effectively destroy the meat for human food purposes. Products from the tanking process are used only for inedible purposes. All tanks and equipment used for the rendering, preparing, or storing of inedible products must be in rooms (or compartments) that are separate from those used for rendering, preparing, or storing edible products.

PRODUCT INSPECTION. Each step in the making of processed meat products is under the continuous scrutiny of food inspectors (Figure 14-7). Examples of such products are sausages, canned meat products, cured and smoked meats, and edible fats. Inspections are made of all the details in the manufacturing processes, to make sure they are carried out in compliance with accepted and approved regulations.

All formulas that are used in processed meat products must be filed with, and approved by, the U.S. Department of Agriculture. Only approved ingredients are permitted for use in processed meat products. The regulations also control the amounts of certain ingredients that are allowed in the product, and the processing procedures used in making

FIGURE 14-7
Inspection is conducted at all stages of processing, in order to assure the consumer that the product has been processed in compliance with U.S. Department of Agriculture approved regulations.

the product. The processor, as well as the inspector, must be thoroughly familiar with and abide by all published regulations governing the manufacturing, storage, and handling of processed meat products. This supervision guards against the adulteration, contamination, and misrepresentation of meat products before they enter the distribution channel to the consumer.

Reinspection is performed in operations such as cutting, boning, trimming, curing, smoking, rendering, canning, sausage manufacturing, packaging, handling, or storing of meats. All products, whether fresh, cured, or otherwise prepared, even though previously inspected and passed, are reinspected as often as necessary in order to determine whether they are still wholesome and fit for human food at the time they leave the plant where the inspection is performed (known as an *official establishment* in the regulations). If the item is found to be unfit for human food upon reinspection, the original inspection mark is removed or defaced and the item is condemned.

Any meat item brought into an official establishment must have been previously inspected by a meat inspector, and must also bear the necessary inspection mark or label. Upon entry into an official establishment, meat items are identified and reinspected. If this reinspection shows that an item is unwholesome, or unfit for human food, the original inspection stamp or label is removed or defaced, and the item is condemned.

If U.S. Department of Agriculture inspectors find unwholesome products bearing an inspection mark outside of an official establishment, they have the authority to condemn them under provisions of the Wholesome Meat Act of 1967. Food and Drug Administration officials, under provisions of the Pure Food, Drug and Cosmetic Act of 1906, also have the authority to condemn foods, in wholesale and retail outlets, that are judged to be unwholesome.

LABORATORY INSPECTION. The U.S. Department of Agriculture has several regional level inspection laboratories that are equipped to make chemical, microbiological, and other technical determinations. Samples of ingredients used in the manufacture or processing of products are submitted to these laboratories for analysis. Likewise, samples of finished products are submitted by inspectors for analysis, to determine whether a particular product is in compliance with the specific regulations governing its production. Products or tissues that are suspected of containing nonapproved ingredients or drug residues are also submitted to these laboratories for analysis. Pathology laboratories in the U.S. Department of Agriculture also assist in the diagnosis of diseased and abnormal tissues sent to them by inspectors.

MARKING AND LABELING OF PRODUCTS. The brands and labels that are applied to carcasses, wholesale cuts, processed meats, edible meat by-products, and containers holding meat, are controlled by the U.S. Department of Agriculture. The inspection legend is applied under the supervision of a meat inspector. No edible product can be legally removed from an official establishment without being marked or labeled in accordance with the regulations.

The inspection legend appears on the label of prepared meat products. In addition to the inspection legend, the label must contain the common or usual name of the product, the name and address of the processor or distributor, and a correct statement of the ingredients used, listed in descending order of their content, if applicable (Figure 14-8). An inspected plant is permitted to use only those labels and markings that have been previously approved by the U.S. Department of Agriculture. Any picture that is used on the label must truly represent the product, and must not convey any false impression of its origin or quality. The

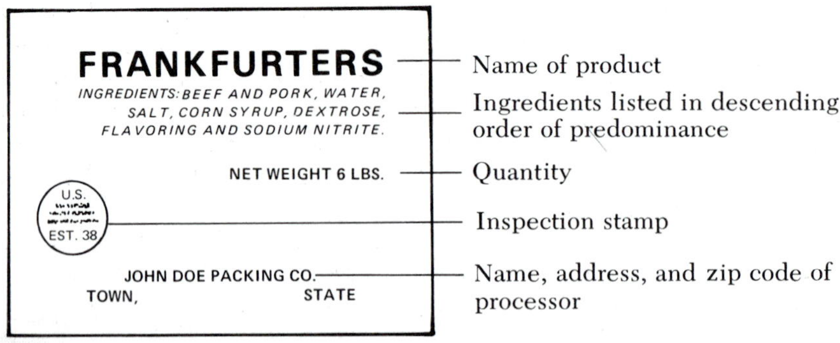

FIGURE 14-8
Sample of a U.S. Department of Agriculture approved label.

label on the product must convey to consumers exactly what they are buying. Any meat products originating from state inspected plants must also bear approved labels. Label requirements are essentially the same as for federally inspected plants, except that the state inspection legend replaces the federal inspection labels.

Kosher Inspection

In addition to federal or state inspection, there is a religious form of inspection. The term *kosher* is derived from a Hebrew word meaning "properly prepared" and, when applied to meat, it means that the product meets the requirements of the Mosaic and Talmudic laws.

Kosher slaughter is performed according to prescribed rabbinical procedures, under the supervision of authorized representatives of the Jewish faith. The actual bleeding, examination for fitness, and the removal of certain organs of the animal is usually done by the "shohet", a specially trained scholar of the dietary laws, who is the slaughterer. If the carcass and parts pass his examination, they are marked in Hebrew with the date of slaughter and the name of the shohet or rabbi.

The kosher inspection does not completely meet the requirements of the Federal meat inspection laws. In plants that are subject to federal or state inspection, another inspection is also done by the federal or state inspectors, in addition to the inspection done by the shohet or rabbi.

Fish, Shellfish, and Seafood Products Inspection

Inspection of fish, shellfish, and seafood products is not mandatory at the present time. However, a voluntary inspection service for these products is made available to processors, by both the Food and Drug Administration and the Department of Commerce. The inspection service of the Food and Drug Administration includes the development of good manufacturing practices regulations for various fish, shellfish, and seafood products that the industry can use as guides in processing, in order to comply with the sanitation requirements of the Federal Food, Drug, and Cosmetic Act.

The Department of Commerce may furnish inspection service to seafood processing plants upon their request, and at their expense. This service is provided on a voluntary basis to plants that have satisfactory facilities and conditions for conducting the service. There are a number of regulations which prescribe various quality and grade classifications for fish, shellfish, and seafood products. Inspection is provided by the

Department of Commerce to assure compliance with the requirements of approved quality or grade classifications. Products packed in plants, operated under the continuous inspection program, may be labeled with official inspection marks of the Department of Commerce.

Currently, there is a growing concern in the United States for the improvement of the legislation that is applicable to the inspection of fish and fishery products. The general idea would be to require a mandatory inspection of these products, which is similar in nature to the programs now applied to meat and poultry products of domestic and foreign origin.

REFERENCES

Brandly, P. J., G. Migaki and K. E. Taylor, *Meat Hygiene* (Lea and Febiger, Philadelphia), 1966.

Collions, F. V., *Meat Inspection* (Rigby Limited, Adelaide), 1966.

"Meat Inspection Regulations," Federal Register, Washington, D.C., Vol. 35, No. 193 (1970).

Levi, S. B. and S. R. Kaplan, *Guide for the Jewish Homemaker* Schocken Books, New York (1964).

Thornton, Horace, *Textbook of Meat Inspection* (Bailliere, Tindall and Cassel, London), 1968.

"U.S. Inspected Meat Packing Plants. A Guide to Construction, Equipment and Layout," U.S. Government Printing Office, Washington, D.C., Agriculture Handbook, No. 191 (1969).

CHAPTER 15

Meat Grading and Standardization

DEFINITION AND PURPOSE OF GRADING

Meat grading and standardization is a procedure by which carcasses, meat, or meat products are segregated on the basis of their expected palatability or yield attributes, or other economically important traits. Grading serves to segregate products into standardized groups with common characteristics, such as appearance, physical properties, or edible portion. For example, carcass grading serves to divide carcasses of a species into groups, each of which have unique characteristics that, to a great extent, will determine its processing, and the form of the finished meat product. Each grade has a specified minimum level for the traits that are used in grading. Each grade also allows a degree of variability above these minimum requirements. The purposes of grading are to facilitate marketing and merchandising through the standardization of products, and by identifying those characteristics of products that are of value to the consumer.

Types of Grades

FEDERAL GRADES. The Federal Meat Grading Service was established by an act of Congress on February 10, 1925. The Bureau of Agricultural Economics was designated as the agency responsible for providing this service. Standards for beef carcass grades were soon published, and the official grading of beef carcasses was started in May,

1927. However, tentative U.S. standards for dressed beef had been formulated in 1916, and these standards had provided the basis for national reporting of dressed beef markets. These tentative dressed beef standards also formed the basis for the first official grading begun in 1927. Tentative standards for pork carcass grades were first published in 1931, and were subsequently made official in 1933. The grading of lamb and mutton was initiated in 1931. Standards and grades for poultry products developed independently of those for beef, pork, and lamb, and were administered in separate bureaus within the Department of Agriculture. However, the historical development of poultry grades paralleled the development of the beef, pork, and lamb grading systems. Grades for poultry meat and carcasses were first proposed in October, 1927. Since their initial institution, the official standards for beef, pork, lamb and mutton, and poultry grades have been amended periodically, both in order to improve the grading systems and to accomplish their intended purposes more efficiently. In 1946, the Agricultural Marketing Act was passed by Congress, reemphasizing and extending the responsibilities and provisions of the Federal Meat Grading Act of 1925.

These federal grades for meat were developed in response to the demands of livestock producers for market reforms. Until the last half of the nineteenth century, both livestock production and the meat industry were highly decentralized; small farmers raised livestock mainly to meet their own needs. But, they also sold some of their livestock to local butchers, who in turn sold meat directly to consumers. Following the Civil War, a widespread economic reorganization occurred in the United States, producing a very complex marketing process for all agricultural products. Factors that contributed to this economic reorganization included the transition from a subsistence agriculture to commercial agriculture, the movement of processing from the farm to factories, the growth of urban markets, the extension of transportation facilities, and the expansion of foreign and domestic trade. Livestock producers did not understand the role of distributors and processors in the new marketing system. In general, they felt that these groups of middlemen were exploiting both the producers and the consumers. Their first effort toward understanding the marketing of meat products was to obtain price information pertaining to the meat trade, so that livestock market conditions could be interpreted. A uniformity of meat classification and terminology was necessary before effective market reporting could occur. Thus, the initial grade standards were specifically designed to allow the establishment of a national price reporting service for livestock.

At present, the grading of beef, pork, lamb, mutton, poultry, and rabbit is administered by the Consumer and Marketing Service of the

U.S. Department of Agriculture. Products or carcasses that are graded must be prepared under the supervision of either federal meat inspection or a state inspection that is its equivalent in standards. Federal grading is a completely voluntary service that is available, upon request, to any company, plant, or individual with a financial interest in the product. In contrast to federal meat inspection, the cost of federal grading is borne by the company or individual who requests it. The charge for the service is based on a set of standard rates that have been established by the Consumer and Marketing Service. In general, two types of services are available; (1) the grading of meat that is for sale in commercial channels on a grade basis, and (2) the examination of meat that is delivered to federal, state, county, or municipal institutions that purchase meat on contract, in order to assure conformance with grade specifications.

There are two general types of federal grades for meat. *Grades for quality* are intended to categorize meat on the basis of its acceptability for consumer cuts. The variables used in assigning quality grades include factors that are related to palatability and the acceptability of meat products to consumers. *Grades for quantity* (yield grades) are designed to categorize carcasses on the basis of an expected yield of trimmed retail cuts, and are established primarily for beef, pork, and lamb carcasses.

BRANDS. Brands are grades which are designed by, and applied to, meat or meat products by the processing industry, including the individual packer, processor, distributor, or retailer.

Since the development of federal grade standards was not supported enthusiastically by the meat packing industry, packers immediately instituted their own branding systems for beef carcasses following the start of federal grading in 1927, especially the large national packers. Many companies still have their own grading or branding system, each with its own unique standards. There was, and still is, no uniformity of standards for a specific product among the various companies, nor is there any uniformity in grade names. Within a few years, a much larger volume of beef was branded than was federally graded. Since then, this situation has become reversed, so that the proportion of fresh meat that is presently being branded is smaller than that which is federally graded. Several factors have contributed to this shift. The retail distribution of meat has become dominated by large national chain retailers; the use of uniform federal grades provides them with increased convenience in their procurement and sales promotion. Temporary compulsory federal grading was instituted during World War II and the Korean conflict.

(A)

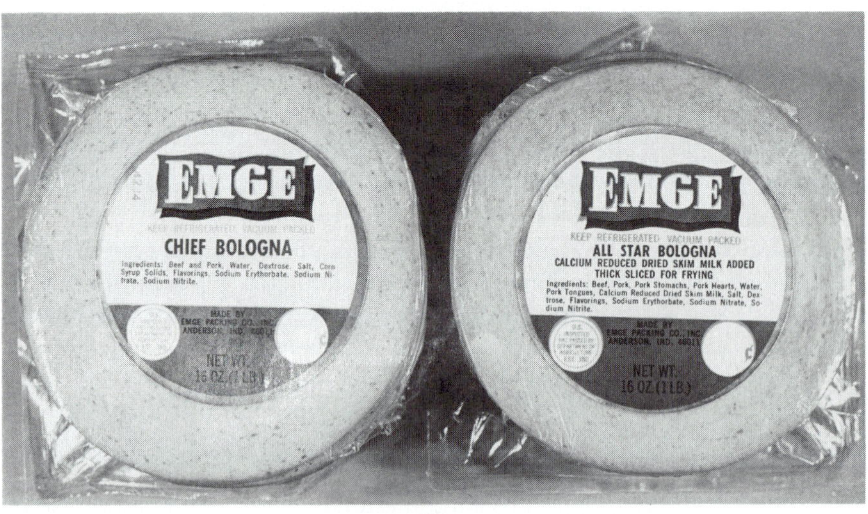

(B)

FIGURE 15-1
Brands used by a meat processor. (A) "Pioneer," "Chief," and "Mild Cure" brands of bacon. (B) "Chief" and "All-Star" brands of sliced bologna. [Courtesy Emge Packing Co., Inc., Fort Branch, Indiana.]

Smaller, regional packers and processors have come to prefer the use of federal grades rather than brands, because they can compete better with large, national packers.

In contrast to the situation with carcasses and fresh meat, brands are widely used for all types of processed meat product. Figure 15-1 illustrates the use of brands on two processed meat products, bacon and bologna. These brands are used to designate differences in sources of ingredients, formulation costs, prestige value conferred on the product by the consumer, and retail price. Brands, in general, play an important role in the merchandising and promotion of processed meat products. Very few federal grades have been established for processed meat products, and they are seldom used.

ADVANTAGES OF GRADES TO THE MEAT INDUSTRY. The purpose behind meat grades and grading is to facilitate the marketing and merchandising of meat products. All other purposes or advantages that can be cited are related in one way or another to some phase of the marketing process, from the initial livestock producer to the ultimate meat consumer.

For the livestock producer, grades form part of the basis on which animals are bred, fed, bought, and sold. Grades have been established for live animals that are designed to correspond to the respective federal grades for carcasses. Producers can estimate, with varying accuracy, the grade of animals they plan to produce, the prices they can expect to pay or receive for those animals, and which market (segment of the consuming public) will buy their production. For the packer, processor, and retailer, grades provide a means for segregating animals, carcasses, and meat into groups that are more uniform as to class, quality, and condition. The use of grades facilitates pricing in the buying and selling of animals, carcasses, and meat products. For consumers, the presence of a grade or brand stamp on a product represents an assurance that the product conforms to some established set of standards. Consumers thus have some assurance that the product meets certain standards which coincide with certain standards that they have set for themselves. This allows an easier selection of those products which will most closely satisfy their desires.

U.S. DEPARTMENT OF AGRICULTURE GRADES FOR QUALITY

For the most part, the quality grades for the various species are intended to classify meat on the basis of palatability traits. However, other factors that have little if any relationship to meat palatability are included, in

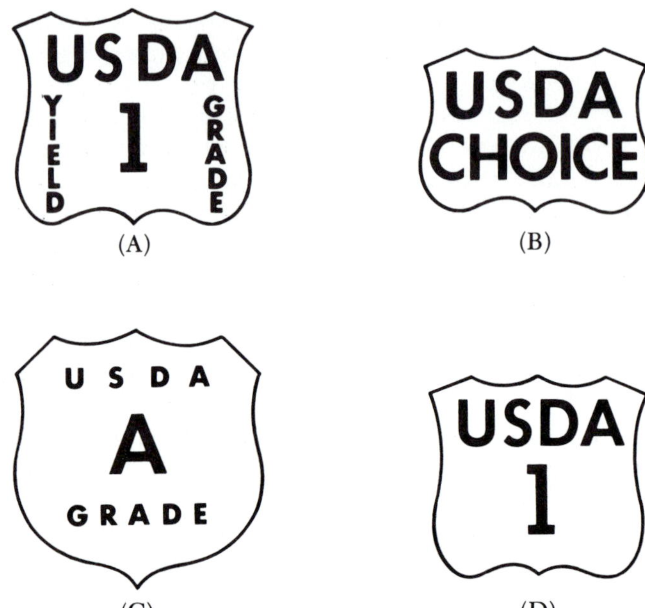

FIGURE 15-2
Identifying marks used for the U.S. Department of Agriculture grades for meat. (A) beef, lamb, and mutton yield grades, (B) beef, lamb, and mutton quality grades, (C) poultry grades and, (D) pork carcass grades.

varying degrees, in all of the quality grade standards. These additional factors are included because they have an effect on acceptability of the product to consumers. These factors include the shape of a carcass or cut, the amount of fleshing and fat cover, its overall appearance, and the presence of physical defects. The identifying marks that are placed on carcasses or packages to indicate U.S. Department of Agriculture quality grades are shown in Figure 15-2.

Factors Used to Establish Quality Grades

KIND AND CLASS. In grading meat, each species is referred to as a *kind*. Each kind of meat is divided into *classes* that are quite similar in physical characteristics. The primary factor in establishing the class of carcasses is sex. In addition, age is very important in establishing the class of poultry, and is also used to some extent in establishing beef, pork, and lamb classes.

MATURITY. As it is used in the grade standards, *maturity* is defined as the physiological age of the animal or bird that produced the carcass. *Physiological age* is an expression of the degree of aging observed in the tissues of an animal. Maturity is one of the more useful indicators of meat palatability. Of all the grade factors that are used, maturity is most closely related to meat tenderness. In general, meat from a physiologically mature animal is less tender than meat from an immature animal. This difference reflects mainly the qualitative changes that occur in an animal's muscle connective tissue (collagen) as it matures. These changes are discussed in Chapters 3 and 4.

In all species, the most useful indicators of carcass maturity are the characteristics of bone and cartilage (Figure 15-3). In young animals, cartilage is abundant in the carcass, particularly the epiphyseal cartilages. As the animal matures this cartilage *ossifies* (is infiltrated with bone salts). Therefore, during aging, a progression from soft cartilage, to hard cartilage, to bone is evident. In birds, the pliability of the breast bone cartilage is evaluated. In beef carcasses, the characteristics of the *buttons*, cartilage caps located on the tips of the dorsal processes of all vertebrae, can be evaluated in any carcass that has been split into right and left sides along the dorsal midline. The shapes and colors of the ribs change during the maturation of all species, and are easily evaluated in the carcass. The presence of a *break joint* is used to differentiate lamb from mutton. The break joint is the epiphyseal cartilage at the distal end of the metacarpal, and is the point at which the front foot is removed from the lamb carcass during slaughter. As the animal matures, this cartilage is converted to bone, and the foot cannot be removed at this location. Since in mature sheep the foot cannot be removed at the break joint, it must be removed at the moveable or *spool joint*. The presence of a spool joint is a criterion for the designation of mutton.

Meat from young animals (veal for example) is very light red in color. As the animal matures, higher myoglobin concentrations accumulate in its muscles, and they become darker red. Therefore, the color changes of lean meat are used, to some extent, in assessing carcass maturity. Changes in the color of lean are more pronounced in species (such as beef) that normally have very red muscles due to their high myoglobin content. In contrast, a less apparent color change occurs in the white breast meat from chickens or turkeys.

The term *texture*, with regard to lean meat, relates to the size of the muscle fiber bundles in the muscle, and to the thickness of the perimysial connective tissues between fiber bundles. Meat with a fine texture has a very smooth, shiny appearance on those surfaces which have been cut across the fibers. This is due directly to the fact that the fiber

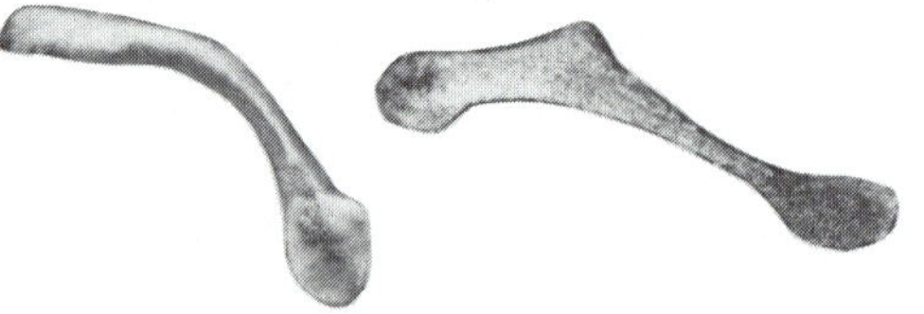

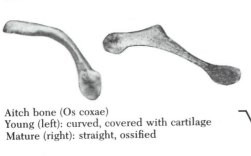

Aitch bone (Os coxae)
Young (left): curved, covered with cartilage
Mature (right): straight, ossified

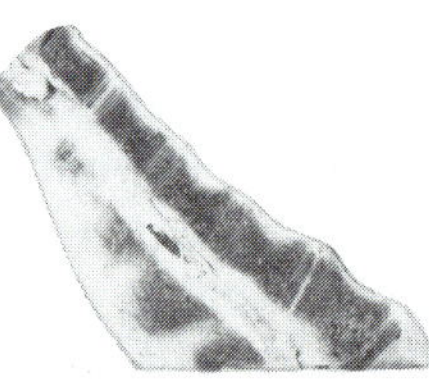

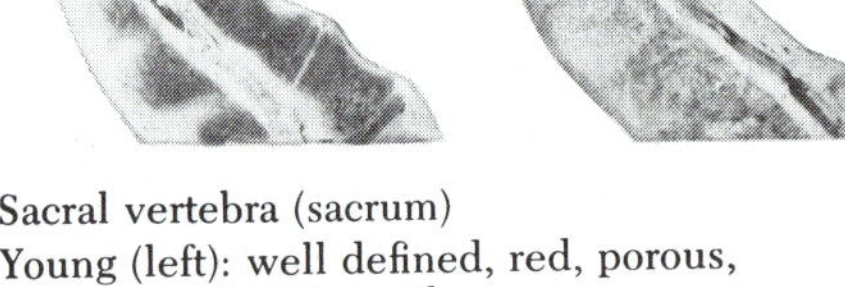

Sacral vertebra (sacrum)
Young (left): well defined, red, porous, interspersed with cartilage
Mature (right): fused, white, free of cartilage

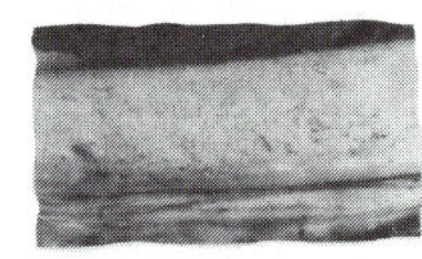

Ribs
Young (left): narrow, tubular, red, porous
Mature (right): wide, flat, white

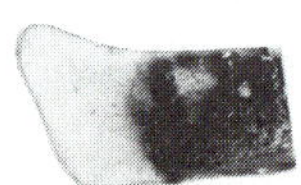

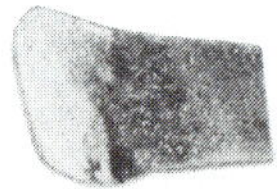

Chine bones (vertical processes of thoracic vertebrae)
Young (left): red, porous bone, terminating in distinct, soft, white cartilage buttons
Mature (right): white, hard bone, terminating in an ossified button, partly or completely indistinguishable from it

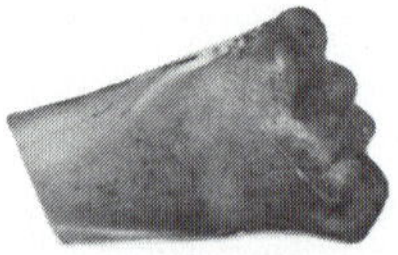

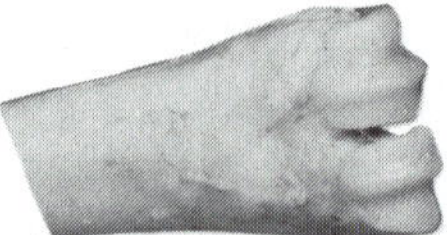

Break (spool) joint (metacarpo-phalangeal joint)
Young (left): red, porous, free of condyles (spools)
Mature (right): white, ossified, with attached condyles

FIGURE 15-3 (*opposite*)
Indices of maturity in meat animal carcasses. A comparison of immature and mature characteristics. [From Briskey, E. J. and R. G. Kauffman, "Quality Characteristics of Muscle as a Food," in *The Science of Meat and Meat Products*, 2nd ed. J. F. Price and B. S. Schweigert, eds. W. H. Freeman and Co., San Francisco. Copyright © 1971.]

bundles are small, and therefore are not readily visible. In coarse textured meat the large fiber bundles are more readily visible, and the cut surface appears dull and slightly rough. As an animal matures, the texture of the lean changes from fine to coarse. In general, a fine textured lean will be slightly more tender than lean with a very coarse texture.

MARBLING. Visible intramuscular fat, located in the perimysial connective tissues between muscle fiber bundles, is called *marbling*. The amount of marbling that is present in meat has been considered an important meat quality characteristic for many years. As stated in Chapter 12, marbling has been widely credited with making meat tender, even though there is little research evidence to indicate that its presence has a strong positive influence on meat tenderness. It is possible that marbling could act as a lubricant during chewing and swallowing, thereby improving the apparent tenderness of inherently tough meat. Marbling probably has a stronger beneficial effect on the juiciness and flavor of meat than on its tenderness. This is due to the melting of the intramuscular fat during cooking, and its release during chewing (along with part of the free water of the meat) giving a sensation of juiciness. A modest amount of marbling, uniformly distributed through the meat (Figure 15-4a), provides for optimum flavor and juiciness. Meat that is nearly devoid of marbling (Figure 15-4b) may be dry and somewhat lacking in flavor, but an excessive amount of marbling (Figure 15-4c) does not give a proportional increase in palatability.

With regard to the establishment of quality grades, ten degrees of marbling are recognized. A file of standard photographs is maintained by the U.S. Department of Agriculture illustrating each degree to insure standard interpretation of degree of marbling. The amount of marbling is subjectively determined based on these photographs, which are of the cross section of the *longissimus* muscle at the 12th rib in beef and lamb, and at the 10th rib in pork. Marbling is a major factor used in establishing beef carcass quality grades. In pork and lamb carcasses, which are normally not ribbed (a cross section of *longissimus* muscle is not available), it is customary to evaluate the amounts of *feathering* and *flank streaking* in the carcass. *Feathering* is a term describing

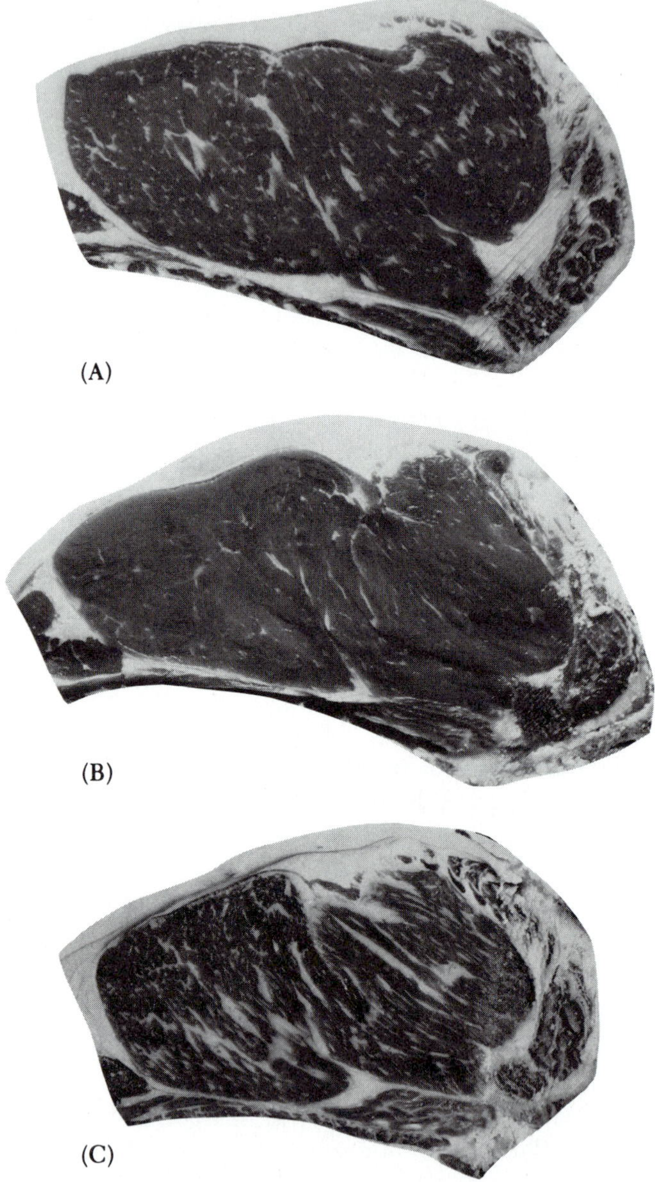

FIGURE 15-4
Marbling in beef *longissimus* muscle. (A) Ideal marbling; a modest amount uniformly distributed and finely dispersed in the muscle. (B) A muscle deficient in marbling. (C) A muscle containing excessive marbling.

the fine streaks of fat visible in the intercostal muscles between the ribs, as viewed from inside the thoracic cavity. *Flank streaking* refers to the streaks of fat located beneath the epimysium of the flank muscles, as viewed from the inside of the abdominal cavity. Both *feathering* and *flank streaking* are used as indicators of marbling, but their relationship to the amount of marbling is not particularly high. Thus, in view of the relatively low relationship of marbling to meat palatability, it is apparent that *feathering* and *flank streaking* are of even more doubtful value.

FIRMNESS. As used in quality grading, the term *firmness* may refer to the firmness of the carcass evaluated in the flank area, or to the firmness of the lean cut surface. Firmness is greatly influenced by the amount of fat that is present. As the carcass is chilled following slaughter, the subcutaneous fat becomes much firmer than the lean. Therefore, a fat carcass will be much firmer than a lean carcass. Similarly, meat that contains a high degree of marbling will be firmer than meat with very little marbling. Firmness makes no direct contribution to meat palatability, but it is a desirable quality trait because it contributes to a better appearance of retail cuts. Thus, it has an influence on the merchandising of meat.

COLOR AND STRUCTURE OF LEAN. A wide variation in the color and structure of the lean (within a species) can develop as a result of the physical and chemical changes that occur during the conversion of muscle to meat. These changes and their results, in terms of meat color and structure, are discussed in Chapter 6. In both color and structure, extremes are undesirable and are discounted in establishing quality grades. At one end of the spectrum of meat color and structure is the very pale colored, soft, exudative lean that is most often observed in pork, but is also seen occasionally in the meat from other species. At the other extreme is very dark colored, firm lean with a dry sticky surface. The dark cutting condition in beef is a typical example of this type of meat (Chapters 6 and 7).

The color and structure of the lean are evaluated in order to establish the federal grade of the meat, since they have an effect on the appearance of meat cuts, which, in turn, affects consumer acceptability. In general, the consumer objects to dark meat (such as dark cutting beef) because a dark color often is associated with meat from old animals, or with meat that has deteriorated. Research evidence indicates that dark cutting beef, or dark, firm, dry pork is the equivalent of normal meat in tenderness, flavor, and juiciness. Consumers generally do not object to the appearance of pale colored meat as severely as they object to that of dark cutting meat. Because of its low water holding capacity and its tendency to lose more juices during cooking, pale, soft, exudative meat might actually be less juicy than meat with a normal color and structure. It might also be slightly more tender, depending somewhat on the method of cookery that is used. However, meat color is much more important from the standpoint of consumer acceptance, than it is in terms of product palatability.

Beef may be graded down as much as one full grade, depending upon the severity of the dark cutting condition. Pale, soft, watery beef muscle is not as severely discounted.

CONFORMATION, FLESHING, AND FINISH. As used in meat grading, *conformation* refers to the proportionate development of the various parts of the carcass or wholesale cut, and to the ratio of meat to bone. Thus, conformation is primarily a function of muscular and skeletal system development, but it is also affected by the amount of fat cover. *Fleshing* (or *muscling*) refers specifically to the development of the skeletal musculature. Conformation and fleshing have no direct effects on the palatability of the meat derived from a carcass. These factors are included in the determination of quality grades, partly because more desirably shaped retail cuts are obtained from carcasses with an acceptable conformation. For example, birds with thinly fleshed thighs and drumsticks, and narrow, tapering, crooked, or concave breasts are not attractive to the consumer making a meat selection. A round, full *longissimus* muscle makes a more attractive rib steak than does a long, narrow muscle.

Finish refers to the amount, character, and distribution of external, internal, and intermuscular fat, either in the carcass or in the wholesale cut. Excess external, internal, and intermuscular fat detracts from the retail cut yield that is obtainable from the carcass. However, a minimum amount of finish is desired for the production of optimum meat quality. The presence of fat beneath the skin of birds serves to baste the meat during cooking, especially in roasting. Thus, this fat helps to prevent dryness in the cooked product. The same advantages can be cited for the presence of external finish on cuts of other kinds of meat. A small amount of external finish on beef and lamb carcasses helps to prevent desiccation and shrinkage of the muscles during post slaughter chilling, aging, merchandising, and cooking. Soft, oily fat is discriminated against in establishing quality grades.

CARCASS DEFECTS. In establishing poultry grades several potential carcass defects are of special importance, and are evaluated. Before a quality grade can be assigned to ready-to-cook poultry, it must be free of protruding pinfeathers. Flesh exposed because of cuts, tears, missing skin, and broken or disjointed bones detracts from the appearance of the bird. The extent of skin discolorations, bruises, and defects resulting from freezing are also considered. The number and severity of such defects that are permitted depends upon their location. Defects on the breast are more objectionable than those on other areas of the carcass.

Establishing Quality Grades

The terminology that is used to designate quality grades varies widely among kinds of meat. There is also considerable variation in grade terminology among classes within a kind of meat. U.S. Department of Agriculture quality grade terms for each kind of meat, and for the various classes within each kind, are given in Table 15-1. It should also be emphasized that not all of the quality grade factors discussed above are used to establish grades for each kind and class of meat. To illustrate this, the specific factors used to establish quality grades for the various kinds of meat are given in Table 15-2.

The beef carcass quality grades can be used to illustrate the combination of various grade factors. Figure 15-5 gives the relationship between degree of marbling, carcass maturity, and quality that are used to estab-

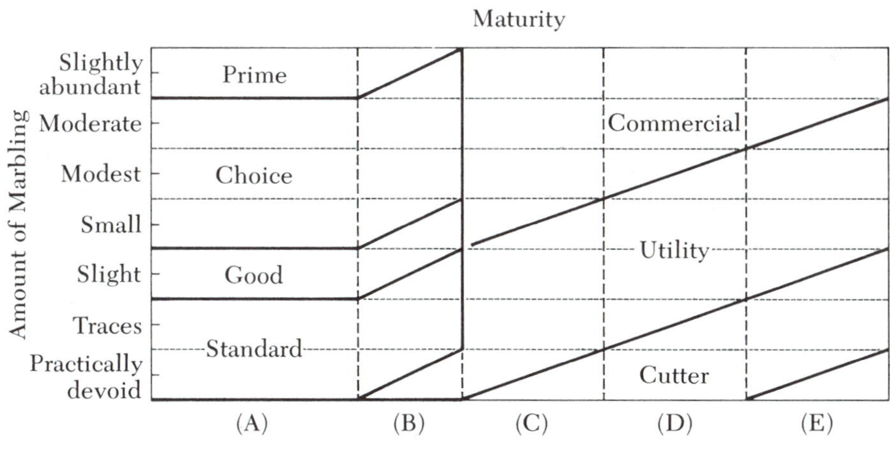

FIGURE 15-5
The relation between marbling, maturity, and quality that is used to establish beef carcass quality grades. Maturity increases from A through E. [Revised official U.S. Standards for Grades of Carcass Beef, adapted from the Federal Register of March 12, 1975.]

TABLE 15-1
United States Department of Agriculture grades for meat

Kind	Class	Grade Names
Beef quality grades	Steer, heifer, cow	Prime, Choice, Good, Standard, Commercial, Utility, Cutter, Canner
	Bullock	Prime, Choice, Good, Standard, Utility
Beef yield grades	All classes	1, 2, 3, 4, 5
Calf quality grades		Prime, Choice, Good, Standard, Utility, Cull
Veal quality grades		Same as Calf
Lamb and mutton quality grades	Lamb, yearling lamb	Prime, Choice, Good, Utility, Cull
	Mutton	Choice, Good, Utility, Cull
Lamb yield grades	All classes	1, 2, 3, 4, 5
Pork carcasses	Barrow, gilt	U.S. No. 1, U.S. No. 2, U.S. No. 3, U.S. No. 4, U.S. Utility
	Sows	U.S. No. 1, U.S. No. 2, U.S. No. 3, Medium, Cull
Chicken	Rock cornish game hens	A, B, C
	Broiler or fryer	A, B, C
	Roaster	A, B, C
	Capon	A, B, C
	Stag	A, B, C
	Cock or rooster	A, B, C
	Hen, stewing chicken, fowl	A, B, C
Turkey	Fryer or roaster	A, B, C
	Young hen	A, B, C
	Young tom	A, B, C
	Yearling hen	A, B, C
	Yearling tom	A, B, C
	Mature or old turkey	A, B, C
Duck	Broiler or fryer	A, B, C
	Mature or old duck	A, B, C
Geese	Young goose	A, B, C
	Mature or old goose	A, B, C
Pigeon	Squab	A, B, C
	Pigeon	A, B, C

TABLE 15-2
Factors used to establish U. S. Department of Agriculture grades for meat

Beef Quality Grades	Beef Yield Grades
Class	Fat thickness
Maturity	Rib eye area
Buttons	Percent kidney, heart, and pelvic fat
Sacral vertebrae	Carcass weight
Color of lean	
Texture of lean	
Marbling	

Lamb Quality Grades	Lamb Yield Grades
Class	Fat thickness
Conformation	Percent kidney and pelvic fat
Maturity	Leg conformation
Shape and color of ribs	
Color of lean	
Feathering	
Flank streaking	
Firmness	

Pork Carcass Grades	Poultry Grades
Class	Class
Backfat thickness	Condition
Muscling	Conformation and fleshing
Carcass length or weight	Finish
	Defects
Color and structure of lean	Pinfeathers
Carcass firmness (of fat)	Exposed flesh
Belly thickness	Discoloration
	Bruises, freezing damage

lish a grade. Note that in each grade, the minimum degree of marbling that is required increases with increasing maturity. Beef carcasses that differ in quality grade are shown in Figure 15-6.

U.S. DEPARTMENT OF AGRICULTURE GRADES FOR CUTABILITY

In the broadest sense, *cutability* is indicative of the proportionate amount of salable retail cuts that can be obtained from a carcass. Beef, pork, and lamb carcass merchandising includes the trimming off of

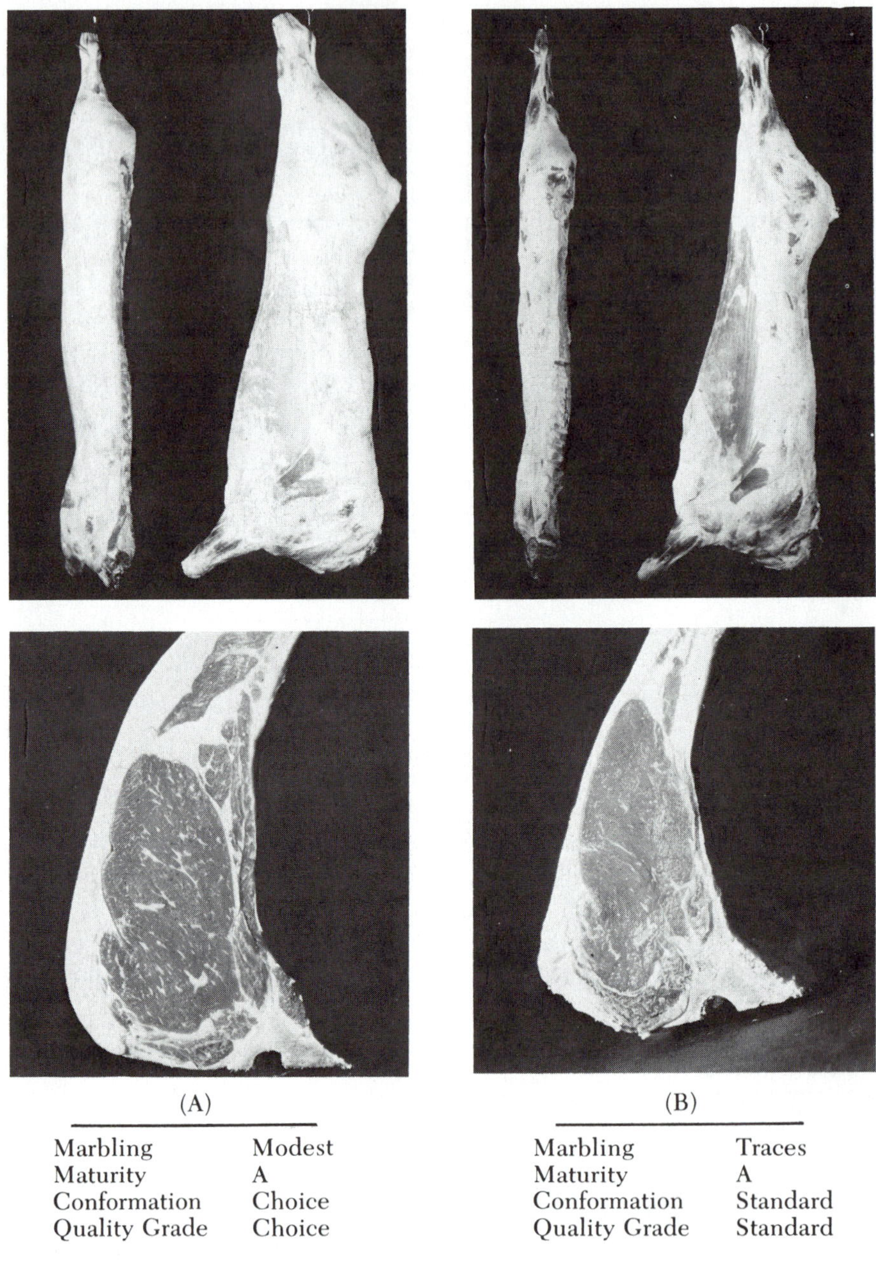

(A)
Marbling Modest
Maturity A
Conformation Choice
Quality Grade Choice
(B)
Marbling Traces
Maturity A
Conformation Standard
Quality Grade Standard

FIGURE 15-6
(A) U.S.D.A. Choice grade beef carcass. (B) U.S.D.A. Standard grade beef carcass. [Courtesy the National Live Stock and Meat Board, Chicago, Illinois.]

excess fat, the removal of many bones, and a subdivision of the carcass into retail cuts (such as steaks, chops, roasts, and ground meat) before their sale to consumers. The percentages of yield in such cuts varies widely among carcasses. Thus, carcasses that have the same quality grade designation may have quite different potential value because of differences in retail cut percentages. Grading systems have been established to segregate carcasses, based on their expected retail yield. For beef and lamb carcasses, these grades are called *yield grades*, and they are a separate and distinct grading system from the quality grades that were previously discussed. Beef yield grades identify carcasses according to differences in their yields of closely trimmed boneless retail cuts from the round, loin, rib, and chuck. Lamb yield grades group carcasses by expected yields of closely trimmed boneless retail cuts from the leg, loin, rack, and shoulder. Only one grading system is established for pork carcasses. It is fundamentally a yield grading system that is based on the expected yield of the four lean cuts (ham, loin, picnic shoulder, and Boston shoulder). Quality is included in pork carcass grades only to the extent that all carcasses are judged to have either an acceptable or an unacceptable quality. Acceptable quality carcasses must meet minimum standards for muscle color and structure, carcass firmness, and fat characteristics. The pork carcass grading system is not available to consumers for their use in selecting pork cuts of different grades, as is done with other species, because quality plays a relatively minor role in establishing carcass grades, compared to cutability.

The yield of retail cuts, as defined in the yield grade standards for beef and lamb carcasses, is popularly known as percent cutability. These percentages are given in Table 15-3. It should be remembered that percent cutability, as defined in the yield grade standards, only predicts the retail yield from the four major wholesale cuts (round, loin, rib and chuck in beef—leg, loin, rack and shoulder in lamb) of the carcass. Total yield of retail cuts that is obtainable from the carcass may be 20 percent or more above the values indicated by the yield grade standards. Similarly, the percentage of lean cuts accounts for only part of the yield from a pork carcass. The belly, which is a very valuable part of the pork carcass, is not included in the values given in Table 15-3.

Factors Used to Establish Cutability Grades

In general, quality grades are best described as subjective grades, and grades for yield or cutability as more objective in nature. Objective measurements are obtained from the carcass and are combined in an equation to arrive at a yield grade, a prediction of the expected retail yield of a carcass. The weight that each factor receives in the equation is based entirely on its value as a predictor of retail yield. Factors that are closely related to the retail yield receive more emphasis in the equation than do factors less closely related to retail yield.

TABLE 15-3
Expected percent cutability for beef and lamb yield grades, and expected percent lean cuts for barrow and gilt pork carcass grades

Beef or lamb yield grade	Percent cutability beef carcasses*	Percent cutability lamb carcasses**
1	52.4–54.6	47.3–49.0
2	50.1–52.3	45.5–47.2
3	47.8–50.0	43.7–45.4
4	45.5–47.7	41.9–43.6
5	43.1–45.4	40.1–41.8

Pork carcass grades	Percent lean cuts***
U. S. No. 1	53.0 and greater
U. S. No. 2	50.0–52.9
U. S. No. 3	47.0–49.9
U. S. No. 4	less than 47.0
U. S. Utility	Unacceptable quality

*Percent closely trimmed boneless retail cuts from the round, loin, rib, and chuck.

**Percent closely trimmed boneless retail cuts from the leg, loin, rack, and shoulder.

***Percent trimmed ham, loin, picnic shoulder, and Boston shoulder.

AMOUNT OF FAT. The amount of external, internal, and intermuscular fat on a carcass has more effect on percent retail yield than any other single factor. Excess finish must be removed in order to prepare attractive, easily merchandisable retail cuts. Considerable time and effort is required to directly measure the amount of excess finish on a carcass. However, some easily obtainable measurements can be made in order to estimate, with reasonable accuracy, the amount of finish on a carcass. These same measurements can also be used to predict cutability.

During the slaughter process, pork carcasses are split into right and left sides by a cut along both the dorsal and ventral midlines of the body. This cut exposes a cross section of each vertebra. The thickness of the layer of *backfat* (the subcutaneous or external fat that covers the pig's back) that is exposed when the carcass is split is a very accurate indicator of the total amount of fat in the carcass. Measurements of backfat thickness are taken opposite the 1st rib, the last rib, and the last lumbar vertebra. The average of these three measurements is recorded as being the *backfat thickness*.

In beef and lamb carcasses, the thickness of external finish over the *longissimus* muscle is measured as an indicator of the total external finish on the carcass. It is the industry practice to separate beef carcass sides into fore and hind quarters by a cut between the 12th and 13th ribs. A cross section of the *longissimus* muscle (rib eye or loin eye) is exposed by this cut and the fat thickness is measured at a point 3/4 of the width of the rib eye from the medial side. In lambs, fat thickness is measured over the midpoint of the *longissimus* muscle between the 12th and 13th ribs. Additional measurements of fat thickness over other areas of the carcass would improve the estimation of external finish, but these are not obtained because the improved accuracy does not offset the time and effort that is required.

Fat that surrounds the kidney, lines the pelvic channel, and surrounds the heart and lungs is left on beef and lamb carcasses during slaughter, and is chilled with the carcasses. This fat is removed and discarded when the carcasses are cut for retail sale. Thus, the amount of these internal fat deposits must be taken into consideration in the establishment of a yield grade. The amount of kidney, pelvic, and heart fat is estimated as a percentage of the carcass weight. An experienced grader can make these estimates with a high degree of accuracy.

MUSCLE DEVELOPMENT. The relative degree of development of the skeletal musculature has an important influence on the percent retail yield. In general, carcasses with less than normal muscle development will have a lower ratio of muscle to bone, and a lower cutability than more muscular carcasses. Several measures of muscle development are used to establish yield grades. The *rib eye area*, in beef carcasses, defined as the cross sectional area of the *longissimus* muscle between the 12th and 13th ribs, is easily obtained, because the side is quartered (ribbed). The *longissimus* muscle area is not used in pork or lamb carcass grades, even though it would contribute somewhat to the prediction of retail yield. This is partly because, in industry practice, pork and lamb carcasses are not ribbed. Instead, six degrees of muscle development, from "very thick" to "very thin," are recognized in pork carcass grading. A specific degree of muscle development is assigned, based on a subjective visual appraisal of the carcass conformation. In lamb carcass yield grading, the subjective conformation score is based on the shape and development of the muscles in the leg region. Measurements of muscle development, particularly the subjective scores for conformation, are not as accurate predictors of retail yield as are fat thickness measurements. Thus, they have less influence on the yield grade which is assigned.

CARCASS SIZE. Carcasses also differ in their percent retail yield due to differences in their degrees of finish and muscle development, in relation to their skeletal size. For example, a large carcass would be expected to have a larger rib eye area than a small carcass. Thus, in order to make accurate predictions of retail yield, the rib eye area must be considered in relation to the carcass size. Similarly, a larger carcass could have greater fat thickness than a smaller carcass, and still have the same expected retail yield. The most widely used indicator of skeletal size is carcass weight. As the carcass weight increases, beef carcasses must have larger rib eye area, in order to maintain the same yield grade. If rib eye area does not increase with the carcass weight, the estimated cutability will be lower. With pork carcasses, the backfat thickness can be greater on heavier carcasses without changing the carcass grade. However, it is general practice to measure carcass length in pork, rather than the carcass weight, as an indicator of skeletal size. The length of a pork carcass is defined as the distance from the anterior point of the aitch (pelvic) bone to the anterior edge of the first rib. The rate of compensation between backfat thickness and carcass length that is applied is 0.1 inch (0.25 cm) of backfat for each 3 inch (7.6 cm) increase in carcass length. Skeletal size measurements are not included in standards for lamb carcass yield grades.

Establishing Yield Grades

As is evident from the discussion of yield grades and cutability, a variety of factors are taken into consideration in order to predict the retail yield. The specific factors used for each kind of carcass are summarized in Table 15-2, and are listed in descending order of importance as to their affect on the percent retail yield. The following equation may be used to establish yield grades of beef carcasses: Beef Carcass Yield Grade = 2.50 + (2.50 × fat thickness, in inches) − (0.32 × rib eye area, in square inches) + (0.20 × percent kidney, heart, and pelvic fat) + (0.0038 ×

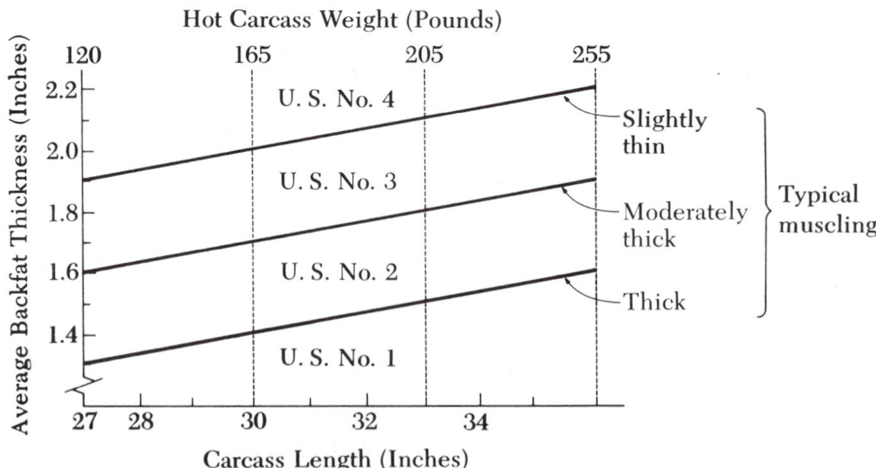

FIGURE 15-7
Relation between backfat thickness, carcass weight or length, and grade, for pork carcasses with muscling that is typical of their degree of fatness.

carcass weight in pounds). The following equation may be used to establish the yield grade of lamb carcasses: Lamb Carcass Yield Grade = 1.66 + (6.66 × fat thickness, in inches) + (0.25 × percent kidney and pelvic fat) − (0.05 × leg conformation score code). The leg conformation score is coded as follows: High Prime = 15, Average Prime = 14, Low Prime = 13, etc. For both beef and lamb, carcasses with calculated grades of 1.0–1.9 are designated as Yield Grade 1, and those with calculated grades of 2.0–2.9 are designated as Yield Grade 2. Yield Grades 3, 4, and 5 are handled in a similar manner. Figure 15-7 gives the relations among the backfat thickness, carcass weight (and length), and the grade of pork carcasses. U.S. No. 1 and No. 3 grades of pork carcasses are illustrated in Figure 15-8.

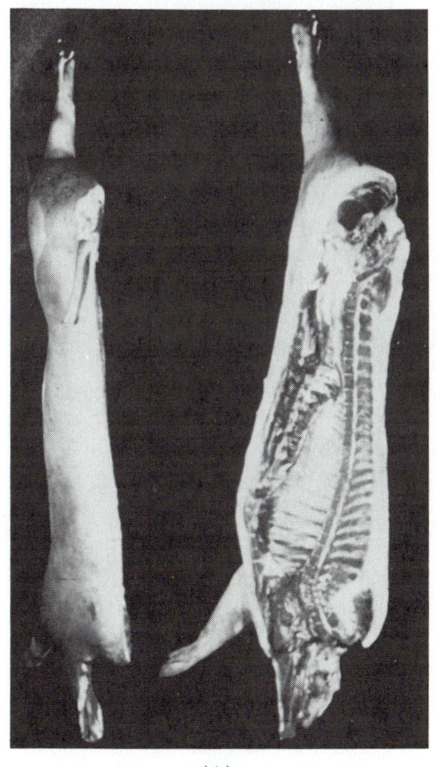

(A)

Backfat thickness	1.3 inches
Length	30.5 inches
Muscling	Thick
Carcass grade	U. S. No. 1

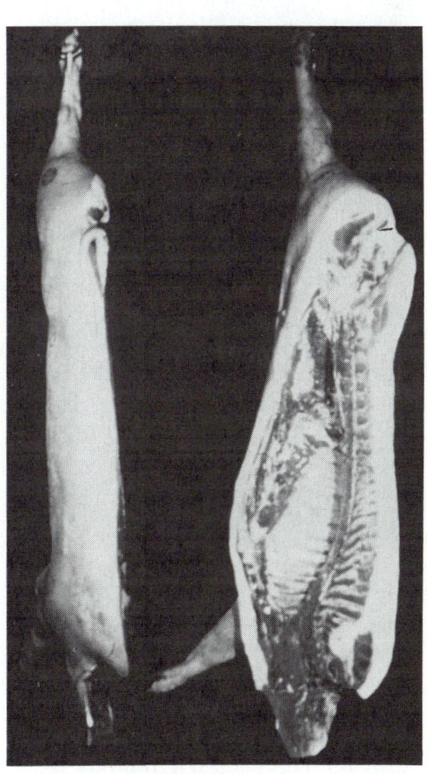

(B)

Backfat thickness	1.9 inches
Length	31.0 inches
Muscling	Slightly thin
Carcass grade	U. S. No. 1

FIGURE 15-8
(A) U.S. No. 1 and, (B) U.S. No. 3 grade pork carcasses.

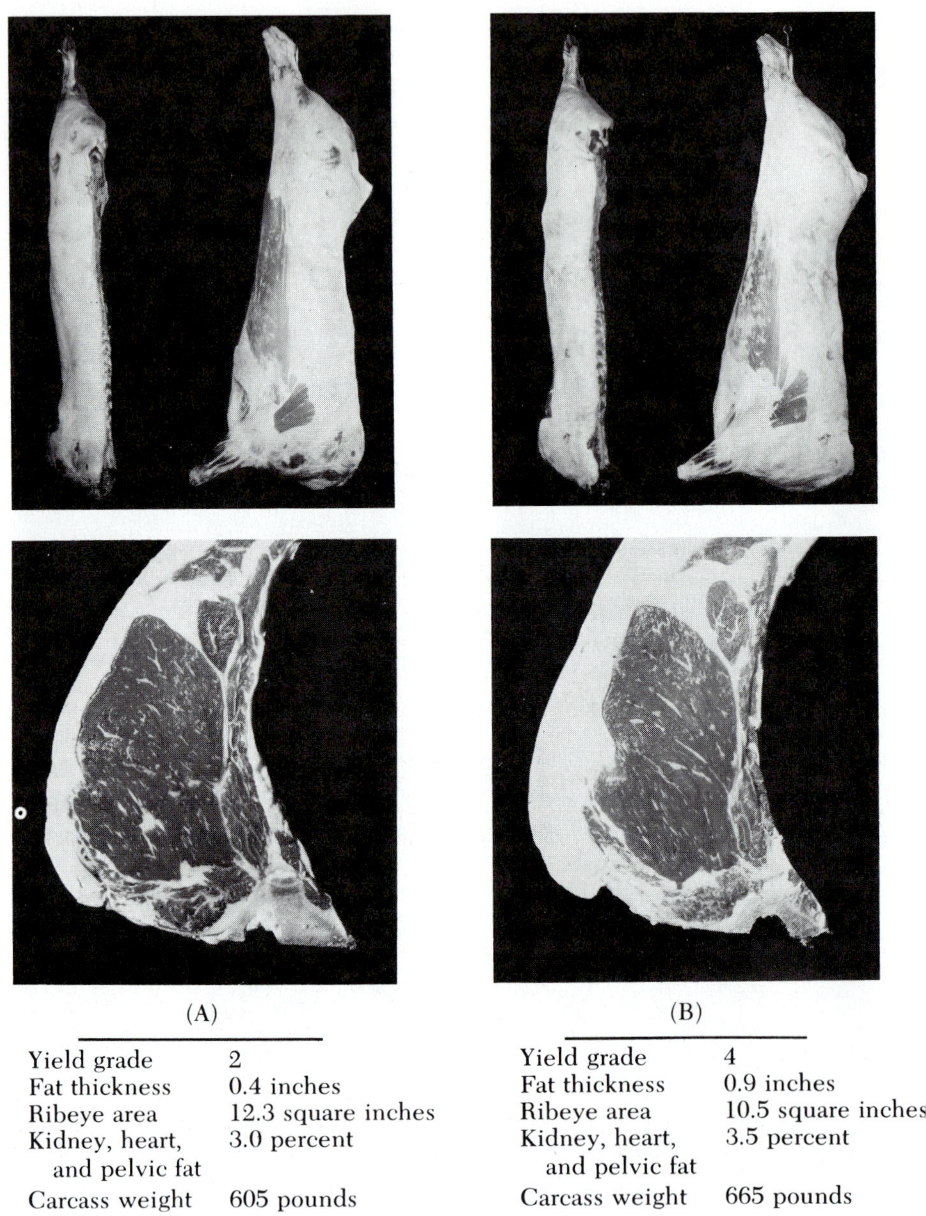

(A)	
Yield grade	2
Fat thickness	0.4 inches
Ribeye area	12.3 square inches
Kidney, heart, and pelvic fat	3.0 percent
Carcass weight	605 pounds

(B)	
Yield grade	4
Fat thickness	0.9 inches
Ribeye area	10.5 square inches
Kidney, heart, and pelvic fat	3.5 percent
Carcass weight	665 pounds

FIGURE 15-9
Examples of beef carcass yield grades. Both carcasses are U.S.D.A. Choice quality grade. (A) Yield Grade 2 and, (B) Yield Grade 4. [Courtesy The National Live Stock and Meat Board, Chicago, Illinois.]

TABLE 15-4
Comparison of yields of retail cuts and retail values between U.S. Department of Agriculture Yield Grade 2 and Yield Grade 4 beef carcasses, each weighing 600 pounds

	Weight of Retail Cuts from a 600 lb. Carcass*			Retail Value	
Closely Trimmed Retail Cuts	Yield Grade 2	Yield Grade 4	Price per Pound**	Yield Grade 2	Yield Grade 4
Rump, boneless	21.0 lbs.	18.6 lbs.	$1.65	$ 34.65	$ 30.69
Inside round	27.0	22.2	1.99	53.73	44.18
Outside round	27.6	25.2	1.89	52.16	47.63
Round tip	15.6	14.4	1.79	27.92	25.78
Sirloin	52.2	47.4	1.89	98.66	89.59
Short loin	31.2	30.0	2.04	63.65	61.20
Rib, short cut (7 in.)	37.2	36.0	1.59	59.15	57.24
Blade chuck	56.4	50.4	1.09	61.48	54.94
Chuck arm, boneless	13.8	11.4	1.49	20.56	16.99
Flank steak	3.0	3.0	1.89	5.67	5.67
Ground beef	141.0	115.8	1.24	174.84	143.59
Fat	76.0	137.4	0.10	7.62	13.74
Bone	59.4	53.4	0.05	2.97	2.67
Kidney	1.8	1.8	0.69	1.24	1.24
TOTALS	600.0	600.0		$716.64	$642.34

*Based on yields presented in Marketing Bulletin No. 45, USDA Yield Grades for Beef, U.S. Department of Agriculture, 1968.
**Prices are representative of retail prices for beef, January, 1974.

In order to emphasize the importance of cutability, the weights of various retail cuts that can be obtained from a Yield Grade 2 *vs* those obtainable from a Yield Grade 4 beef carcass, both of which weigh 600 pounds, together with the retail value of these cuts, are presented in Table 15-4. Both carcasses are U.S. Department of Agriculture Choice quality grade. Excluding bone and fat trim, 55.2 pounds more salable retail cuts are obtained from the Yield Grade 2 carcass, and its retail value is $72.30 greater. The retail values that are given include the wholesale cost of the carcasses plus the retailer's markup for operating expenses and profit. Figure 15-9 illustrates typical examples of Yield Grade 2 and Yield Grade 4 beef carcasses.

REFERENCES

Kiehl, E. R. and V. J. Rhodes, "Historical Development of Beef Quality and Grading Standards," University of Missouri Agricultural Experiment Station, Columbia, Missouri, Research Bulletin 728 (1960).

National Live Stock and Meat Board, Meat Evaluation Handbook. 36 South Wabash Avenue, Chicago, Illinois, 1969.

United States Department of Agriculture, "Official United States Standards for Grades of Carcass Beef," Agriculture Marketing Service (1973).

United States Department of Agriculture, "Official United States Standards for Grades of Lamb, Yearling Mutton and Mutton Carcasses," Agricultural Marketing Service, Service and Regulatory Announcements No. 123 (1960).

United States Department of Agriculture, "USDA Yield Grades for Beef," Agricultural Marketing Service, Marketing Bulletin No. 45 (1968).

United States Department of Agriculture, "USDA Yield Grades for Lamb," Consumer and Marketing Service, Marketing Bulletin No. 49 (1970).

United States Department of Agriculture, "USDA Grades for Pork Carcasses," Consumer and Marketing Service, Marketing Bulletin No. 52 (1970).

United States Department of Agriculture, Poultry Grading Manual. Consumer and Marketing Service, Poultry Division, Washington, D.C. Agriculture Handbook No. 31, 1971.

CHAPTER 16

By-products of the Meat Industry

Animal by-products include everything of economic value, other than the carcass, obtained from an animal during slaughter and processing. These products are classified as either being edible or inedible, based upon whether or not they are intended for human food.

The meat packing industry has long been noted for its efficiency in the processing and utilization of by-products. The ingenuity of this industry in devising new products and for discovering new uses for by-products has given rise to the expression that "the packer uses everything but the squeal." Although the value of by-products is only a small fraction of the current live animal value, they are of considerable economic importance to the entire livestock and meat industry. Their value influences the amount the livestock producer receives for animals, and the return on the live animal investment by the processor. Over the past several years, the value of by-products has declined with respect to the value of the live animal. This relative decline in value has been due largely to technological progress in producing competitive products from nonanimal sources. For example, various synthetic materials have been developed to make many items which were once made of leather. Synthetic fibers, instead of wool, are use to make clothing. In many food products, vegetable fats and oils have replaced animal fats. Synthetic detergents have replaced soaps, which were once made entirely from animal fats. Many other examples could be cited to illustrate the competition from nonanimal products.

TABLE 16-1
Annual by-product raw materials available in 1967 in the United States

	Cattle	Calves	Lambs and sheep	Hogs
Number of animals slaughtered, million	34	6	13	85
Liveweight, billion kg	16	0.6	0.7	9.4
Edible organs and trimmings at packer level, percent of liveweight	4	4	4	7
Total inedible products at packer level, percent of liveweight	39	40	37	40
Bones and fat at retail level, percent of liveweight	12	5	16	1
Total by-products all levels, percent of liveweight	55	49	57	48

SOURCE: Little, Arthur D. Inc., "Opportunities for Use of Meat By-Products in Human and Animal Foods," Report to Iowa Development Commission (1969).

TABLE 16-2
Approximate yield of various items obtained from meat animals

Item	Steer	Lamb	Pig
Grade	Choice	Choice	U.S. No. 1
Live weight, kg	455	45	100
Dressed carcass, kg	273	23	70
Retail cuts, kg	190	16	56
By-products, kg			
Hide or pelt	36	7	–
Edible fats	50	4	16
Variety meats	17	1	4
Blood	18	2	4
Inedible fats, bone, and meat scrap	80	10	8
Unaccounted items (stomach contents, shrink, etc.)	64	5	12

Enormous quantities of by-products are generated at each step of the meat processing and distribution system, as shown in Table 16-1. Yields of by-products vary greatly, depending upon the method of processing and the weight, grade, sex, and species of animal (Table 16-2). For example, whether beef tallow is saved and processed for edible or inedible use markedly influences its price. The value of by-products is greater when these products are processed immediately after slaughter. The great the distance from the slaughtering plant or the longer the elapsed time after slaughter, the greater the probability of edible materials being degraded to inedible products. In the case of inedible materials such as fats, the grade may be lowered because of free fatty acid buildup and the development of rancidity.

In addition to the monetary value that is derived from processed by-products, the conversion of inedible parts of the animal into useful products performs a very important function from a sanitary standpoint. All inedible parts, unless processed or disposed of in a proper manner, would accumulate and decompose, causing undesirable conditions in the surrounding environment.

EDIBLE MEAT BY-PRODUCTS

Numerous edible meat by-products are obtained during the slaughter and processing of meat animals. Some of the more common items are listed in Table 16-3. Items, such as tongue, brains, sweetbreads, heart, liver, and kidney are called variety meats. They are excellent sources of many essential nutrients that are required in the human diet, and offer many interesting and appetizing variations for the consumer. For example, liver is the best known and most widely served of all the variety meats. It ranks among the better food sources for vitamin A and iron.

The number of edible by-products, and the extent to which these items are processed for edible uses, varies among processors. Unless the processor has a potential market for an item, it may not be economically feasible to process it for edible use. Without a potential market, it might become more profitable to process the item for an inedible use.

In recent years there has been a decline in the general acceptance of variety meats. This may be due to the image, held by some people, that variety meats are foods for poor people who cannot afford steak. However, various ethnic groups have traditionally been large consumers of variety meats. Variety meats are also popular menu items in many of the finer restaurants. They could potentially become a pleasurable source of food for large segments of the population who do not now eat them.

TABLE 16-3
Edible usage of by-products obtained from meat animals

Raw by-product	Principal use
Brains	Variety meat
Heart	Variety meat
Kidneys	Variety meat
Liver	Variety meat
Spleen (melt)	Variety meat
Sweetbreads	Variety meat
Tongue	Variety meat
Oxtails	Soup stock
Cheek and head trimmings	Sausage ingredient
Beef extract	Soups and bouillion
Blood	Sausage component
Stomach	
(a) Suckling calves	Rennet for cheese making
(b) Pork	Sausage container, sausage ingredient
(c) Beef (1st and 2nd)	Sausage ingredient, variety meat (tripe)
Bones	Gelatin for confectioneries, ice cream, and jellied food products
Fats	
(a) Cattle, calves, lambs and sheep	Shortening Candies, chewing gum
(b) Pork	Shortening (lard)
Intestines, small	Sausage casings
Intestines, large (pork)	Variety meat (chitterlings)
Intestines, large	Sausage casings
Esophagus (weasand)	Sausage ingredient
Pork skins	Gelatin for confectioneries, ice cream, and jellied food products; french fried pork skins
Calf skin trimmings	Gelatin for confectioneries, ice cream, and jellied food products

Almost all of the edible by-products, especially some of the variety meats, are more perishable than the carcass. Liver is one of the most perishable of these items. Therefore, these items must be chilled quickly after slaughter, and be processed or moved quickly into the retail trade. Variety meats can be merchandised as fresh or frozen items, or used in making other processed items.

The following brief discussion will illustrate the source and use of many edible by-products. (For detailed descriptions of the methods used for processing these various items, it is suggested that readers refer to the references listed at the end of this chapter.)

Beef extract is a very useful beef by-product. It consists of the concentrated cooking water from the heating of beef during the canning process. This liquid is usually concentrated by evaporation under vacuum to about a 20 percent moisture content, and sets up, upon cooling, as a pasty solid. It is a major ingredient in bouillion cubes and bouillion type broth, and is used as a flavoring agent in gravies. It is not produced to any extent in this country, but is made primarily in South America in the course of producing canned beef items.

The testicles of lambs and calves are sold fresh or frozen. When cooked they are commonly known as "fries" or "mountain oysters."

Blood is used as a component of blood sausages, and special hygienic precautions must be taken when it is collected for edible use. Recently, ultrafiltration processes have been used to recover proteins from both the plasma and cellular components of blood. These blood proteins have many potential uses as a binder material in sausage, and in the manufacture of other food products.

Tripe is obtained from the first (rumen) and second (reticulum) stomach compartments of cattle. It is used in some sausages, and is also sold to the consumer as a variety meat. Pig stomachs are also processed for use as an ingredient in some sausages. The Pennsylvania Dutch use the cleaned stomach of the pig as a container for a delectable meat dish called "filled pig stomach," or "hog maw."

Chitterlings are made from the thoroughly cleaned and cooked intestines of the pig, and are sold as a variety meat.

Cheek and head trimmings are commonly used in the manufacture of various sausages.

Brains, which are regarded as a delicacy by some people, are sold to the consumer either chilled, frozen, or canned.

Beef tongues are used fresh, cured and smoked, canned, or as an ingredient in sausage items. Sheep and pig tongues can be used fresh, but they are most commonly used in making other processed meat items.

Kidneys are used fresh, or they can be frozen for the meat trade. However, they are more commonly used as an ingredient in pet food.

Sweetbreads are the thymus glands obtained from the necks of calves or young cattle. These are used fresh or frozen.

Hearts are used fresh, frozen, or as an ingredient in processed luncheon meats.

Liver is used fresh, frozen, or as an ingredient in liver sausages or other processed luncheon meats. Comparatively, more beef liver is used fresh, whereas more pork liver is used in making sausages. However, kosher style liver sausages include only beef or lamb liver.

Oxtails are used mainly for making soups.

Spleens can be used for retail sale as "melt," or for manufacturing pet and mink foods.

Gelatin is made by heating collagen rich connective tissues in hot water. Tissues such as pork skins, calfskins, and bone are used to make gelatin. Gelatin is used in the making of other processed meats; and it is widely used in gelatin desserts, consommes, marshmallows, candies, bakery products, various dairy products, and ice cream. The pharmaceutical industry uses gelatin in making items such as capsules, ointments, cosmetics, and emulsions. Gelatin is also used in the manufacture of photographic films and paper, and in the paper and textile industries.

Historically, animal intestines have been used as edible sausage casings. However, as was pointed out in Chapter 9, most sausages are currently processed in synthetic casings that are made from cellulose. Edible regenerated collagen casings, which have characteristics of both synthetic and natural casings, are being produced commercially. Animal hides are the source of the collagen used in these casings.

Fatty tissues (adipose tissue), obtained from various parts of meat animals, are processed into edible fats. These fats are *rendered* (separated from their supporting tissues) from fatty tissues by a variety of heat treatments. Lard and rendered pork fat are obtained from the fatty tissues of pork, and are used as *shortenings.* The fatty tissues of beef yield edible tallow that is used in making shortenings. Shortenings are used for frying and tenderizing foods, particularly baked foods, by interposing fat throughout the food in such a manner that the protein and carbohydrates do not cook into a continuous hardened mass. If fatty tissues are processed at low temperatures, and the fat is removed by centrifugation, the remaining protein residue can be used as an ingredient in processed luncheon meats. If the fat is extracted at higher temperatures, the protein residue, known as cracklings, is used in animal feeds.

Any discussion of the potential uses of edible by-products is necessarily incomplete, because the creativity and ingenuity of the meat processor is continually generating more practical uses for these items. The unique properties and nutritional value of these meat by-products should be sufficient encouragement for the meat industry to continue making concerted efforts at creating new and useful products.

INEDIBLE MEAT BY-PRODUCTS

A list of some of the more important inedible by-products that are obtained from meat animals is presented in Table 16-4. The use of some of these processed by-products are varied, and are almost unlimited in number. New uses are continually being found for these products, while other nonmeat products are replacing existing meat by-products.

TABLE 16-4
Inedible by-products obtained from meat animals, and their major usages

Raw by-product	Processed by-product	Principal use
Hide (cattle and calves)	Leather	Numerous leather goods
	Glue	Paper boxes, sandpaper, plywood, sizing
	Hair	Felts, plaster binder, upholstery
Pork skins	Tanned skin	Leather goods
Pelts	Wool	Textiles
	Skin	Leather goods
	Lanolin	Ointments
Fats		
(cattle, calves, lambs, and sheep)	Inedible tallow	Industrial oils, lubricants, soap, glycerin
(cattle, calves, lambs, sheep, and hogs)	Tankage	Livestock and poultry feeds
	Cracklings	
	Stick	
(hogs)	Grease	Industrial oils
		Animal feeds
		Soap
Bones	Dry bone	Glue
		Hardening steel
		Refining sugar (bone charcoal)
	Bone meal	Animal feed
		Fertilizer
	Blood albumen	Leather preparations, textile sizing
Cattle feet	Neatsfoot stock	Fine lubricants
	Neatsfoot oil	Leather preparations
Glands	Pharmaceuticals	Medicines
	Enzyme preparations	Industrial uses
Lungs (all species)		Pet foods

Hides, Skins and Pelts

CLASSIFICATION. Cattle hides and skins are classified on the basis of hide weight, sex, maturity, the presence and location of brands, and the method of curing. Hides and skins are technically divided into *hides*, *kips*, and *skins*. These three classes are based on the weight of the clean hide (Table 16-5). Hides come from large and mature animals. Kips are the skins of immature animals, and skins are from small animals, such as calves. The presence of brands burned into the hide affects its value.

TABLE 16-5
Classification of cattle hides

Origin	Weight* (Kilograms)	Classification
Unborn calf		Slunk skin
Calf	Less than 4	Light calf skin
	4–7	Heavy calf skin
	7–11	Kip skin
	11–14	Overweight kip skin
Cow	14–16	Light cow hide
	Greater than 24	Heavy cow hide
Steer	Less than 22	Extra light steer hide
	22–26	Light steer hide
	Greater than 26	Heavy steer hide
Bull	27–54+	Bull hide
Stag		Accepted as steer or bull hide depending upon characteristics

SOURCE: American Meat Institute Foundation, *The Science of Meat and Meat Products* (W. H. Freeman, San Francisco), 1960.
*Directly from the animal.

TABLE 16-6
Classification of sheep pelts

Classification	Subclass Number	Wool length (centimeters)
Shearlings	1	1.25–2.5
Shearlings	2	0.63–1.25
Shearlings	3	0.31–0.63
Shearlings	4	0.31
Fall clips	–	2.5–3.75
Wool pelts	–	3.75

SOURCE: American Meat Institute Foundation, *The Science of Meat and Meat Products* (W. H. Freeman, San Francisco), 1960.

Consequently, branded hides are known as Colorado or Texas hides, and unbranded hides are known as natives. When hides are taken off by skilled workmen in packing plants, they are known as standard packer hides. Hides that are taken off by unskilled workmen are referred to as country hides.

The skins and accompanying wool from sheep are known as pelts. Sheep pelts are classified on the basis of wool length, as shown in Table 16-6. After the wool is removed from pelts, they are classified as skins.

It is a common practice to leave the skin on pork carcasses until after they have been chilled. Most pigskins come from the skin that has been removed from the belly and backfat after the carcass has been chilled and cut into wholesale cuts.

PROCESSING. After cattle hides and skins are removed from the carcass, it is a common practice to trim off the ears, lips, and tail. Most trimmed hides are preserved either by curing with solid salt or with brine. The purpose of curing is simply to preserve the hides against bacterial decomposition until they can be processed by the tannery. The process of curing with salt consists of spreading salt over the flesh side, and stacking the hides in a pack for about 30 days or more.

The process of brine curing has become quite common in recent years. In this process the fresh hides are washed to remove dirt and manure, trimmed of excess fat and flesh, and then submerged in vats of saturated brine. In order to facilitate curing, paddle wheels in the vats keep the hides moving in the brine. The curing process requires about 24 hours. After the hides are removed from the vat they are drained and packed for shipment to the tannery.

The removal of hair from cattle hides and skins is usually done at the tannery by soaking them in a lime water solution. The actual tanning processes are rather complex, and their specifics depend upon the ultimate use of the finished product. Basically, leather is made by either of two processes; *vegetable tanning* or *chrome tanning.* The vegetable tanning process consists of suspending the hides and skins in a vat containing a solution prepared from tannin bearing woods or barks. In the chrome process, a solution of basic chromium sulphate is used. The tannin or chromium permeates each collagen fiber of the skin and converts it into leather. After tanning, the leather is subjected to additional operations to attain various final characteristics. These processes include dying, lubrication, filling, impregnation, flexing, surface coating, embossing, and polishing.

Sheep pelts must be cooled after removal from the animal. Then they may be salted to preserve the skins, depending upon the time that will elapse before further processing. Next, the fresh or salted pelts are thoroughly washed in cold water, and then the excess water is removed in a centrifugal machine. The wool is removed from the skin by painting the flesh side with a paste containing sodium sulfide and lime, and allowing the pelts to hang for about 24 hours. These reagents loosen the wool roots so that the wool can be easily pulled away from the skin. The pulled wool is sorted and graded as it is removed. The skins are then treated with an alkaline sulfide solution that dissolves any residual wool. Following this treatment, the chemicals are removed by a thorough washing. Next, the skins are treated with ammonium salts and pancreatic enzymes to remove extraneous protein material. They are washed once again, and then are preserved by being soaked in a solution of salt and sulfuric acid. After draining, the skins are graded and are ready for tanning.

Calf skins, as well as pig skins, are generally removed from the carcass after it has been chilled. They are then cured, either in dry salt or in brine, to preserve them until they are processed by the tannery.

WOOL AND HAIR. Although most wool produced in this country is shorn from sheep, pulled wool is an important by-product of the packing industry. As the wool is pulled from each sheep pelt by the handful, it is graded. Although various grading systems may be used, the wool is generally graded on the basis of its cleanliness, length, and fineness of the wool fiber. After removal and grading, the wool is washed, dried, and shipped to processing mills. Pulled wool can be used in making blankets, felt, carpets, and fabrics; in other words, for any product in which shorn wool can be used. Fabrics can be made entirely with wool, or the wool can be blended with cotton or synthetic fibers.

The processing of wool results in other by-products of commercial value. One of the best known of these is lanolin, a refined form of wool grease, which is recovered in the washing process. Lanolin is the base from which many ointments and cosmetics are made.

Cattle hair and hog hair have been extensively used for furniture upholstery, carpet paddings, insulation, and brushes. However, these uses for hair are steadily declining, due to a strong competition by synthetic materials. Because of this situation, the disposal of waste hair is becoming an increasing problem.

Tallows and Greases

Inedible fats are classified as tallow or grease mainly on the basis of their *titer*, which is the congealing or solidification point of the fatty acids in the fat. Any fat having a titer above 40°C is classed as a tallow. Fat with a titer below 40°C is classed as a grease. Most cattle and sheep fats are tallow; pork fat is classified as grease.

The sources of inedible tallows and greases include animals that die in transit, diseased and condemned animals and parts, and waste fat and trimmings from retail meat markets, hotels, and restaurants. However, a large portion of the inedible fat produced by packinghouses comes from edible fats that had a potential market as such.

Most inedible fats are processed by the *dry rendering process*. The fatty tissues are ground and placed into a horizontal, steam-jacketed cylinder equipped with a set of internal rotating blades. Rendering may be accomplished at atmospheric pressure, at an elevated pressure, or under a partial vacuum. The fat cells are ruptured, and the melted fat is released from the supporting tissues. When sufficient moisture has cooked out, the mixture is filtered or strained to remove the cracklings from the rendered tallow or grease.

Many uses have been found for the products obtained from inedible tallows and greases. These fats can be split, by the action of acids or bases, into glycerine and fatty acids. Glycerine is used in the manufacture of pharmaceuticals, explosives, cosmetics, transparent wrapping materials, paints, and many other products. Likewise, fatty acids have many industrial uses, such as in the manufacture of soaps and detergents, wetting agents, in insecticides and herbicides, cutting oils, paints, lubricants, and even as an additive to asphalt.

The mixed feed industry uses large quantities of stabilized inedible tallows and greases. Fats make feedstuffs less dusty, more palatable, and facilitates the pelleting process. They also add energy to the feed. Fats used in animal feeds are stabilized against rancidity by the addition of antioxidants, such as butylated hydroxyanisole (BHA) and butylated hydroxytoluene (BHT), during the chilling that follows the rendering process.

Animal Feeds and Fertilizers

Most large packing plants have facilities for rendering inedible materials produced during their slaughter operations. Independent rendering plants collect and process materials from slaughter and processing plants that do not have their own rendering facilities. The end products of the rendering operations are fats and proteinaceous materials. After they are separated from the fats, the proteinaceous materials are dried and ground. These protein concentrates are quite valuable, and are used as protein supplements in feeds for pigs, chickens, and pets.

The pet food industry has grown very rapidly, and uses large quantities of select meat by-products. Such items as liver, spleen, lung, meat meal, horse meat, and cereal products are used in making dry, semimoist, and canned cat and dog food.

Dried blood (blood meal) is made by coagulating fresh blood with steam, draining off the liquid, and drying the coagulum. This dried blood is a rich source of protein, and it is used as an ingredient in animal feeds. Blood meal is used rather extensively in the formulation of feed for commercial fish operations. Meat meal (proteinaceous materials from the inedible rendering process) and organs, such as liver, are also used for fish foods.

Steamed bone meal is made by cooking bones with steam, under a high pressure, in order to remove any fat and meat that may be left on them. The dried bone is then ground up and used as a calcium and phosphorus supplement in animal feeds.

The use of animal by-products in fertilizers is limited almost entirely to the manufacture of speciality fertilizers for home gardening use, which represents a very small proportion of fertilizer production.

Glue

Chemically and physically, glue and gelatin are very much alike. The raw materials from which glue is made include the skins or hides, connective tissues, cartilage, and bones of cattle and calves. Glue is extracted from these materials by successive heatings in water under specific temperature conditions. Cooking in water converts the collagen in these materials to gelatin. The extracts are concentrated, dried, and ground up. Glue has many and varied uses in the woodworking, paper, and textile industries, and its manufacture dates back more than 3000 years to the cabinet makers of Egypt. Blood albumen (obtained from blood plasma after the red cells have been removed) is now used to make an adhesive, almost all of which is used in manufacturing plywood and wood veneers.

Pharmaceuticals

The endocrine glands in the animal body secrete hormones that exert specific effects on the physiological functions of the body. These substances, when extracted from the endocrine glands, have great value in treating disorders and diseases of both humans and animals.

Some of the endocrine glands are removed during the slaughter process, and are immediately chilled and frozen. The frozen glands are shipped to pharmaceutical plants where extraction, concentration, and purification processes are performed to prepare the final products for use in the medical profession.

The following subsections list examples of the animal tissues and organs that yield useful pharmaceutical products. The location of the various endocrine organs is illustrated in Figure 4-13.

GLANDS

ADRENAL. *Epinephrine* is extracted from the adrenal medulla and adrinocortical extract from the adrenal cortex of the adrenal glands of cattle, hogs, and sheep. Epinephrine is used to stimulate the heart, and as a remedy in treating bronchial asthma. The cortical extract is a source of several steroid hormones that are used to treat adrenal hormone deficiencies.

OVARIES. Bovine ovaries yield *estrogens,* which are used in the treatment of menopausal syndromes, and progesterone, which is used to prevent abortion.

PANCREAS. The pancreas yields *insulin*, which is used as a palliative for the treatment of diabetes. *Trypsin* is also obtained from the pancreas, and is used to liquify and remove necrotic, abscessed, or infected tissue. *Chymotrypsin* from the pancreas is also used to remove dead tissue in wounds and lesions in order to promote healing.

PARATHYROID. *Parathyroid hormone* extract, from beef parathyroids, is used to prevent large scale muscular rigidity and tremors (tetany) in humans whose glands have been removed or function improperly.

PITUITARY. A number of hormones, including the *adrenocorticotropic hormone* (*ACTH*), is produced by, and can be obtained from, the anterior lobe of the pituitary of beef, sheep, and pigs. This hormone stimulates the adrenals, and is used to treat several disorders associated with a deficiency of this hormone, as well as many other diseases and conditions, including arthritis, acute rheumatic fever, and numerous inflammations.

TESTES. *Hyaluronidase*, obtained from bull testes, is used as a "spreading factor" in combination with other drugs. This enzyme has the ability to penetrate body cells very easily, and thus it can increase the rate of distribution of an administered drug, and intensify its effect.

THYROID. Dessicated thyroid, and thyroid extract, from both cattle and hogs are used to treat humans having a deficiency of hormones produced by this gland. *Thyroxine* and *calcitonin* are the major hormones that are obtained from the thyroid.

TISSUES AND ORGANS

BLOOD. In addition to its industrial uses, blood yields many pharmaceutical products. Purified bovine *albumen* is used as a reagent in testing for the presence of Rh factor in human blood, as a stabilizer for vaccines and other sensitive biological products, in antibiotic sensitivity tests, and in microbiological culture media. Blood is also a source for amino acids that are used in the intravenous feeding of hospital patients.

BONE. Purified bone meal is used as a source of calcium and phosphorus in pediatric foods.

INTESTINES. The small intestine of sheep is made into surgical ligatures for suturing internal incisions or wounds. This product consists mainly of collagen which the enzymes of the body will subsequently digest.

LIVER. *Liver extracts* are used in treating pernicious anemia. However, since vitamin B_{12} was isolated from liver, and synthesized, the use of liver extract for the treatment of this anemia has declined. *Bile extract*, obtained from the bile of cattle, is used to increase the secretory activity of the liver. Bile extract can also be used to make *cortisone*, an adrenalcortical steroid hormone with anti-inflammatory properties similar to those of ACTH.

LUNGS. *Heparin* is obtained from lungs or liver, and is used as an anticoagulant to prevent blood clots.

SPINAL CORD. The spinal cord from cattle is a source of *cholesterol*. The principal use of cholesterol is in the preparation of vitamin D.

STOMACH. *Rennet*, from the stomach of calves, is used to curdle milk in the cheese making process. This enzyme can also be added to the diet of infants to aid in their digestion of milk. *Mucin* is obtained from pig stomachs, and is used in the treatment of ulcers. *Pepsin* is also obtained from pig stomachs, and was used at one time as an aid to digestion.

This list of preparations obtained from animal tissues, although incomplete, will serve to illustrate the importance of animals and the meat packing industry to our health and well-being.

REFERENCES

American Meat Institute Committee on Textbooks, *Beef, Veal and Lamb Operations* (Institute of Meat Packing, University of Chicago), 4th Rev. ed., 1945.

American Meat Institute Committee on Textbooks, *By-Products of the Meat Packing Industry* (Institute of Meat Packing, University of Chicago), Rev. ed., 1953.

American Meat Institute Committee on Textbooks, *Pork Operations* (Institute of Meat Packing, University of Chicago), 6th Rev. ed., 1957.

Clemen, Rudolf A., *By-Products in the Packing Industry* (The University of Chicago Press), 1927.

Little, Arthur D., Inc., "Opportunities for Use of Meat By-Products in Human and Animal Foods," Report to Iowa Development Corporation, 1969.

Price, J. F. and B. S. Schweigert, *The Science of Meat and Meat Products* (W. H. Freeman and Company, San Francisco), 1971.

CHAPTER 17

Meat Identification

This final chapter is devoted to the identification of wholesale and retail cuts of beef, veal, pork, lamb, and retail parts of chicken. The skeletal diagrams, sketches, and photos of retail cuts that are presented here will enable students to identify them in relation to their anatomical location. In the identification of cuts, the skeletal and muscle structures, cut size, and the color of the lean are the main factors to be considered.

A standardized cutting of, and nomenclature for, wholesale and retail cuts are essential prerequisites for good merchandising and the proper utilization of meat. Since the various cuts differ in composition and tenderness, the lean cuts are separated from the fat cuts, the more tender from the less tender, and the thick from the thin cuts. The more tender cuts come from the loin and rib, which contain muscles that were used for support. By comparison, cuts from the leg, containing muscles that were used for locomotion, are less tender. The belly of the pork carcass is a fat cut, for example, and is separated from the lean cuts—ham, loin, picnic shoulder, and Boston style shoulder.

A distinguishing characteristic among beef, veal, pork, and lamb is the color of the lean. Usually beef is a bright cherry red, pork is light grayish pink, veal is brownish pink, and lamb is a light to brick red. Poultry meat is usually a grey-white to dull red, but its color is rarely used as a means of identification.

Skeletal diagrams of beef, pork, and lamb carcasses are presented in Figure 17-1. The anatomical and common names for the major bones

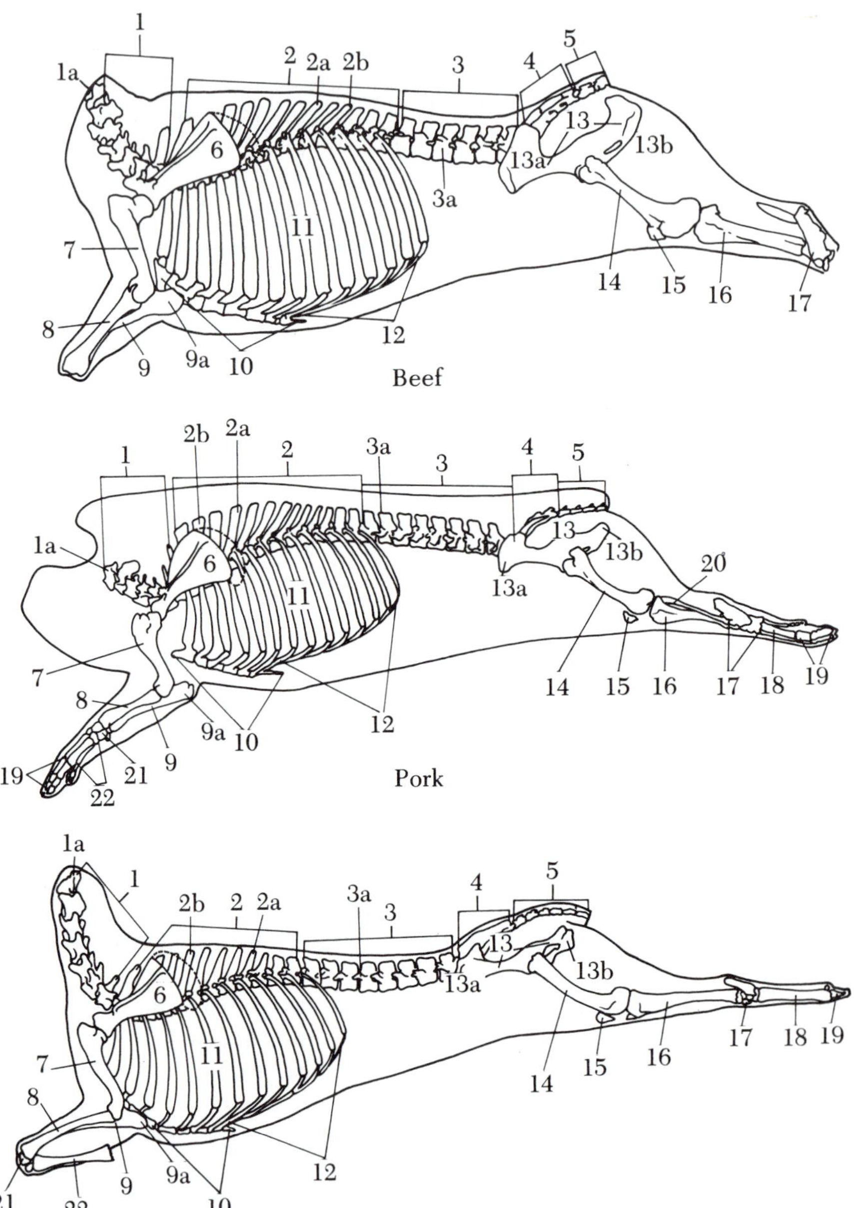

FIGURE 17-1
Beef, pork, and lamb skeletal diagrams. The skeletal anatomies of veal and mutton are the same as those of beef and lamb, respectively. See Table 17-1 for anatomical and common terms used for bones. [Adapted from Lessons on Meat, courtesy National Live Stock and Meat Board.]

are presented in Table 17-1. The bone names are the same for the three species, but the styles of dressing results in more bones being removed from the fore and hind legs of the beef carcass than from lamb and pork carcasses. Also, the fore and hind feet are removed from lamb, but not from pork, during slaughter. There are some differences among the three

TABLE 17-1
Anatomical and Common Terms Used For Bones From the Beef, Pork and Lamb Carcass*

Bone number (See Figure 17-1)	Anatomical term	Common term
1	Cervical vertebrae	Neck bone
1a	First cervical vertebrae	Atlas
2	Thoracic vertebrae	Back bone
2a	Spinous process	Feather bone
2b	Cartilage of spinous process	Button
3	Lumbar vertebrae	Back bone
3a	Spinous process	Finger bone
4	Sacral vertebrae	Back bone
5	Coccygeal vertebrae	Tail bone
6	Scapula	Blade bone
7	Humerus	Arm bone
8	Radius	Foreshank bone
9	Ulna	Foreshank bone
9a	Olecranon process	Elbow bone
10	Sternum	Breast bone
11	Ribs	Ribs
12	Costal cartilages	Rib cartilages
13	Pubis	Pelvic bone
13a	Ilium	Hip bone or pin bone
13b	Ischium	Aitch bone
14	Femur	Leg (round) bone
15	Patella	Knee cap
16	Tibia	Hind shank bone
17	Tarsal bones	Hock bones
18	Metatarsal	Hind shank bone
19	Phalangeal bone(s)	Hind shank bone (lamb), foot bones (pork)
20	Fibula	Hind shank bone
21	Carpal bones	Shank bones (lamb), front foot bones (pork)
22	Metacarpal bone(s)	Shank bones (lamb), front foot bones (pork)

*Bones of veal and mutton carcasses are the same as those for beef and lamb, respectively.

species in the number of bones. For example, beef and lamb each have 13 thoracic vertebrae, but their number is variable in pork, usually being between 13–15. Lamb and pork have seven lumbar vertebrae, and beef have six. Pigs have a tibia and fibula, but cattle and sheep have only a tibia.

The names of retail cuts frequently are related to and are named for bone structures. These include rib roast, rib chops, rib steak, blade roasts, arm roasts, arm chops, short ribs, neck bones, spareribs, riblets, T-bone, wedge bone sirloin, flat bone sirloin, and pin bone sirloin.

A skeletal diagram of the chicken is presented in Figure 17-2. Many chicken bones have the same names as those of beef, pork, and lamb. However, when they are compared to the skeletons of mammals, the vertebral trunk of the chicken and other birds is seen to have several fused vertebrae. The last four thoracic, three lumbar, two sacral, and the five coccygeal vertebrae are fused together to form a rigid bone called the *lumbosacral.* This bone gives rigidity to the body in flight, and enables the bird to walk with two legs while maintaining its body in a horizontal position. The strength of the flexible back of a mammal is produced by the combination of an articulated backbone with a complex system of attached muscles. In birds, the backbone is rigid and the muscular system connected to it is extremely reduced in size compared to that of pigs, lambs, and cattle. The sternum (keel bone) of the chicken is the largest bone in its body, and is not segmented as in mammals. The two clavicular bones are fused with each other to form the forked clavicle, commonly known as the wishbone. Mammals always have seven cervical vertebra, but the chicken has thirteen. The chicken has seven thoracic vertebrae, four of which are fused together. Cattle and sheep have thirteen thoracic vertebrae, and pigs have a variable number.

Certain locations on the skeleton, as illustrated in Figure 17-3, are used as a guide in making wholesale cuts. For example, the chuck of beef and shoulder of lamb are removed from the wholesale rib, between the fifth and sixth ribs. The Boston style shoulder, and picnic shoulder, of pork are separated from the loin and belly between the second and third ribs.

Since the beef carcass is so large, it is impractical to transport it in one piece from the packer to the wholesaler and/or retailer. Consequently, the beef carcass is split into sides, and the sides are cut into fore- and hindquarters. Currently, approximately half of the beef destined for the retail, and hotel, restaurant, and institutional trade (HRI) is processed by the packer into *primal* or *subprimal cuts.* A primal cut is a major wholesale cut, such as beef rib, and a subprimal cut is a large cut, such as a rib roast, which is prepared from a primal cut. These cuts are usually vacuum packaged, boxed, and shipped as "boxed beef." Veal and lamb are small enough to be handled without being cut into sides.

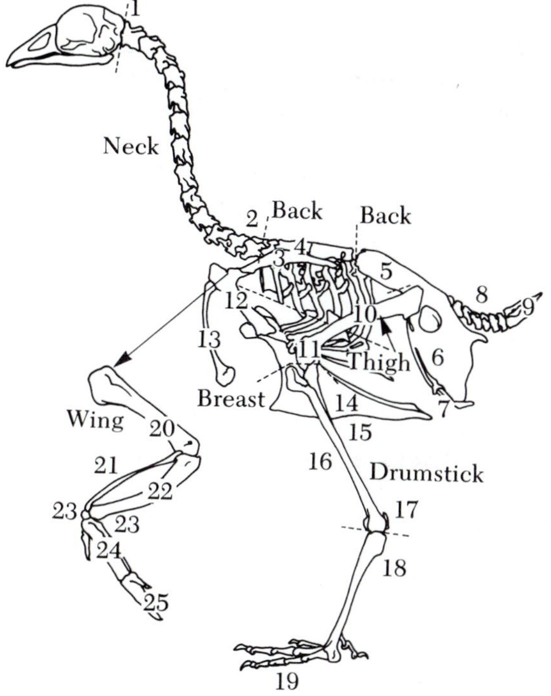

FIGURE 17-2
Skeleton of the chicken with parts and bones identified. The parts are named on the illustration, and dashed lines indicate locations where the parts are cut. Bones are identified with numbers, as follows: (1) first cervical vertebra (atlas), (2) last cervical vertebra, (3) scapula, (4) fused thoracic vertebrae, (5) ilium, (6) ischium, (7) pubis, (8) coccygeal vertebrae, (9) pygostyle, (10) femur, (11) patella, (12) coracoid, (13) clavicle, (14) sternum, (15) fibula, (16) tibia, (17) sesamoid, (18) metatarsus, (19) bones of toes (phalanges), (20) humerus, (21) radius, (22) ulna, (23) carpal bones, (24) metacarpus and, (25) finger bones (phalanges).

Sometimes these carcasses are cut into fore- and hindsaddles, or even into wholesale cuts to facilitate their merchandising. Pork carcasses are cut into sides and are small enough to be transported easily as such, but generally they are not. This is because many pork products, such as cured and smoked hams, bacon, sausage, and lard are processed at the packing plant. Therefore, retailers usually receive pork in the form of wholesale or even as retail cuts. The approximate yields of wholesale cuts from beef, veal, lamb, and pork carcasses are given in Table 17-2.

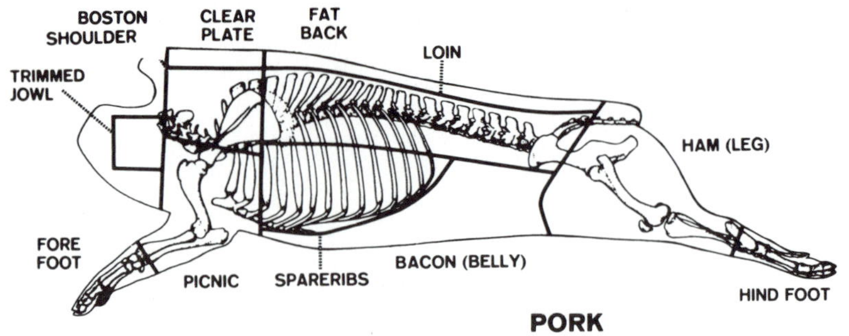
BOSTON SHOULDER
CLEAR PLATE
FAT BACK
LOIN
TRIMMED JOWL
HAM (LEG)
FORE FOOT
PICNIC
SPARERIBS
BACON (BELLY)
HIND FOOT
PORK

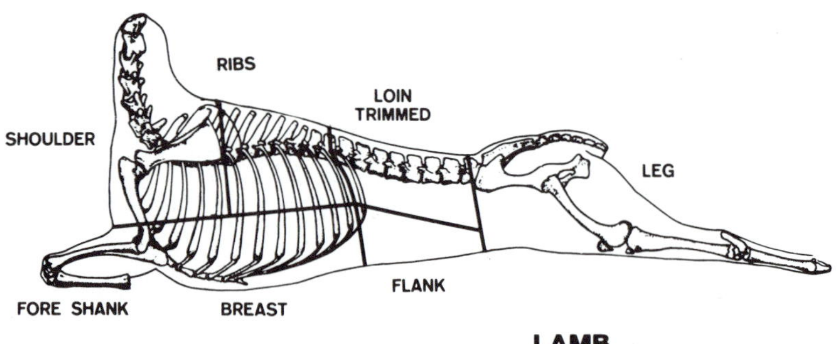
RIBS
LOIN TRIMMED
SHOULDER
LEG
FORE SHANK
BREAST
FLANK
LAMB

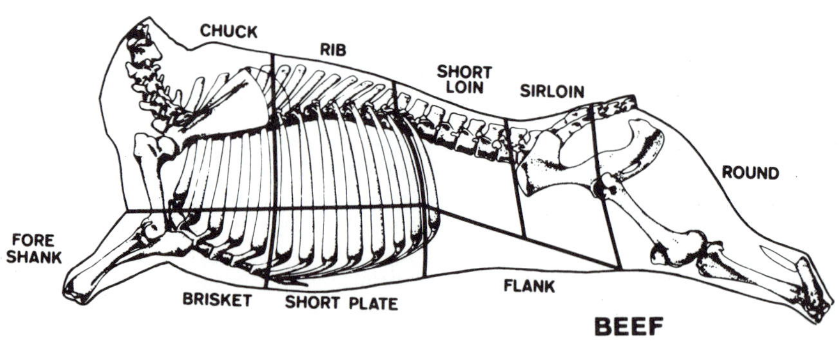
CHUCK
RIB
SHORT LOIN
SIRLOIN
ROUND
FORE SHANK
BRISKET
SHORT PLATE
FLANK
BEEF

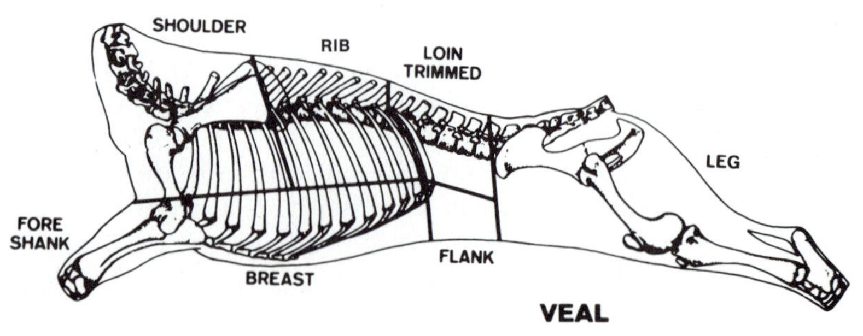
SHOULDER
RIB
LOIN TRIMMED
LEG
FORE SHANK
BREAST
FLANK
VEAL

TABLE 17-2
Approximate Yields of Wholesale Cuts From Beef, Veal, Lamb and Pork Carcasses*

Wholesale cut	Percent	Wholesale cut	Percent
	BEEF		
Hindquarter (48 percent)		Forequarter (52 percent)	
Round	23	Chuck	26
Loin	17	Rib	9
Flank	5	Shank	4
Kidney, kidney fat, and hanging tender	3	Brisket	5
		Short plate	8
	VEAL		
Shoulder	28.0		
Rib	7.3		
Loin	7.7		
Leg	34.0		
Shank	3.8		
Breast and flank	13.4		
Kidney and kidney fat	5.8		
	LAMB		
Hindsaddle (50 percent)		Foresaddle (50 percent)	
Leg	39	Shoulder	26
Loin	7	Rib	9
Flank	2	Shank	5
Kidney and kidney fat	2	Breast	10
	PORK**		
Ham	21		
Loin	18		
Boston shoulder	6.6		
Picnic shoulder	8.8		
Belly	17.3		
Spareribs	3.8		
Jowl	3.0		
Feet, tail, and neckbones	6.0		
Fat back, clear plate, and other fat trim	11.2		
Lean trim	4.3		

*No allowance for cutting shrink.
**Packer style, dressed without leaf fat.

FIGURE 17-3 (*opposite*)
Diagrams of beef, veal, pork, and lamb carcasses illustrating wholesale cuts in relation to the skeleton. [Courtesy National Live Stock and Meat Board.]

In some instances, wholesale cuts are cut in a different manner than that illustrated in Figure 17-3. This is done for a specific purpose. For example, the beef chuck may be cut to contain the arm and be called an arm chuck. Many of these special cuts are boneless or partially boneless. Such wholesale cuts are processed into special roasts, steaks, or chops for the HRI trade. We will not go into a detailed discussion of the nomenclature and specifications for these cuts. However, the discussion of muscle structure and (to some extent) bone structure that is presented later in this chapter can serve as a guide to the identification of HRI cuts and their anatomical locations. References, listed at the end of this chapter, contain the specifications and directions for making these cuts.

Chicken is sold to the consumer as a whole bird, in halves or quarters, or as cut up parts. When quartered, the cut is made where the front and hind portion of the back are separated (Figure 17-2). The hindquarter (dark meat quarter) contains the drumstick, thigh, and hind portion of the back. The forequarter (white meat quarter) contains the breast, wing, and the front portion of the back.

Many retail cuts have the same names as the wholesale cut from which they come. These include the names of most of the retail cuts illustrated in Figures 17-5 through 17-15. Since the name of the retail cut is the same as the wholesale cut, this latter name is a clue both to the name of the retail cut and to its anatomical location.

Whether a retail cut is called a roast or steak is based mainly on the thickness of the cut and the recommended cookery method. Roasts are generally thicker than steaks, and are especially thicker than chops. Steaks and roasts are obtained from beef. Roasts, steaks, and chops are cut from pork and veal. Roasts and chops are cut from lamb.

Considerable fat and bone may be removed from cuts during processing in order to meet consumer preference. If cuts contain at least part of the bones that are shown in Figure 17-4, they should be easy to identify. There are essentially seven basic retail cuts of meat: the blade, arm, rib, loin, sirloin, leg, and breast cuts. The retail cuts from these seven basic areas are common to beef, veal, pork, and lamb. Among these four meats, the breast cuts will probably vary most, due to retail cutting methods. For example, in lamb and veal the breast is cut as one piece, or converted into a number of retail cuts. In beef, the breast is cut into two pieces, the brisket and short plate; in turn, these are made into smaller retail cuts. The flanks of beef, veal, and lamb usually are processed into ground meat. The breast and flank of pork are processed as one piece (belly) into bacon, after the rib bones (spareribs) are removed.

When retail cuts are processed as boneless cuts, the recognition of muscle structure becomes more important as a means of identification.

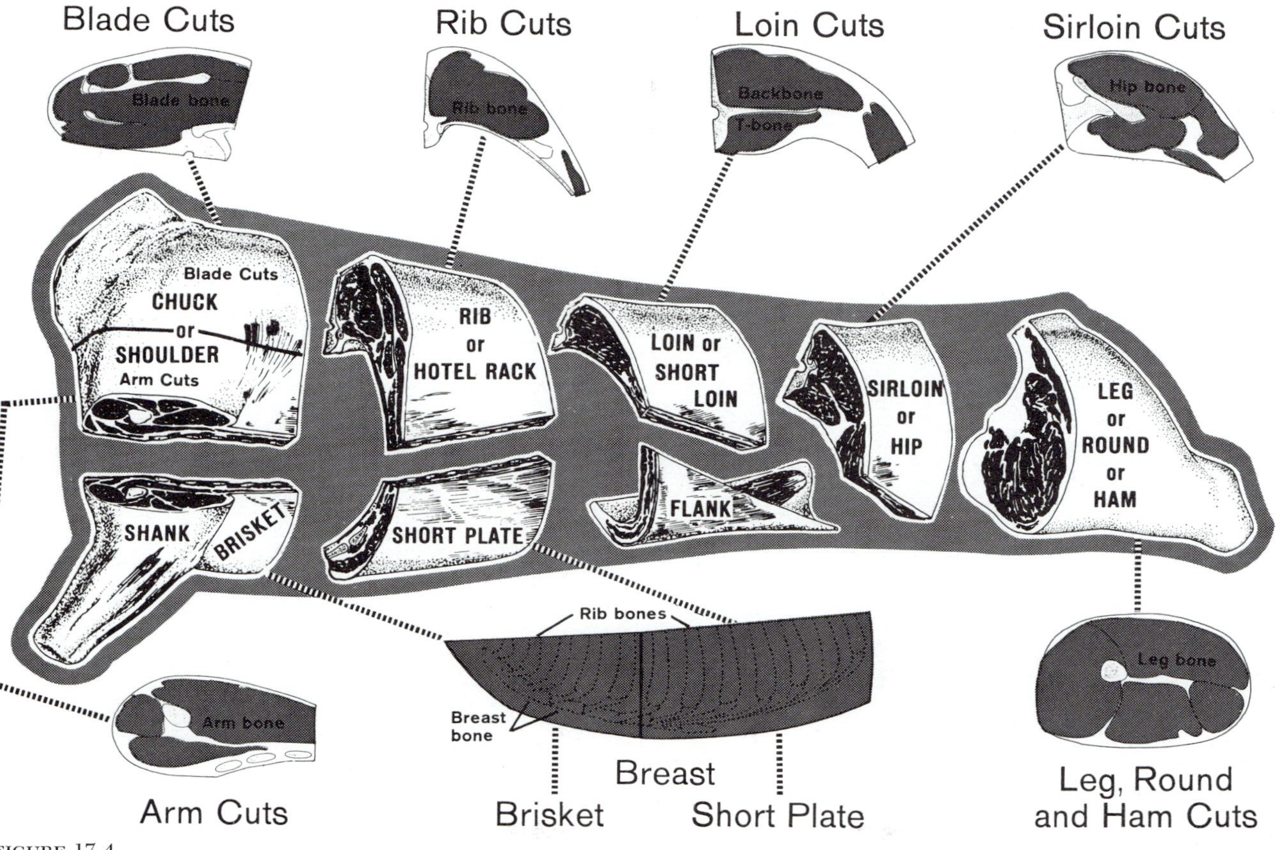

FIGURE 17-4
The seven basic retail cuts of meat obtained from beef, veal, pork, and lamb. A side of beef is used as an example. [Courtesy National Live Stock and Meat Board, Chicago, Illinois.]

Photographs of the most commonly made boneless retail cuts are presented in Figures 17-5 through 17-11, along with the bone-in cuts from the seven basic areas.

Retail cuts from the breast area are presented in Figure 17-5. The sternum, ribs, and rib cartilages of the breast are shown in the small diagram, and in the shaded areas of the carcass diagrams. The main characteristics to observe in the diagrams and photographs are: (1) the similarity in contour of the beef boneless brisket, veal breast, lamb breast, and pork spareribs, (2) the similarity of bones in pork spareribs and veal and lamb breasts, (3) the alternating layers of fat and lean in sliced bacon, beef short ribs, and lamb riblets, and (4) the diaphragm muscle in lamb breast and pork spareribs, and the absence of this muscle in veal breast, indicating that it was removed.

Shoulder arm cuts are illustrated in Figure 17-6. These cuts contain the humerus and cross-sections of the rib bones. The main features of these cuts to observe in the diagrams and photographs are: (1) the similarity in bone and muscle structure of beef arm steak, veal and pork steaks, and lamb arm chop, (2) the shape of the bone-in cuts, and the muscle structure and differences in size of the respective cuts, and (3) the absence of rib bones in beef and pork arm steaks, and the presence of rib bones in the beef chuck cross rib pot-roast, and beef chuck short ribs.

Most of the shoulder blade cuts can be identified by the blade bone (Figure 17-7). The scapula may be flat, or it may have a "ridge" resembling the figure seven, depending on whether the cut comes from the center or the posterior portion of the bone. Beef, veal, and lamb blade cuts may also contain rib bones. The pork shoulder blade steak contains no ribs because they are removed with the neck bones. When the beef blade chuck is boned, several cuts can be made. The three most common ones are pictured. Each has a characteristic muscle structure, which aids in its identification. The whole beef mock tender has a shape similar to that of the whole beef tenderloin.

The identifying bone characteristics of retail cuts from the rib area are the rib bone, and the thoracic vertebra, consisting of the chine bone and feather bone (Figure 17-8). When all of the bones are removed, cuts from the rib are easy to recognize by the appearance of the ribeye muscle, which extends along the backbone in the rib and loin.

Bone-in cuts from the short loin contain only the lumbar vertebra, except for the beef top loin steaks from the anterior position, which contain one rib (Figure 17-9). The lumbar vertebra, which is cut in half when the carcass is split into halves, consists of the spinous process, chine bone, and finger bone, and forms a characteristic T-shape.

Two muscles in particular, the *longissimus* (loin eye) and the *psoas* (tenderloin), assist in identifying either bone-in or boneless cuts from the short loin. The size of the tenderloin muscle in the beef porterhouse steak is the distinguishing characteristic between this steak and the beef T-bone steak. The center of the tenderloin muscle must be 1.25 inches (3.2 cm) in diameter, or more, to be called a porterhouse, and 0.5–1.25 inches (1.5–3.2 cm) in diameter to be called a T-bone. The tenderloin is a long tapering muscle that extends the full length of the loin (short loin and sirloin). The diameter of this muscle is greatest in the sirloin. Beef boneless top loin steaks are commonly known as Kansas City steak, New York cut, and strip steak.

The shapes of bones in cuts from the sirloin area vary, as is illustrated in Figure 17-10. Upon close examination of the skeletal diagrams, it should be evident that the cut may contain lumbar or sacral vertebrae, and a portion of the ilium. The shape of the ilium bone in the cut determines whether the steak is a wedge, round, flat, or pinbone sirloin. Veal steaks, and lamb and pork chops from the sirloin area, are identified only as sirloin steaks, or sirloin chops. The beef flat bone sirloin steak is prepared by cutting through the connective tissue that joins the ilium and sacral vertebra, and removing the latter.

The beef sirloin, when processed into boneless cuts, yields the boneless sirloin (top sirloin) and tenderloin. These cuts can be identified by muscle structure, as illustrated by the photographs in the bottom portion of Figure 17-10.

Only a cross-section of the femur bone is present in cuts from the center portion of the leg, round, or ham (Figure 17-11). This bone is round and could be confused only with the humerus from the arm cuts. However, the distinctive muscle structure of the round is very different from that of the arm cuts.

The beef round is often divided into top round, bottom round, and eye of round steaks. Their individual muscle structures are their identifying characteristics. The round tip or "sirloin tip" is a separate cut in beef, but it is part of the leg or round steaks in lamb, veal, and pork.

Summaries of the more common retail cuts of beef, veal, pork, and lamb are presented in Figures 17-12 through 17-15, respectively. Recommended methods of cookery are included for the various cuts. If each cut is cooked and served properly, then the sacrifice of the animal shall not have been for naught.

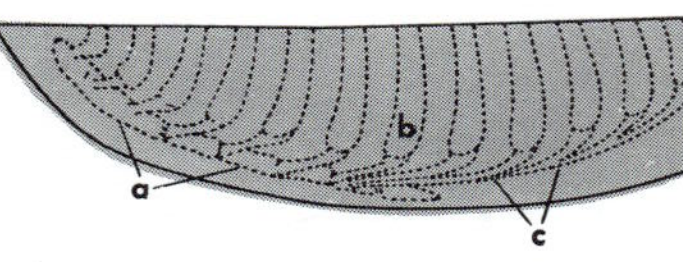

(a) Sternum
(b) Rib
(c) Costal cartilages

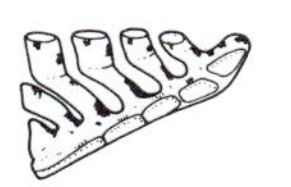

Breast and Rib Bones

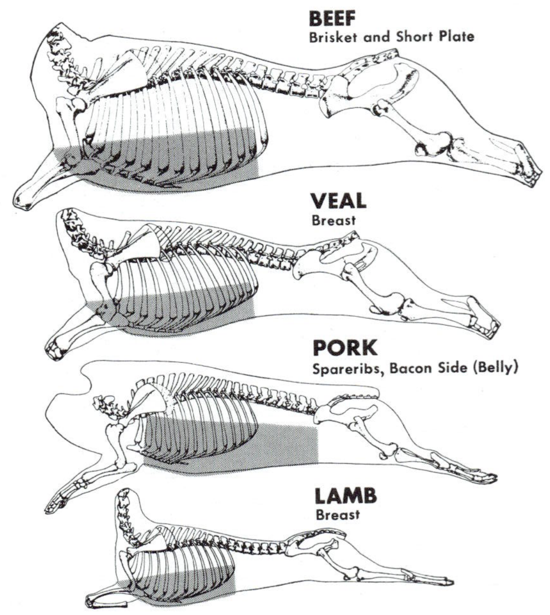

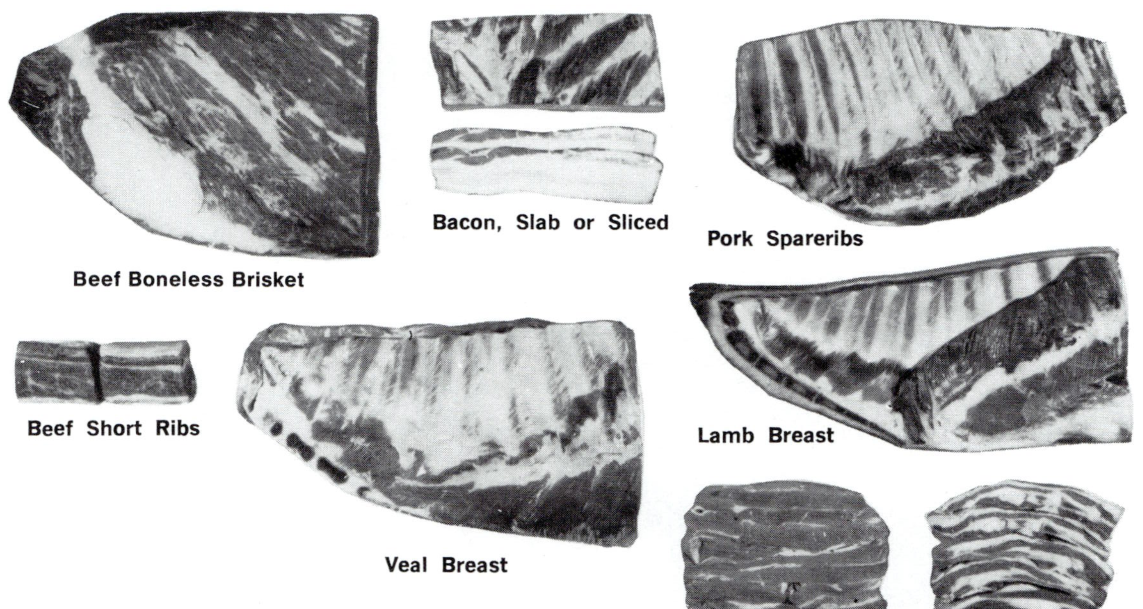

FIGURE 17-5
Retail cuts from the breast, brisket, and short plate. [Courtesy National Live Stock and Meat Board, Chicago, Illinois.]

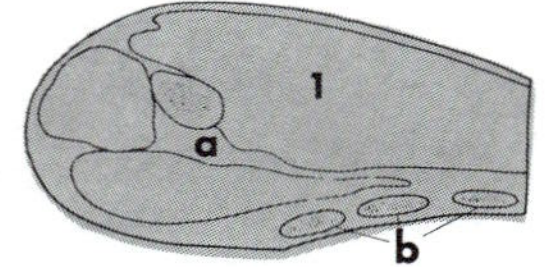

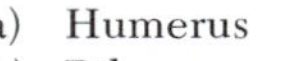

Arm Bone

(a) Humerus
(b) Ribs
(1) *Triceps brachii* and other muscles (Beef boneless shoulder steaks and pot-roasts are cut from this section.)

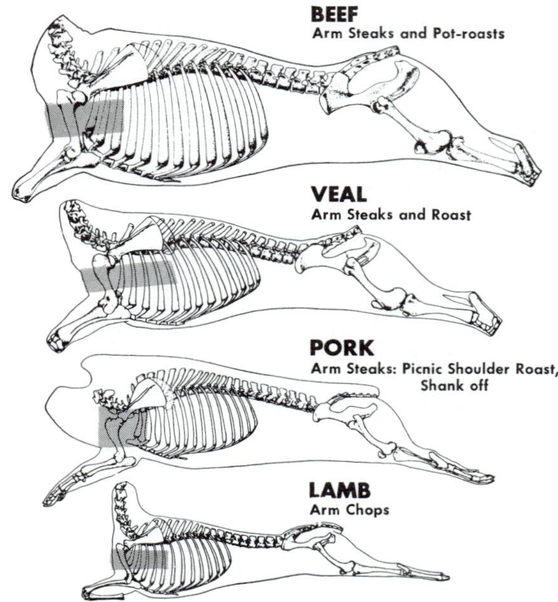

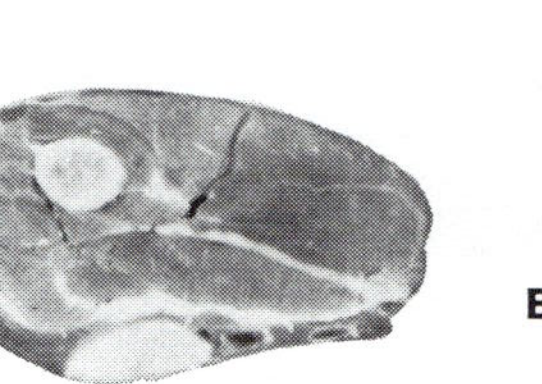

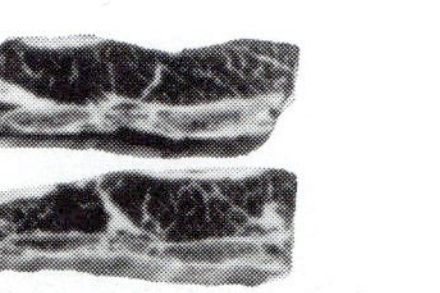

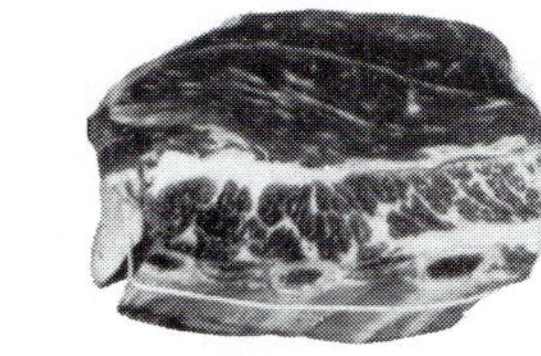

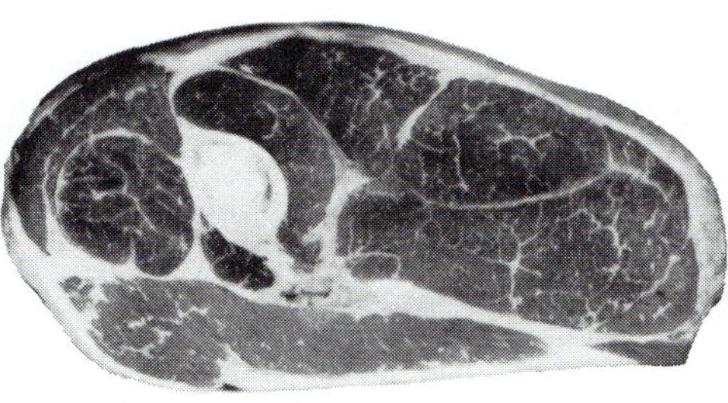

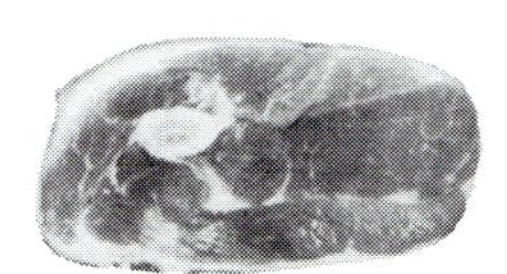

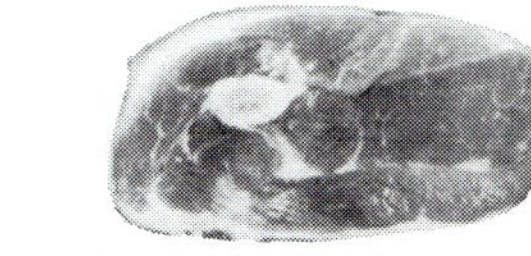

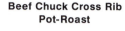

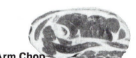

FIGURE 17-6
Retail cuts from the shoulder (chuck) arm. [Courtesy National Live Stock and Meat Board, Chicago, Illinois.]

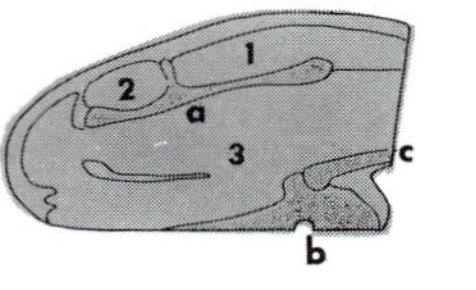

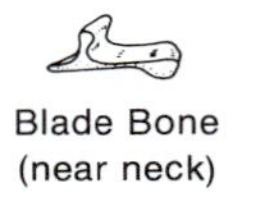

Blade Bone (near neck)

Blade Bone (center cuts)

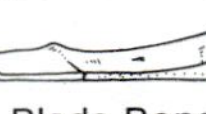

Blade Bone (near rib)

(a) Scapula
(b) Thoracic vertebra
(c) Rib bone
(1) *Infraspinatus* (Boneless top blade steak)
(2) *Supraspinatus* (Beef mock tender)
(3) Several muscles including *longissimus* (Boneless eye roast)

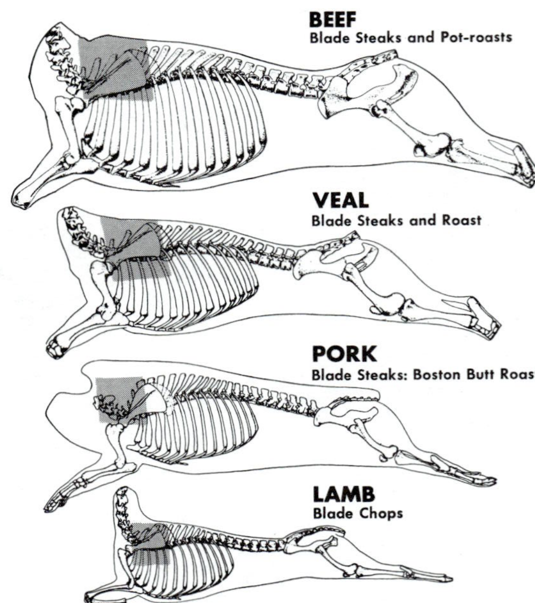

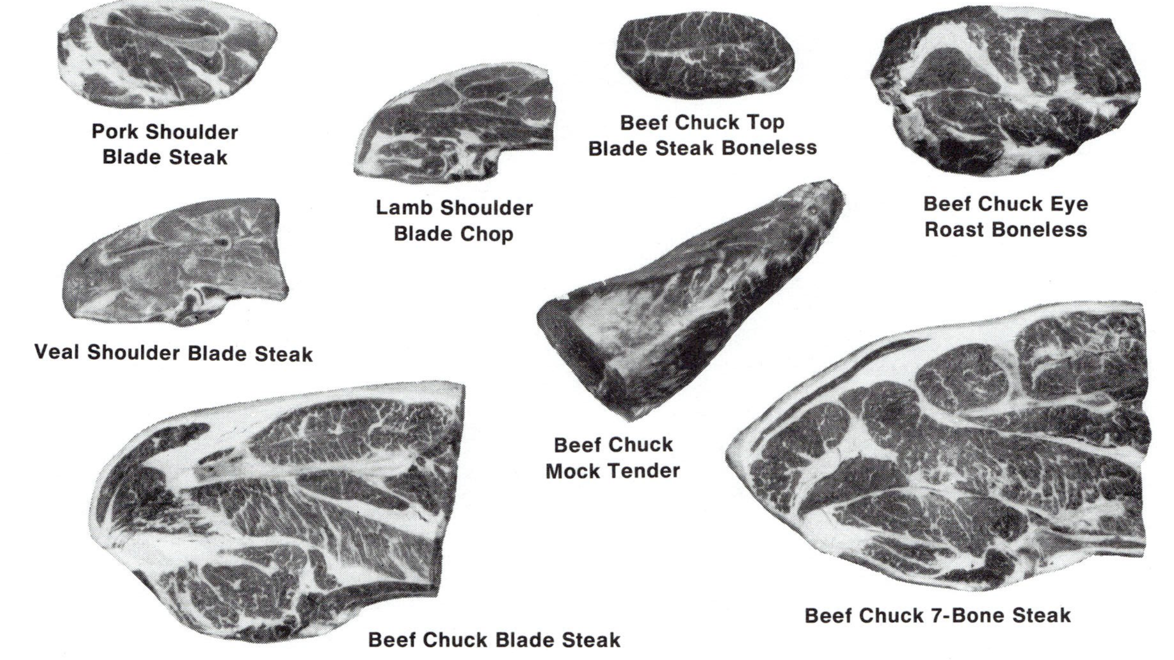

FIGURE 17-7
Retail cuts from the shoulder blade. [Courtesy National Live Stock and Meat Board, Chicago, Illinois.]

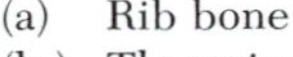

(a) Rib bone
(b_1) Thoracic vertebra (spinous process)
(b_2) Thoracic vertebra (chine bone)
(1) *Longissimus* (Ribeye)

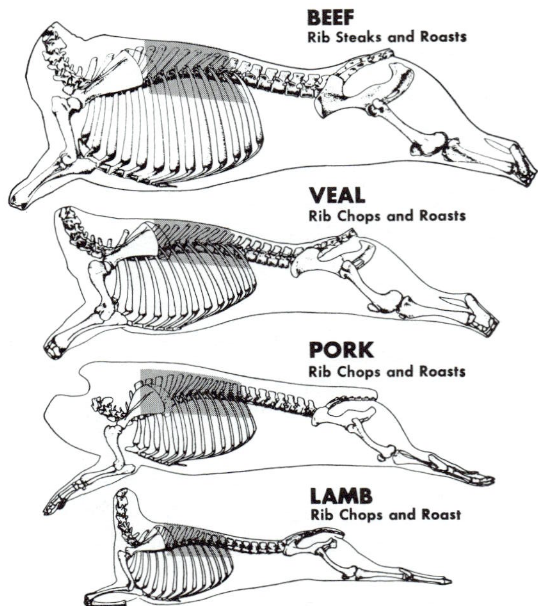

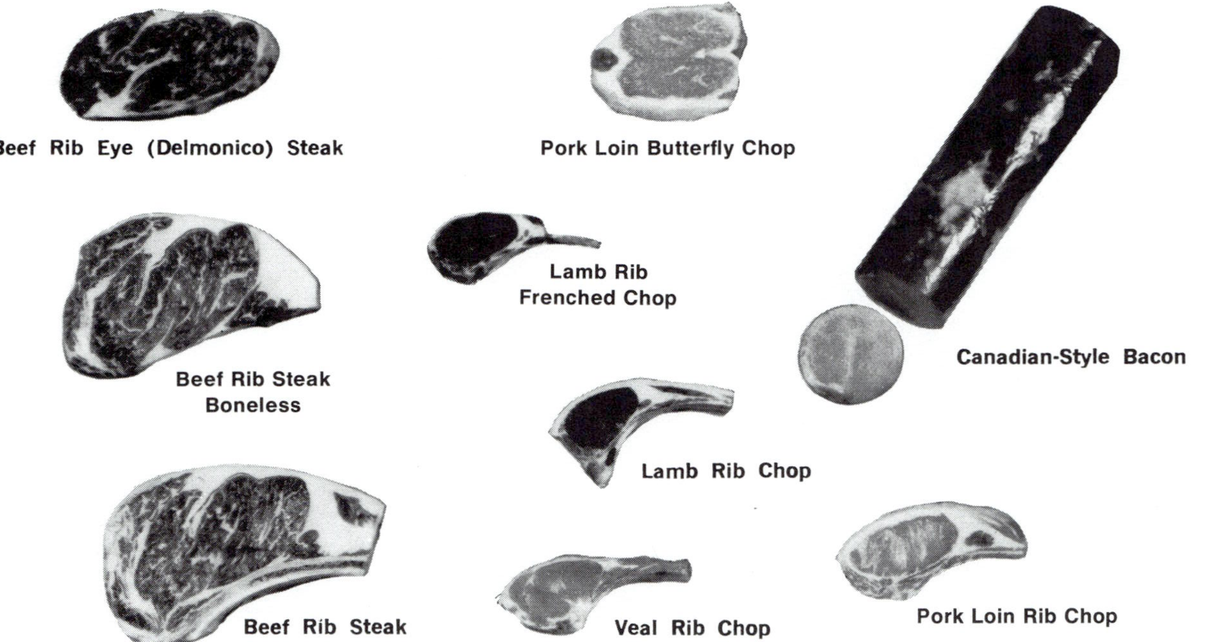

FIGURE 17-8
Retail cuts from the rib. [Courtesy National Live Stock and Meat Board, Chicago, Illinois.]

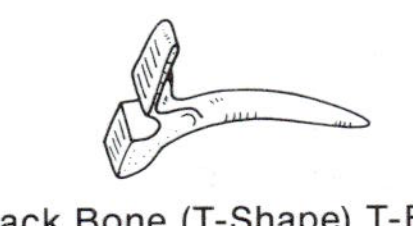

Back Bone (T-Shape) T-Bone

(a,b) Lumbar vertebra
(c) Lumbar vertebra (finger bone)
(1) *Longissimus*
(2) *Psoas* (tenderloin)
(3) *Obliquus abdominis internus* and *obliquus abdominus externus* muscles (flank muscles)

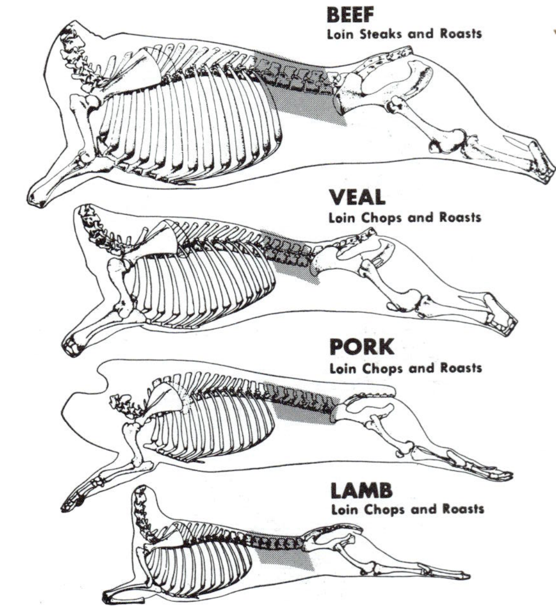

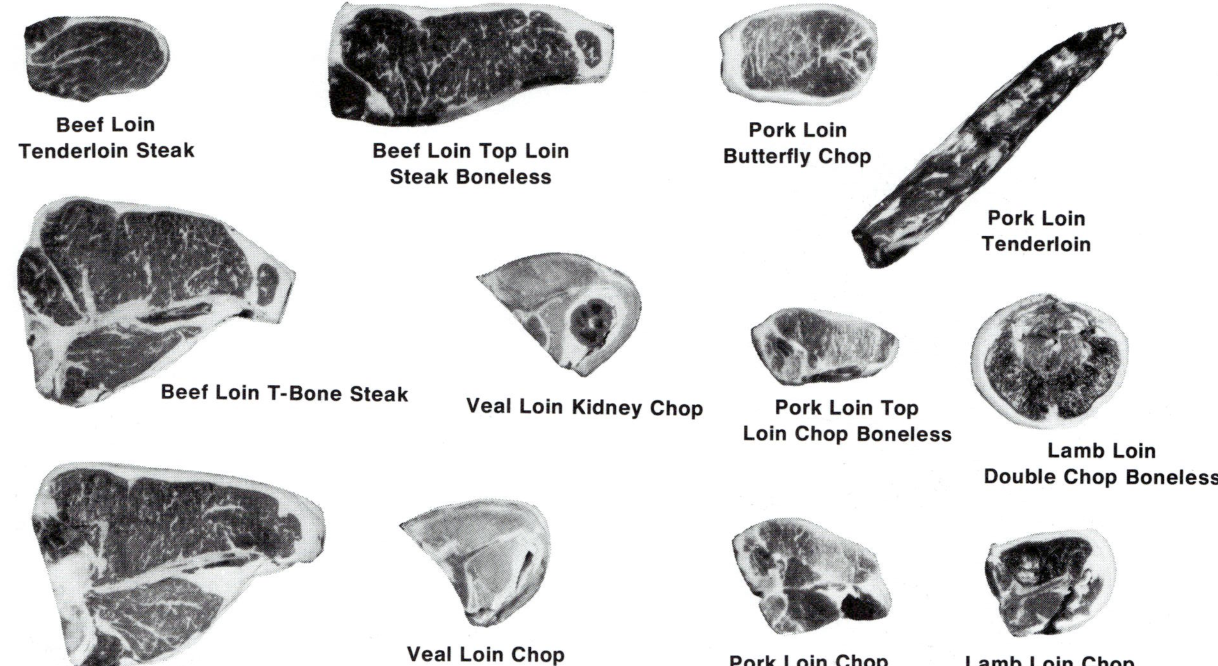

FIGURE 17-9
Retail cuts from the loin (short loin). [Courtesy National Live Stock and Meat Board, Chicago, Illinois.]

(a) Sacral vertebra
(b) Ilium
(1) *Gluteus medius* (Top sirloin)
(2) *Psoas* (Tenderloin)
(3) *Tensor fascia lata* (Sirloin tip)

Wedge Bone
(near round)

Flat Bone
(center cuts)

Pin Bone
(near short loin)

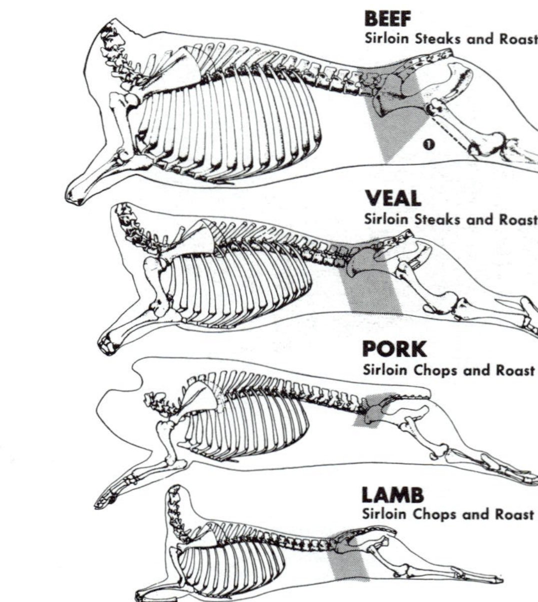

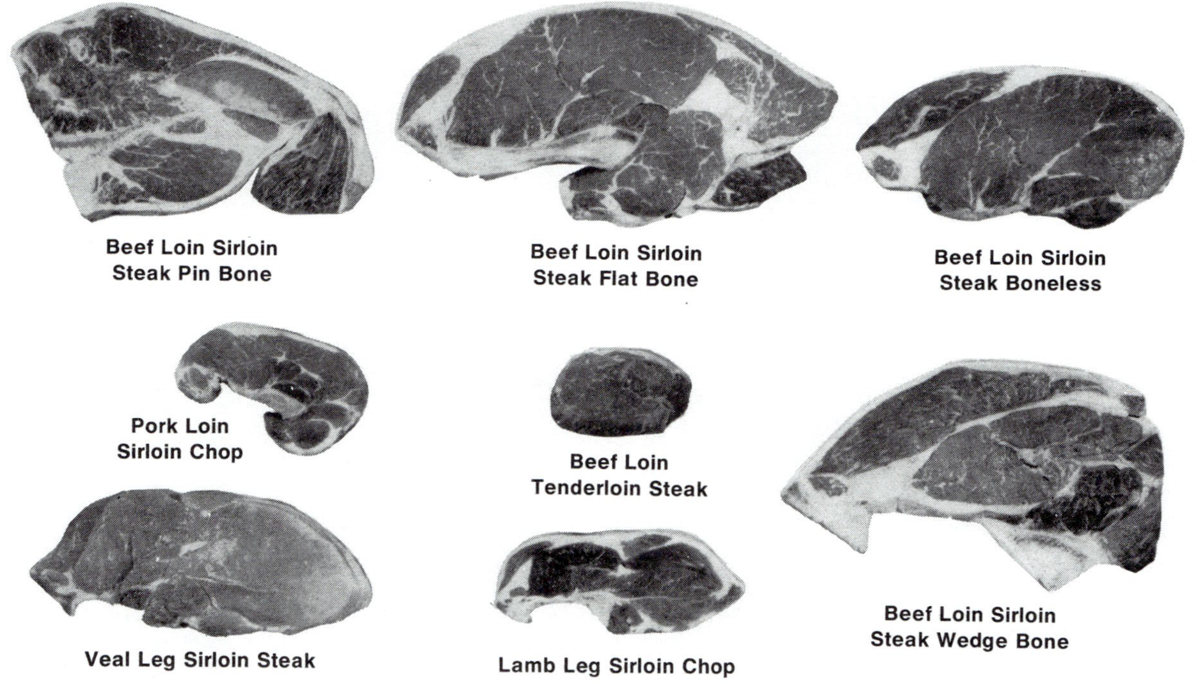

FIGURE 17-10
Retail cuts from the sirloin. [Courtesy National Live Stock and Meat Board, Chicago, Illinois.]

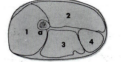

Leg or Round Bone

(a) Femur
(1) *Rectus femoris* and other muscles (Sirloin tip)
(2) *Semimembranosus* & *adductor* (Top round)
(3) *Biceps femoris* (Bottom round)
(4) *Semitendinosus* (Eye of round)

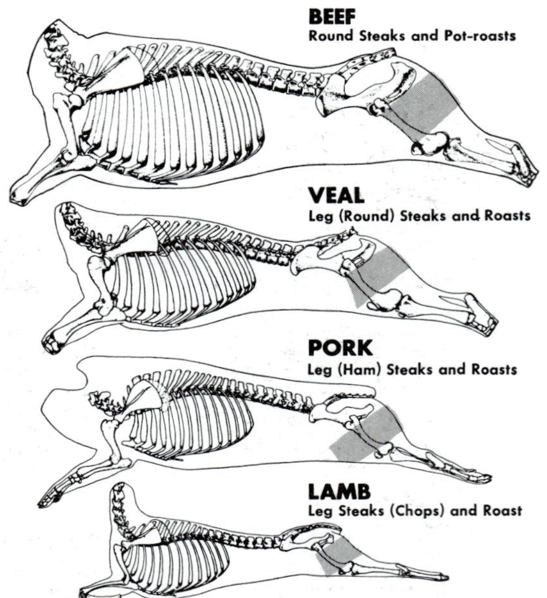

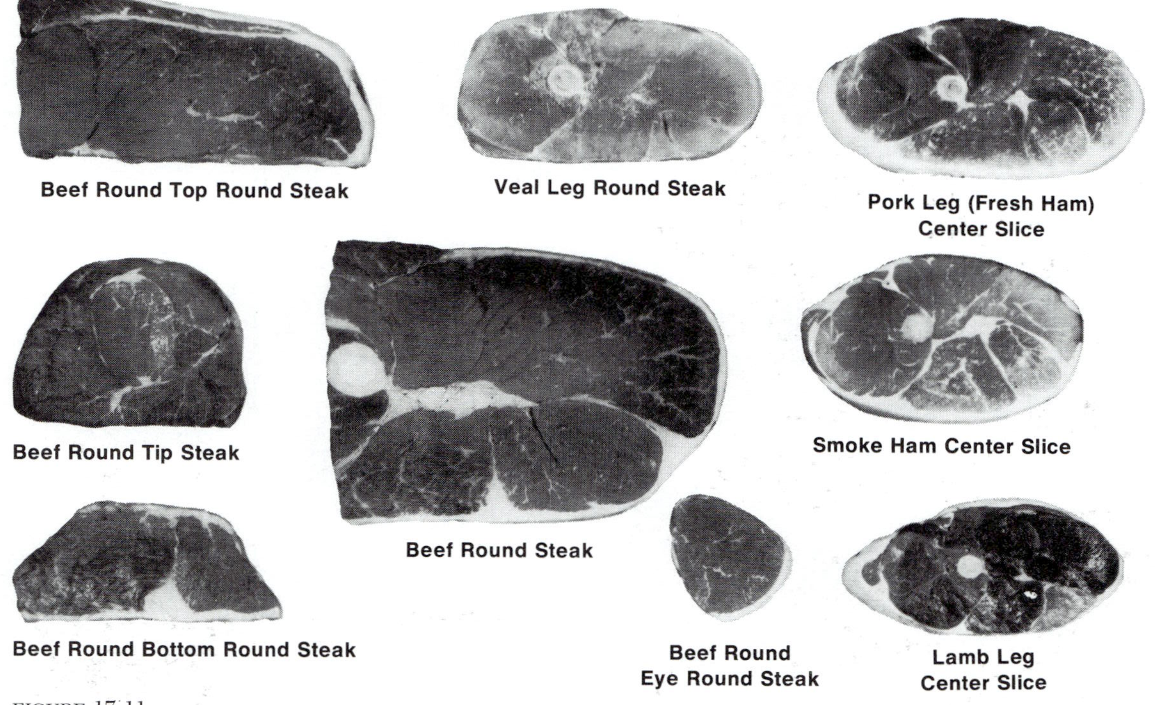

FIGURE 17-11
Retail cuts from the leg, round, or ham. [Courtesy National Live Stock and Meat Board, Chicago, Illinois.]

FIGURE 17-12
Retail cuts of beef that are obtained from the various wholesale cuts, and recommended methods of cookery. [Courtesy National Live Stock and Meat Board, Chicago, Illinois.]

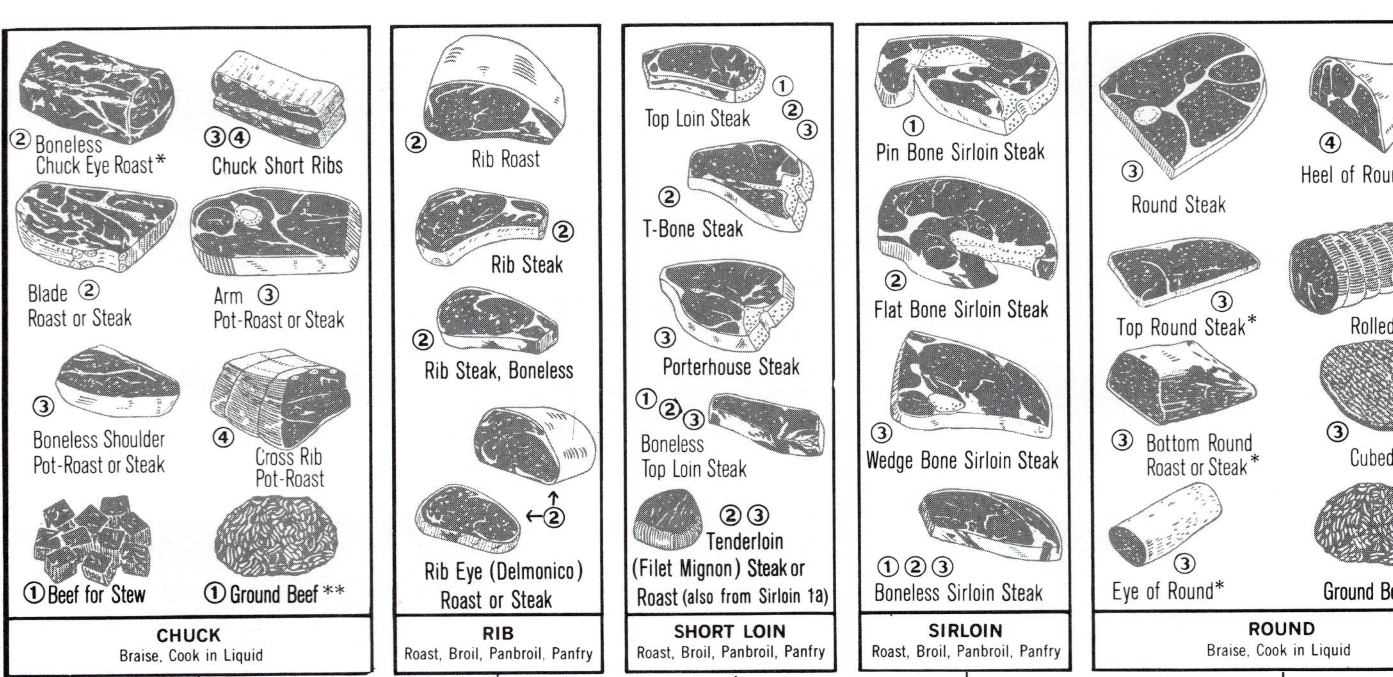

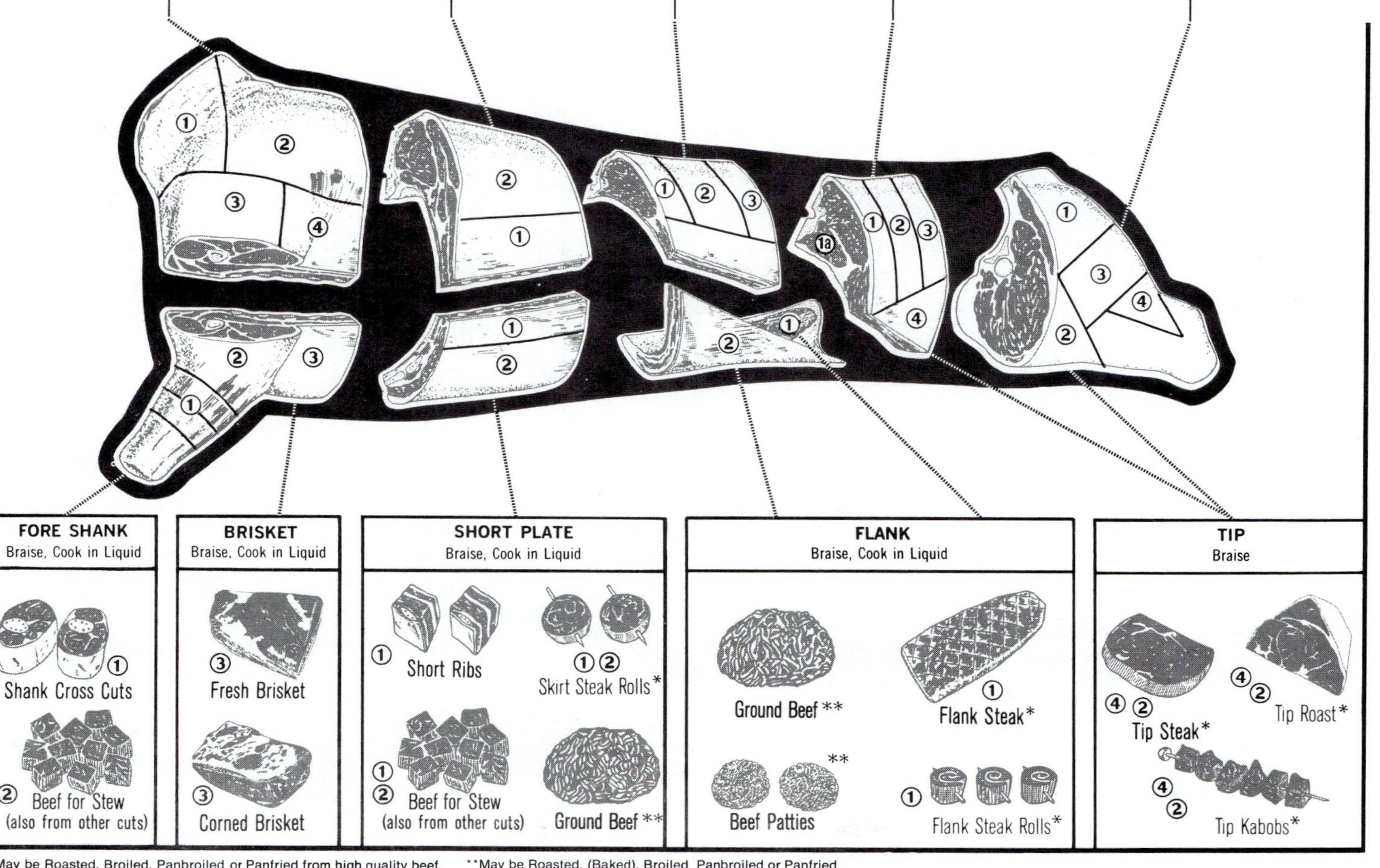

*May be Roasted, Broiled, Panbroiled or Panfried from high quality beef. **May be Roasted, (Baked), Broiled Panbroiled or Panfried.

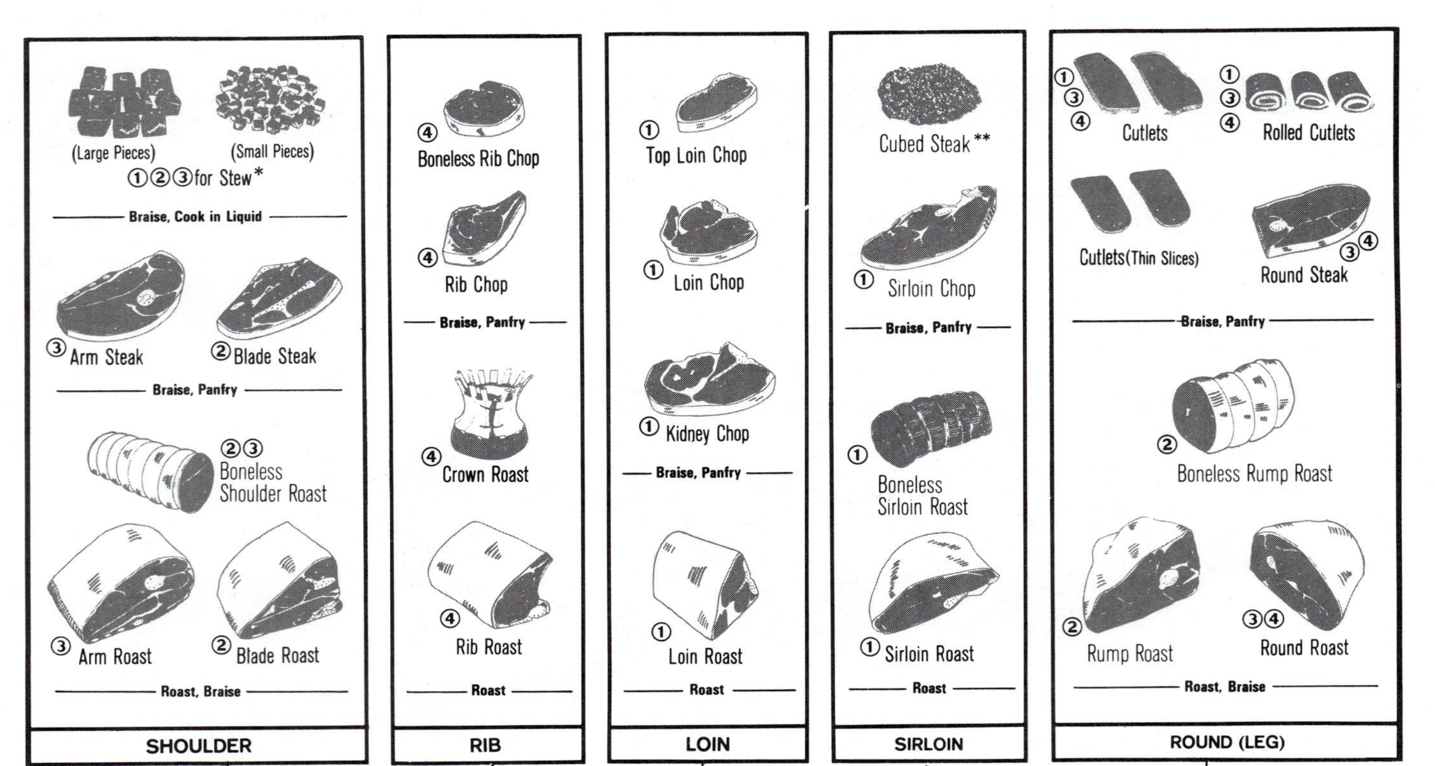

(Large Pieces)
(Small Pieces)
①②③for Stew*
Braise, Cook in Liquid
③Arm Steak
②Blade Steak
Braise, Panfry
②③ Boneless Shoulder Roast
③ Arm Roast
② Blade Roast
Roast, Braise
SHOULDER
④ Boneless Rib Chop
④ Rib Chop
Braise, Panfry
④ Crown Roast
④ Rib Roast
Roast
RIB
① Top Loin Chop
① Loin Chop
① Kidney Chop
Braise, Panfry
① Loin Roast
Roast
LOIN
Cubed Steak **
① Sirloin Chop
Braise, Panfry
① Boneless Sirloin Roast
① Sirloin Roast
Roast
SIRLOIN
①③④ Cutlets
①③④ Rolled Cutlets
Cutlets(Thin Slices)
③④ Round Steak
Braise, Panfry
② Boneless Rump Roast
② Rump Roast
③④ Round Roast
Roast, Braise
ROUND (LEG)

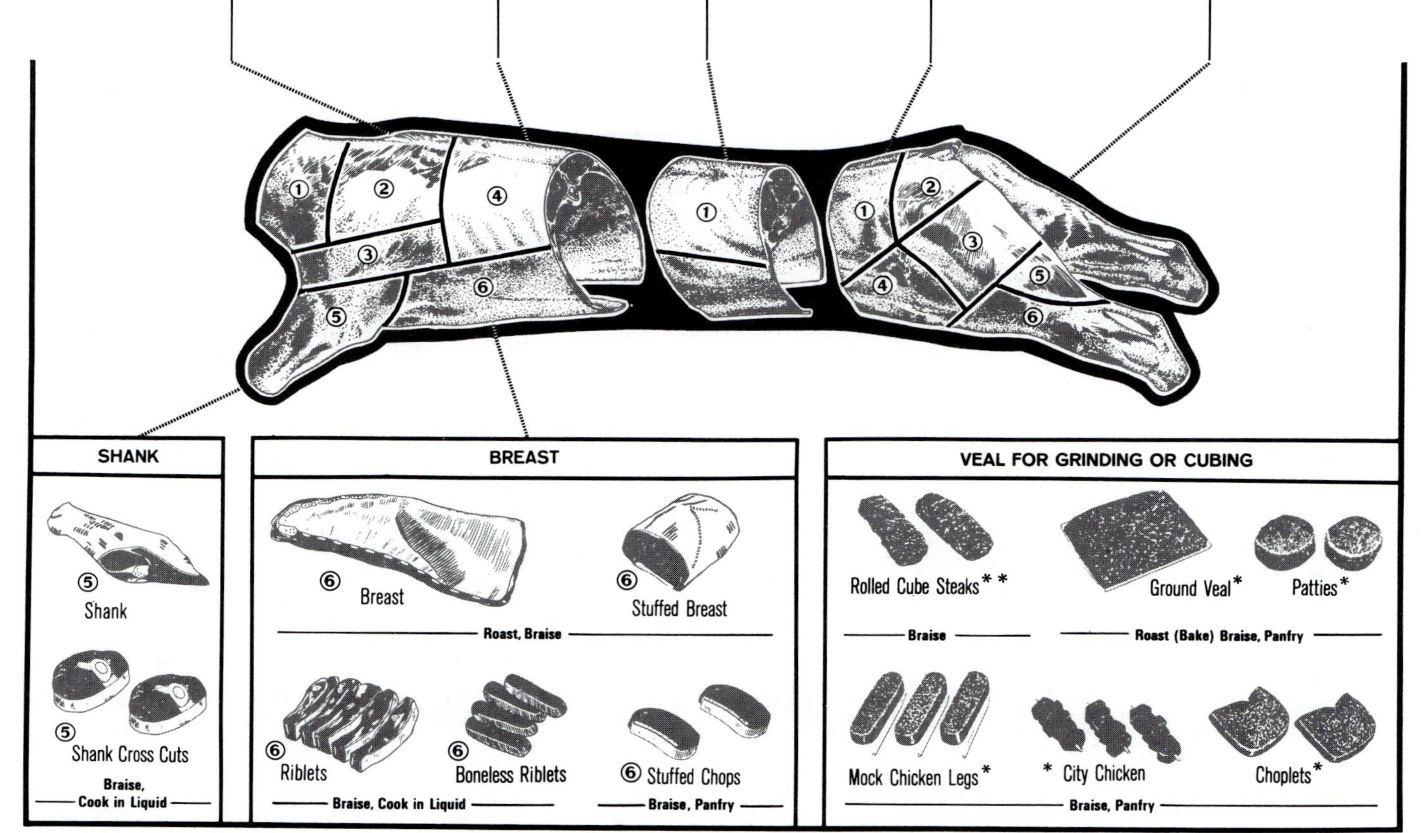

FIGURE 17-13
Retail cuts of veal that are obtained from the various wholesale cuts, and recommended methods of cookery. [Courtesy National Live Stock and Meat Board, Chicago, Illinois.]

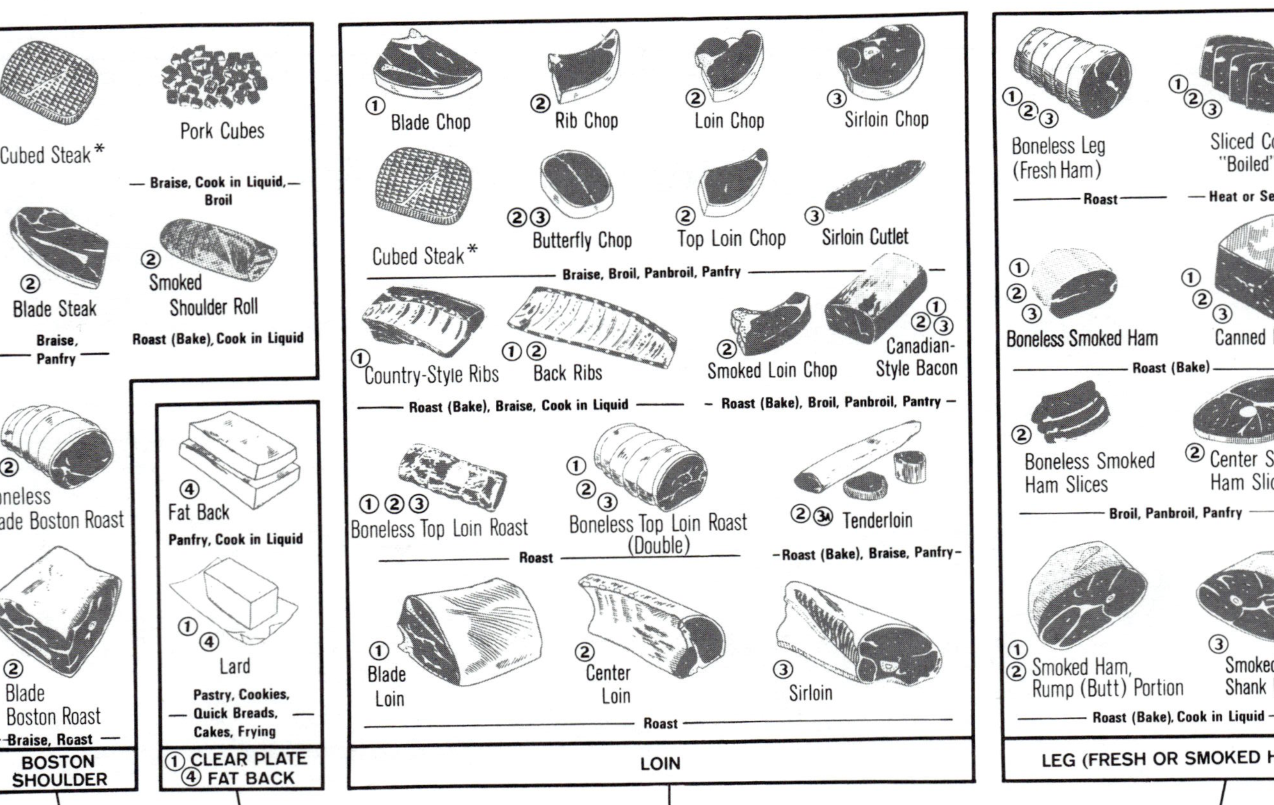

Cubed Steak*
Pork Cubes
— Braise, Cook in Liquid, Broil —
② Blade Steak
② Smoked Shoulder Roll
— Braise, Panfry —
Roast (Bake), Cook in Liquid
② Boneless Blade Boston Roast
② Blade Boston Roast
— Braise, Roast —
BOSTON SHOULDER
④ Fat Back
Panfry, Cook in Liquid
① ④ Lard
— Pastry, Cookies, Quick Breads, Cakes, Frying —
① CLEAR PLATE
④ FAT BACK
① Blade Chop
② Rib Chop
② Loin Chop
③ Sirloin Chop
Cubed Steak*
②③ Butterfly Chop
② Top Loin Chop
③ Sirloin Cutlet
— Braise, Broil, Panbroil, Panfry —
① Country-Style Ribs
①② Back Ribs
② Smoked Loin Chop
①②③ Canadian-Style Bacon
— Roast (Bake), Braise, Cook in Liquid —
– Roast (Bake), Broil, Panbroil, Pantry –
①②③ Boneless Top Loin Roast
①②③ Boneless Top Loin Roast (Double)
②③④ Tenderloin
— Roast —
-Roast (Bake), Braise, Panfry-
① Blade Loin
② Center Loin
③ Sirloin
— Roast —
LOIN
①②③ Boneless Leg (Fresh Ham)
①②③ Sliced Cooked "Boiled" Ham
— Roast —
— Heat or Serve Cold —
①②③ Boneless Smoked Ham
①②③ Canned Ham
— Roast (Bake) —
② Boneless Smoked Ham Slices
② Center Smoked Ham Slice
— Broil, Panbroil, Panfry —
①② Smoked Ham, Rump (Butt) Portion
③ Smoked Ham, Shank Portion
— Roast (Bake), Cook in Liquid —
LEG (FRESH OR SMOKED HAM)

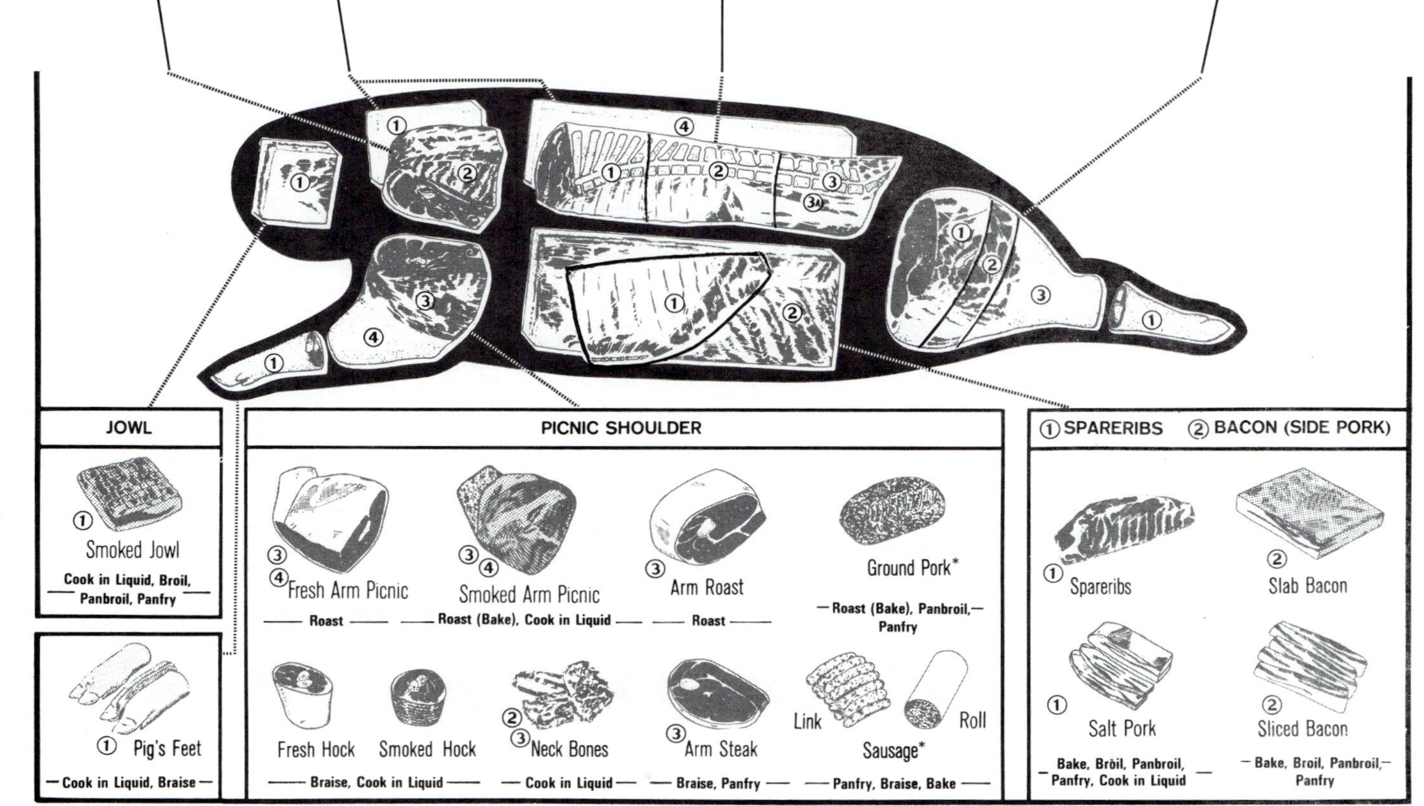

FIGURE 17-14
Retail cuts of pork that are obtained from the various wholesale cuts, and recommended methods of cookery. [Courtesy National Live Stock and Meat Board, Chicago, Illinois.]

SHOULDER
Cubes for Kabobs**
Broil
② ③ Boneless Shoulder
② ③ Cushion Shoulder
② ③ Square Shoulder
Roast
② Boneless Blade Chops (Saratoga)
② Blade Chop
③ Arm Chop
Broil, Panbroil, Panfry
NECK
① Neck Slices
Braise
RIB
① Frenched Rib Chops
① Rib Chops*
Broil, Panbroil, Panfry
① Crown Roast
① Rib Roast
Roast
LOIN
① Loin Chops
① Boneless Double Loin Chop
Broil, Panbroil, Panfry
① Boneless Double Loin Roast
① Loin Roast
Roast
SIRLOIN
① Sirloin Chop
Broil, Panbroil, Panfry
① Boneless Sirloin Roast
① Sirloin Roast
Roast
LEG
② ③ Leg Chop (Steak)
Broil, Panbroil, Panfry
① ② ③ ④ Combination Leg
② ③ Center Leg
① ② ③ ④ Rolled Leg
② ③ ④ American Leg
① ② Sirloin Half of Leg
③ ④ Shank Half of Leg
① ② ③ ④ French-Style Leg
② ③ ④ French-Style Leg, Sirloin Off
Roast

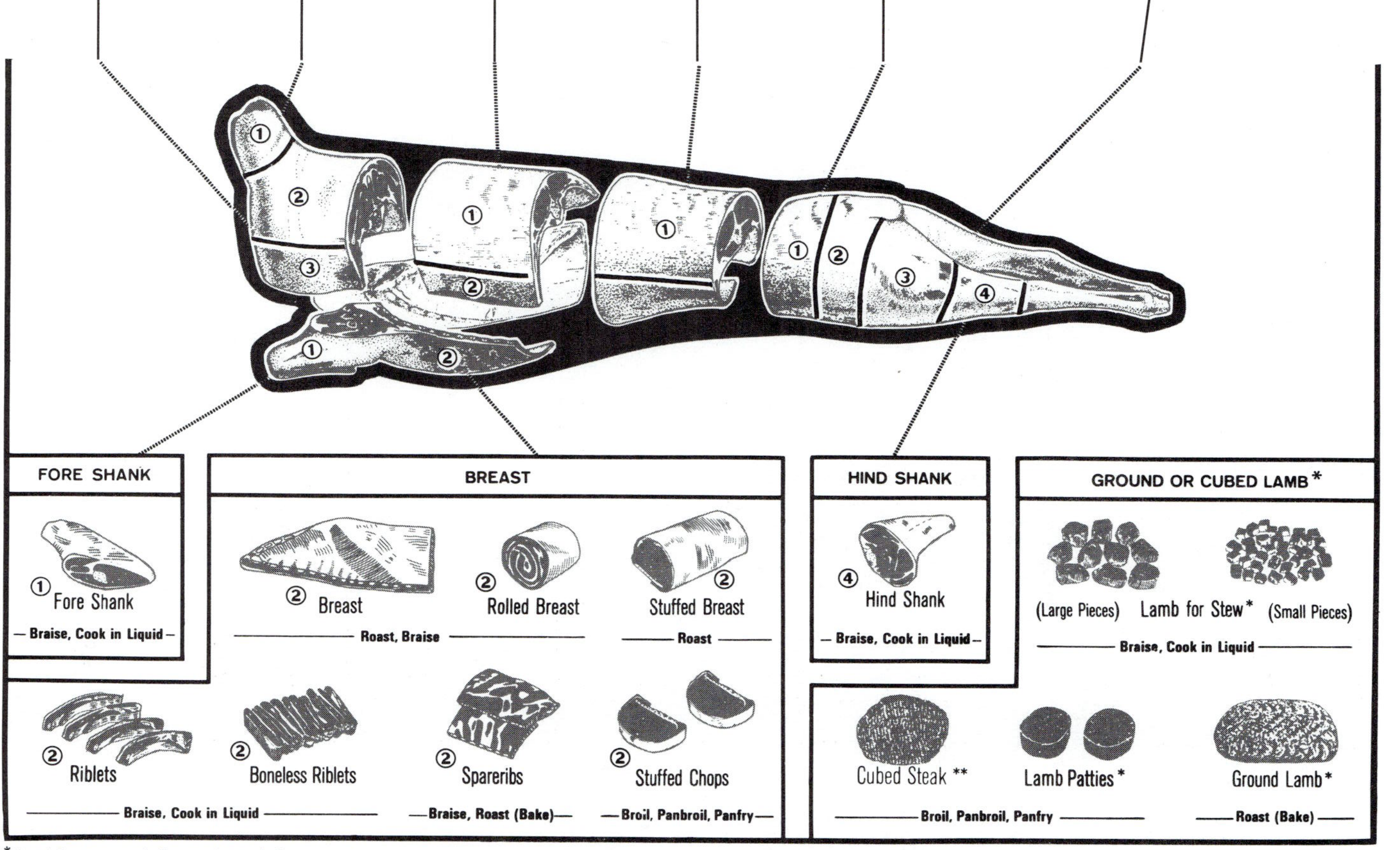

FIGURE 17-15
Retail cuts of lamb that are obtained from the various wholesale cuts, and recommended methods of cookery. [Courtesy National Live Stock and Meat Board, Chicago, Illinois.]

REFERENCES

National Association of Meat Purveyors, Meat Buyers Guide to Portion Control Meats. Chicago, Illinois, 1967.

National Association of Meat Purveyors, Meat Buyers Guide to Standardized Meat Cuts. Chicago, Illinois, 1961.

National Live Stock and Meat Board, Lessons on Meat. Chicago, Illinois, 1973.

National Live Stock and Meat Board, Uniform Retail Meat Identity Standards. Chicago, Illinois, 1973.

United States Department of Agriculture, Institutional Meat Purchase Specifications, "Series 100 (Fresh Beef); 200 (Lamb and Mutton); 300 (Veal and Calf); 400 (Fresh Pork); 500 (Cured and Smoked and Fully Cooked Pork Products); 600 (Cured, Dried and Smoked Beef Products); 800 (Sausage Products); and 1000 (Portion-cut Meat Products)," Consumer and Marketing Service, Livestock Division, Washington, D.C.

Index

NOTE: Page numbers in *italics* refer to tables or illustrations.